VOLUME FIFTY SIX

Advances in
ECOLOGICAL RESEARCH

Networks of Invasion: A Synthesis
of Concepts

ADVANCES IN ECOLOGICAL RESEARCH

Series Editors

DAVID A. BOHAN
Directeur de Recherche
UMR 1347 Agroécologie
AgroSup/UB/INRA
Pôle GESTAD, Dijon, France

ALEX J. DUMBRELL
School of Biological Sciences
University of Essex
Wivenhoe Park, Colchester
Essex, United Kingdom

Advances in
ECOLOGICAL RESEARCH

Networks of Invasion: A Synthesis of Concepts

Edited by

DAVID A. BOHAN
Directeur de Recherche
UMR 1347 Agroécologie
AgroSup/UB/INRA
Pôle GESTAD, Dijon, France

ALEX J. DUMBRELL
School of Biological Sciences
University of Essex
Wivenhoe Park, Colchester, Essex,
United Kingdom

FRANÇOIS MASSOL
UMR 8198 Evo-Eco-Paleo
CNRS, Université de Lille
Lille, France

ACADEMIC PRESS

An imprint of Elsevier
elsevier.com

Academic Press is an imprint of Elsevier
The Boulevard, Langford Lane, Kidlington, Oxford OX5 1GB, United Kingdom
125 London Wall, London, EC2Y 5AS, United Kingdom
50 Hampshire Street, 5th Floor, Cambridge, MA 02139, United States
525 B Street, Suite 1800, San Diego, CA 92101-4495, United States

First edition 2017

Notices
Knowledge and best practice in this field are constantly changing. As new research and experience broaden our understanding, changes in research methods, professional practices, or medical treatment may become necessary.

Practitioners and researchers must always rely on their own experience and knowledge in evaluating and using any information, methods, compounds, or experiments described herein. In using such information or methods they should be mindful of their own safety and the safety of others, including parties for whom they have a professional responsibility.

To the fullest extent of the law, neither the Publisher nor the authors, contributors, or editors, assume any liability for any injury and/or damage to persons or property as a matter of products liability, negligence or otherwise, or from any use or operation of any methods, products, instructions, or ideas contained in the material herein.

ISBN: 978-0-12-804338-7
ISSN: 0065-2504

For information on all Academic Press publications
visit our website at https://www.elsevier.com/

Working together
to grow libraries in
developing countries

www.elsevier.com • www.bookaid.org

Publisher: Zoe Kruze
Acquisition Editor: Alex White
Editorial Project Manager: Helene Kabes
Production Project Manager: Magesh Kumar Mahalingam
Cover Designer: Mark Rogers

Typeset by SPi Global, India

CONTENTS

CONTRIBUTORS

O. Anneville
UMR 42 CARRTEL, INRA-Université Savoie Mont Blanc, Thonon-les-Bains, France

T.J. Bartley
University of Guelph, Guelph, ON, Canada

D.A. Bohan
UMR1347 Agroécologie, AgroSup/UB/INRA, Pôle Gestion des Adventices, Dijon Cedex, France

J.R. Boutain
Botanical Research Institute of Texas, Fort Worth, TX, United States

V. Calcagno
Université Côte d'Azur, CNRS, INRA, ISA, France

K. Cazelles
Université du Québec à Rimouski, Rimouski; Quebec Center for Biodiversity Science, Montréal, QC, Canada; UMR MARBEC (MARine Biodiversity, Exploitation and Conservation), Université de Montpellier, Montpellier, France

E. Chapuis
UMR 3P PVBMT, CIRAD-Université de la Réunion, Saint Pierre; UMR IPME, IRD-Université de Montpellier-CIRAD, Montpellier, France

R.I. Colautti
Queen's University, Kingston, ON, Canada

P. David
Centre d'Ecologie Fonctionnelle et Evolutive, UMR 5175, CNRS-Université de Montpellier-UMIII-EPHE, Montpellier, France

I. Domaizon
INRA, Université de Savoie Mont Blanc, UMR CARRTEL, Thonon les Bains, France

M. Dubart
CNRS, Université de Lille-Sciences et Technologies, UMR 8198 Evo-Eco-Paleo, SPICI Group, Villeneuve d'Ascq; Institut des Sciences de l'Évolution, Université de Montpellier, CNRS, IRD, EPHE, CC065, Montpellier, France

P.-F. Duyck
CIRAD, UMR PVBMT, 97410 Saint Pierre, La Réunion, France

C. Fontaine
Muséum National d'Histoire Naturelle—CESCO, UMR 7204 MNHN-CNRS-UPMC, Paris, France

D. Gravel
Université du Québec à Rimouski, Rimouski; Quebec Center for Biodiversity Science, Montréal; Faculté des Sciences, Université de Sherbrooke, Sherbrooke, QC, Canada

C. Jacquet
Université du Québec à Rimouski, Rimouski; Quebec Center for Biodiversity Science, Montréal, QC, Canada; UMR MARBEC (MARine Biodiversity, Exploitation and Conservation), Université de Montpellier, Montpellier, France

S. Kamenova
University of Guelph, Guelph, ON, Canada

S. Kéfi
Institut des Sciences de l'Évolution, Université de Montpellier, CNRS, IRD, EPHE, CC065, Montpellier, France

I. Le Viol
Muséum National d'Histoire Naturelle—CESCO, UMR 7204 MNHN-CNRS-UPMC, Paris, France

A. Lemainque
Commissariat à l'Energie Atomique et aux Energies Alternatives (CEA), Institut de Génomique (IG), Genoscope, Evry, France

N. Loeuille
Institute of Ecology and Environmental Sciences UMR 7618, Sorbonne Universités-UPMC–CNRS-IRD-INRA-Université Paris Diderot-UPEC, Paris, France

N.D. Martinez
Pacific Ecoinformatics and Computational Ecology Lab, Berkeley, CA; University of Arizona, Tucson, AZ, United States

F. Massol
CNRS, Université de Lille, UMR 8198 Evo-Eco-Paleo, SPICI Group, Lille, France

G. Mollot
SupAgro, UMR CBGP (INRA/IRD/CIRAD/Montpellier SupAgro), Montferrier-sur-Lez; CESAB-FRB, Immeuble Henri Poincaré, Aix en Provence, France

J.H. Pantel
Centre d'Ecologie Fonctionnelle et Evolutive, UMR 5175, CNRS-Université de Montpellier-UMIII-EPHE, Montpellier; Centre for Ecological Analysis and Synthesis, Foundation for Research on Biodiversity, Bâtiment Henri Poincaré, Rue Louis-Philibert, 13100 Aix-en-Provence, France

M.-E. Perga
INRA, Université de Savoie Mont Blanc, UMR CARRTEL, Thonon les Bains, France

V. Ravigné
CIRAD, UMR PVBMT Pôle de Protection des Plantes, Saint-Pierre, Réunion, France

T.N. Romanuk
Dalhousie University, Halifax, NS, Canada; Pacific Ecoinformatics and Computational Ecology Lab, Berkeley, CA, United States

E. Thébault
Institute of Ecology and Environmental Sciences UMR 7618, Sorbonne Universités-UPMC-CNRS-IRD-INRA-Université Paris Diderot-UPEC, Paris, France

P. Tixier
CIRAD, UR GECO, Montpellier Cedex, France; CATIE, Cartago, Costa Rica

F.S. Valdovinos
University of Arizona, Tucson, AZ, United States

Y. Zhou
Pacific Ecoinformatics and Computational Ecology Lab, Berkeley, CA; School of Public Policy, University of Maryland, College Park, MD, United States

PREFACE

Biological invasions are considered one of the pre-eminent sources of disturbance of ecosystems, causing marked loss of biodiversity at large spatial scales and concomitant changes in function and ecosystem service provision (Ehrenfeld, 2010; Murphy and Romanuk, 2014; Vilà et al., 2006). The past half century has seen increased concern about the impacts of biological invasions (Gurevitch and Padilla, 2004), from local species extinction to changes in ecosystem functioning, as well as effects on genetic diversity or facilitation of further invasions through invasion meltdown (Murphy and Romanuk, 2014; Paolucci et al., 2013; Simberloff et al., 2013; Strayer, 2012; Vilà et al., 2011). In order to prevent invasions, much effort has also been devoted towards understanding conditions favouring the success of invasions (Beisner et al., 2006; Pyšek et al., 2010; Romanuk et al., 2009). In spite of numerous empirical and theoretical developments, understanding, predicting and managing species invasions remain one of the main issues of ecology today, with potentially massive costs now and in the years to come (Bradshaw et al., 2016).

Invasion biology, born as a new discipline more than half a century ago (Elton, 1958), is only now coming of age. Despite an accumulation of detailed case studies, a growing theoretical corpus and connections made to many areas of ecology, evolutionary biology and economy, there remain some key gaps in the study of invasions. One is the need for a more integrative, systems-based view of an invasive species encountering an established, recipient community, with an emphasis placed on the complexity of reciprocal relationships, including the realization that the latter is itself a complex entity. The research that results will likely be neither a simple, linear addition of a species to a community nor the summation of all pairwise interactions between the constituent species.

Until recently, relatively little investigation of species invasions has been done using systems- or network-based approaches (but see Romanuk et al., 2009), despite growing evidence that consideration of the invader alone, or the species it appears to displace, is not enough. In biocontrol, whereby species have been deliberately introduced to counteract the effects of particular pest species, there is ample evidence that the insertion of a novel species into an established community can markedly modify community structure and functioning in ways that are often unexpected (Lombaert et al., 2010;

Roy et al., 2011). The results of such introductions range from mere persistence of the biocontrol agent, to successful introduction and control effect, through to wholesale switching to other unintended prey species and the loss of these from the system instead (Clarke et al., 1984; Phillips et al., 2006). These findings in biocontrol, when placed alongside others from 'natural' invasion events, argue for the effects of invading species ramifying across whole ecosystems and that systems- or network-based approaches would have great power to resolve mechanisms and understand invasion.

'Networks of invasion', understood both through biological invasions into ecological interaction networks and through the web-like nature of invasion patterns, spatially and in terms of sequential invasions, are the subject of two thematic issues of *Advances in Ecological Research*. These deal, in turn, with papers synthesizing concepts in invasion ecology and the empirical evidence and case studies of invasions viewed with a network perspective. This first volume opens with a set of papers, both empirical and theoretical, which set out the general Perspectives in Invasion Ecology, when viewed from the network standpoint. These papers first synthesize known impacts of invaders on food webs, with both a general review focusing on the 'insertion point' of invaders within food webs (David et al., Chapter 1) and a meta-analysis (Mollot et al., Chapter 2) which detail how often and under which conditions species invasions lead to species losses. Chapter 3 (Kamenova et al.) introduces current tools—genetic, isotopic, participatory, model based, etc.—available to monitor, detect and predict invasions in ecological networks. The following two chapters (Massol et al., Chapter 4; Romanuk et al., Chapter 5) present two different models aimed at elucidating the effects of food web topology on invasion probability and impact. While Chapter 4 develops a spatial food web model based on MacArthur and Wilson's (1963) theory of island biogeography to understand the effects of food web properties on sequences of invasions, Chapter 5 focuses on what makes a food web robust to species invasions, elaborating on the niche model of Williams and Martinez (2000) to explore whether quantitative and qualitative robustness are traded off along the food web connectance gradient. This first volume closes with a general perspective papers (Pantel et al., Chapter 6) which focuses on 14 different questions for future studies on invasions in ecological networks.

The second volume consists of a series of papers that detail invasions of particular types of invaders (e.g. parasites) or into particular types of ecosystems (e.g. freshwater ecosystems). It begins with Médoc et al., which synthesizes predictions of impacts of parasite invasions in food webs. The following chapter (Jackson et al., Chapter 2) focuses on shifts in freshwater food web topologies due to species invasions and emphasizes the possibility of trophic link disruption in such situations. Chapter 3 (Amsellem et al.) then synthesizes current evidence and hypotheses regarding the effects of micro-organisms on species invasions, from the classic enemy release hypothesis to spillback and spillover effects linked to the co-invasion of organisms with their symbionts. Plant–pollinator interactions and the topology of the asso-ciated networks under a regime of massive species introductions are the focus of Chapter 4 (Geslin et al.), which detail the potential impacts of such massive management of both plants and insects on nearby, non-managed ecosystems. The volume then ends with Chapter 5, by Murall et al., with a change of scale to describe invasions in microbial networks within hosts, emphasizing what is already known regarding mammal microbiota and suggesting ways in which interdisciplinary dialogue between ecologists, microbiologists and physiologists might pave the way for a better under-standing of problems such as the emergence of antibiotic resistance.

These two volumes present a snapshot of the current state of the art in the study of invasion ecology, emphasizing species interactions and using network-based approaches. These 11 chapters reveal that there is great value in using networks to study invasion, both from a conceptual viewpoint and in practical terms to measure, predict and monitor the impacts of invasions on ecosystems. When taken together, the work presented here suggests that building a richer understanding of invasion, which networks afford, could be an important step forward in developing predictive approaches to managing or preventing invasions. We hope that the variety of approaches and ques-tions to networks of invasion contained within these pages provide a fruitful and stimulating framework for future studies and research programmes aimed at tackling invasions.

F. MASSOL
P. DAVID
D.A. BOHAN

ACKNOWLEDGEMENTS

This series of papers came, in part, out of the COREIDS project, cofinanced by TOTAL and the Fondation pour la Recherche sur la Biodiversité (FRB), at the Centre for the Synthesis and Analysis of Biodiversity (CESAB) in Aix-en-Provence, France. The COREIDS project has the goals of unifying and analyzing databases on species invasions into sensitive communities and zones using network ecological approaches, with the aim of identifying the generic processes and mechanisms of invasion that might be used in invasion prediction and management.

REFERENCES

Beisner, B.E., et al., 2006. Environmental productivity and biodiversity effects on invertebrate community invasibility. Biol. Invasions 8, 655–664.

Bradshaw, C.J.A., et al., 2016. Massive yet grossly underestimated global costs of invasive insects. Nat. Commun. 7, 12986.

Clarke, B., et al., 1984. The extinction of endemic species by a program of biological control. Pac. Sci. 38, 97–104.

Ehrenfeld, J.G., 2010. Ecosystem consequences of biological invasions. Annu. Rev. Ecol. Evol. Syst. 41, 59–80.

Elton, C.S., 1958. The Ecology of Invasions by Animals and Plants. Methuen & Co Ltd., London, UK.

Gurevitch, J., Padilla, D.K., 2004. Are invasive species a major cause of extinctions? Trends Ecol. Evol. 19, 470–474.

Lombaert, E., et al., 2010. Bridgehead effect in the worldwide invasion of the biocontrol Harlequin Ladybird. PLoS One 5, e9743.

MacArthur, R.H., Wilson, E.O., 1963. An equilibrium theory of insular zoogeography. Evolution 17, 373–387.

Murphy, G.E.P., Romanuk, T.N., 2014. A meta-analysis of declines in local species richness from human disturbances. Ecol. Evol. 4, 91–103.

Paolucci, E.M., et al., 2013. Origin matters: alien consumers inflict greater damage on prey populations than do native consumers. Divers. Distrib. 19, 988–995.

Phillips, B.L., et al., 2006. Invasion and the evolution of speed in toads. Nature 439, 803.

Pyšek, P., et al., 2010. Disentangling the role of environmental and human pressures on biological invasions across Europe. Proc. Natl. Acad. Sci. U.S.A. 107, 12157–12162.

Romanuk, T.N., et al., 2009. Predicting invasion success in complex ecological networks. Philos. Trans. R. Soc. B 364, 1743–1754.

Roy, H.E., et al., 2011. Living with the enemy: parasites and pathogens of the ladybird Harmonia axyridis. Biocontrol 56, 663–679.

Simberloff, D., et al., 2013. Impacts of biological invasions: what's what and the way forward. Trends Ecol. Evol. 28, 58–66.

Strayer, D.L., 2012. Eight questions about invasions and ecosystem functioning. Ecol. Lett. 15, 1199–1210.

Vilà, M., et al., 2006. Local and regional assessments of the impacts of plant invaders on vegetation structure and soil properties of Mediterranean islands. J. Biogeogr. 33, 853–861.

Vilà, M., et al., 2011. Ecological impacts of invasive alien plants: a meta-analysis of their effects on species, communities and ecosystems. Ecol. Lett. 14, 702–708.

Williams, R.J., Martinez, N.D., 2000. Simple rules yield complex food webs. Nature 404, 180–183.

CHAPTER ONE

Impacts of Invasive Species on Food Webs: A Review of Empirical Data

P. David*,1, E. Thébault†, O. Anneville‡, P.-F. Duyck§, E. Chapuis¶,‖, N. Loeuille†

*Centre d'Ecologie Fonctionnelle et Evolutive, UMR 5175, CNRS-Université de Montpellier-UMIII-EPHE, Montpellier, France
†Institute of Ecology and Environmental Sciences UMR 7618, Sorbonne Universités-UPMC-CNRS-IRD-INRA-Université Paris Diderot-UPEC, Paris, France
‡UMR 42 CARRTEL, INRA-Université Savoie Mont Blanc, Thonon-les-Bains, France
§CIRAD, UMR PVBMT, 97410 Saint Pierre, La Réunion, France
¶UMR 3P PVBMT, CIRAD-Université de la Réunion, Saint Pierre, France
‖UMR IPME, IRD-Université de Montpellier-CIRAD, Montpellier, France
1Corresponding author: e-mail address: patrice.david@cefe.cnrs.fr

Contents

Advances in Ecological Research, Volume 56
ISSN 0065-2504
http://dx.doi.org/10.1016/bs.aecr.2016.10.001

Abstract

We review empirical studies on how bioinvasions alter food webs and how a food-web perspective may change their prediction and management. Predation is found to underlie the most spectacular damage in invaded systems, sometimes cascading down to primary producers. Indirect trophic effects (exploitative and apparent competition) also affect native species, but rarely provoke extinctions, while invaders often have positive bottom-up effects on higher trophic levels. As a result of these trophic interactions, and of nontrophic ones such as mutualisms or ecosystem engineering, invasions can profoundly modify the structure of the entire food web. While few studies have been undertaken at this scale, those that have highlight how network properties such as species richness, phenotypic diversity, and functional diversity, limit the likelihood and impacts of invasions by saturating niche space. Vulnerable communities have unsaturated niche space mainly because of evolutionary history in isolation (islands), dispersal limitation, or anthropogenic disturbance. Evolution also modulates the insertion of invaders into a food web. Exotics and natives are evolutionarily new to one another, and invasion tends to retain alien species that happen to have advantage over residents in trophic interactions. Resident species, therefore, often rapidly evolve traits to better tolerate or exploit invaders—a process that may eventually restore more balanced food webs and prevent extinctions. We discuss how network-based principles might guide management policies to better live with invaders, rather than to undertake the daunting (and often illusory) task of eradicating them one by one.

1. INTRODUCTION

Over the past half century there has been increased concern about the impacts of biological invasions (Gurevitch and Padilla, 2004). Examples abound of invasions studies detailing spectacular impacts of invasions on whole ecosystems. The impacts are diverse and, depending on the variables considered, not necessarily negative (Gallardo et al., 2016; Jeschke et al., 2014). For example, while introduced filter-feeding species deplete planktonic communities, they improve water quality and create substrates that favour benthic invertebrates and macrophytes (Higgins and Vander Zanden, 2009). Invasive species must also exploit the resources available in the recipient ecosystem, and thus establish trophic interactions with the resident species. Among all various ways to describe an ecosystem, we believe that networks, and especially food webs, which describe the trophic links among species in a system (Montoya et al., 2006), are particularly interesting for studying the impacts of invasions. Indeed, focusing on interspecific interactions is necessary to understand how impacts might (or not) propagate to the entire system. Food webs themselves are simplifications of the ecology of systems; they usually ignore spatial subdivisions, nontrophic interactions (allelopathy, agonistic interference, creation of habitat, etc.) and mutualisms. Yet they can be rich enough to capture the information necessary to interpret and/or predict impacts of invasive species. For this review, our objective is to provide an overview of what is known about the impact of invasive species on food webs. We particularly focus on how these impacts are transmitted through trophic interactions, and when/where they are more likely to occur. Our review is divided into four sections. The first section defines the key terms used in this chapter; the second reviews studies of trophic interactions between one invasive species and one or a few resident species; the third section looks at more integrative studies focusing on how invasions may alter the whole food web and the properties that make a food web more or less robust to invasions. Finally, the fourth section explores how food web "thinking" may help in the management of biological invasions.

2. DEFINITIONS AND LIMITS

Most important definitions have been collected in Glossary; corresponding words are highlighted in bold in the text.

2.1 Invasive Species: An "Anthropocentric Concept"

There are many definitions of an **invasive species**, which use different combinations of criteria based on origin, demography, and impact (Blackburn et al., 2011; Gurevitch et al., 2011). For example, a species is sometimes called invasive when, regardless of its origin, its growth is not strongly regulated, resulting in monoculture or community dominance. In aquatic ecosystems, bloom-forming cyanobacteria could be considered invasive, by this definition, as they proliferate because of human-induced changes in abiotic conditions that provide them with a competitive advantage (Carey et al., 2012). However, the term "invasive" is more often used for nonnative species whose expansion can cause economic or environmental harm and/or has negative effects on public health (Definition approved by the Invasive Species Advisory Committee, April 27, 2006). The difficulty with this definition resides in both the "nonnative" and "harmfulness" notions. Regarding the former, changes in geographic range may reflect different processes. Although species invasions are often considered to be an anthropogenic disturbance linked to growing commercial transportation and habitat anthropization (Murphy and Romanuk, 2014), species distributions have always changed in time and colonization is also a natural component of natural systems such as metapopulations (Hanski, 1999; Levins, 1969) and metacommunities (Leibold et al., 2004). Consequently, whether a species is considered native depends on the time of introduction. For this reason, in this chapter, we will distinguish between "**native species**" (present in the community since prehuman times) and **resident species**, which include ex-invasives that have become established locally. It should be noted that the notion of "resident" is relative to a particular time and invader; when contrasting "invasive" and "resident", the residents are all species that were established prior to the arrival of a particular invasive.

The impact of an introduction can be difficult to evaluate and predict. Despite the development of tools such a bioeconomic modelling frameworks (Leung et al., 2002), impact assessments are strongly influenced by human expectations of ecosystem services, that depend on the stakeholder point of view or sector of society, and on the spatial and temporal scales considered. The definition of "invasive species" based on harmfulness thus appears "anthropocentric" and subjective despite its use in a policy context. As a consequence, for the purpose of this review, we have chosen not to include the notion of "harmfulness" in our definition of invasive species. **In the following we adopt the convention of using "invasive species" to characterize any nonresident species, intentionally or accidentally introduced, that is able to maintain, spread, and reproduce in the new habitat** (Blackburn et al., 2011).

When introduced, a species may persist only if it is able to pass through environmental and biotic **filters**. Environmental filters include all the abiotic conditions that determine the range of physicochemical properties, often called the fundamental niche, that make a new habitat suitable for a species to complete its life cycle. As an example, temperature is a major environmental factor constraining the distribution of organisms (Gunderson and Leal, 2016). Biotic filters include the level and availability of resources, competition, and natural enemies, which define the realized niche. Species that have been able to pass through these filters may not come alone. There are numerous documented examples of parallel invasions, frequently of species from the same area of origin (Ricciardi and MacIsaac, 2000).

2.2 Measuring Impacts on Food Webs: Objects of Study and Methodology

There have been two ways to envisage the impact of an invasive species on a resident community; one—by far the most common—is through a simple, reductionist, species-centred approach, while the other is a holistic, **food web**-based approach. **Food webs** are complex structures, which can be summarized as graphs with species or groups of species as nodes connected by trophic links ("is eaten by" or, sometimes, "is parasitized by"), possibly with some quantification of the intensity of transfer of energy or matter through these links.

The idea of the species–centred approach is to study just the trophic links through which an invasive species A influences a resident species B, focusing on particular species known to have changed in frequency, abundance, or diet after invasion. Experimental approaches of this kind include comparisons of diet and abundance of focal species before and after and invasion into the community, and manipulative addition or depletion of populations in controlled environments or in situ. Such approaches can be complemented by a dynamical model of the abundances and interactions among the studied species (e.g. Lotka–Volterra-type models) to identify key mechanisms (Courchamp et al., 2000). Fig. 1 illustrates the fact that trophic links may be direct (bottom-up or top-down effects, one step away from the invasive species in the food web) or indirect, with one or more intermediate steps

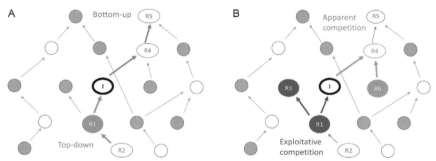

Fig. 1 An idealized food web to illustrate direct and indirect trophic links through which an invasive species (I) can affect resident species (R). *Arrows* go from prey (or host) to predator (or parasite). *Open circles* represent species that are expected to increase upon arrival of the invader, *filled circles* represent species that are expected to decrease. Types of interactions are as follows, and for clarity vertical effects are highlighted in (A), while horizontal effects are highlighted in (B) (although they exist simultaneously). Vertical effects include direct top-down (R1) and bottom-up (R4) effects and indirect top-down (R2) and bottom-up (R5) trophic cascades. Horizontal effects are only indirect and include exploitative (R3) and apparent (R6) competition. We highlighted only one- and two-step impacts of invasive species but effects may propagate further (*blue circles*). The positive or negative direction of the impact will depend on the relative positions of I and R in the food web. Considering the trajectory from I to R, if it includes an even number of steps opposite to *arrows*, the impact will be positive; if the number of such steps is odd, it will be negative (parity effects). Note that real food webs are much more connected and complex than this idealization; often, more than one trajectory links together the invasive species to a particular resident (e.g. intraguild predation, change in trophic position with ontogeny), and the prediction of effects becomes very complex.

involving other species than the invasive and the resident (cascades, **exploitative competition**, and **apparent competition**). The complexity of most real food webs makes it difficult to predict how the impacts of an invasive species may propagate more than two steps away from it, as the number of connected nodes rapidly increases with additional steps. Thus, in practise, the species–centred approach is usually limited to the study of resident species at two steps or less from the invasive (red circles in Fig. 1).

The principle of the food-web-centred approach is to compare food webs before and after invasion, or to compare the position of invasive species to that of residents in a food web. Research on ecological networks has underlined the role of some general characteristics of food webs such as the distribution of the interaction strength, diversity, or **connectance**, in controlling the productivity and stability of food webs (Ings et al., 2009). Sophisticated algorithms are now available to measure such variables, and undertake cross-network comparisons. The limiting step remains the acquisition of data; a challenging and resource consuming task, which explains the small number of empirical studies of invasion into food webs. In addition the available networks have some strong biases. For instance, plant–insect–parasitoid networks are relatively short and, being based on the sampling of vegetation and the animals found within it, they tend to ignore those herbivores and predators that do not permanently reside on plants (Carvalheiro et al., 2008; Heleno et al., 2009). Other food webs include more functional groups but changes after invasion are inferred mostly from species abundance, combined with expert knowledge on feeding habits; thereby assuming that changes in species diet follow changes in abundance, as expected from optimal-foraging theory (Charnov, 1976). This is a valuable first step and most available data are of this kind. A few observational studies go further and track changes in species diet through isotopes analysis, fatty acid, or stomach contents (Deudero et al., 2014; Vander Zanden et al., 1999). Although trophic characterization is challenging, recent molecular methods will provide unprecedented opportunities for this kind of sampling (Kamenova et al., 2017; Mollot et al., 2014; Pompanon et al., 2012; Vacher et al., 2016).

3. LOCAL EFFECTS: EFFECT OF INVADERS AT ONE OR TWO STEPS OF DISTANCE

Introduced species inevitably create new trophic links, because they eat and/or are eaten by resident species. Through these links, they affect the demography and abundance of species around them and these effects may propagate at two or more steps of distance in the network (e.g. to the prey of their prey, the predator of their predator). Based on this simple premise, there is an abundant literature examining the impacts of an invasive species on one or more resident species. These studies examine local impacts on the resident food web. Changes in the overall structure of food webs are not the focus, and there is no attempt to exhaustively describe them. In this section, we review these approaches using case studies to highlight the common themes and to answer the question "what makes the interaction between invasive and resident different from similar interactions among residents"?

HIGHLIGHTS

— *Impacts of invaders on a few interacting species in the recipient ecosystem are more often studied than impacts at the whole food web scale.*

— *Direct predation underlies the most spectacular impacts, and invasive predators can drive resident species to extinction (Section 3.1). Competition (Section 3.3) and apparent competition (Section 3.4) with invasive species lead to declines in resident species, but rarely provoke extinctions.*

— *Invaders are often an important new resource benefitting the higher trophic levels in the recipient ecosystems, but this positive effect may be opposed or reversed by the decline in local prey due to competition with the invaders (Section 3.2).*

— *Predation or competition between invaders and residents are often asymmetrical, with more negative impacts of invaders on residents than the reverse (Sections 3.1 and 3.2). This situation puts a selective pressure on resident species which often evolve traits to better tolerate or exploit the invasive species.*

— *Introduced and resident species have no coevolutionary history, which may result in extreme reciprocal impacts one way or the other. Asymmetry in impacts, in favour of invaders over residents, most likely results from a filtering process whereby only introduced species that overall benefit from interactions with residents become successful invaders (Section 3.1).*

3.1 Top-Down Effects

3.1.1 *Invasive Predators May Have Large Impacts on Resident Species*

Direct predation is the predominant mechanism by which invaders can dramatically decrease populations of indigenous species or even cause their extinction (Bruno et al., 2005). The most spectacular cases come from insular ecosystems; classical examples include the snake *Boiga irregularis*, introduced accidentally in the Guam island, where it quickly devastated forest bird populations (Savidge, 1987) or the predatory snail *Euglandina rosea*, introduced in Moorea (Pacific) to control populations of the introduced snail *Achatina*, but that instead wiped out endemic *Partula* snails (Clarke et al., 1984). Invasion of lakes by large fish species that become top predators provide other examples: catch rates of local fishes were six times lower after the invasion by smallmouth bass (*Micropterus dolomieu*) and/or rock bass (*Ambloplites rupestris*) in Canadian lakes (Vander Zanden et al., 1999, see also Arthington, 1991). Planktonic communities of lakes previously devoid of fish undergo similarly large changes after planktivorous fish are introduced (Reissig et al., 2006). In some cases, although indigenous prey do not go extinct, invasive predators displace them from their usual range or habitat; thus, the predatory cladoceran *Bythotrephes longimanus* causes some copepod taxa to escape to deeper and colder water layers in lake Michigan (Bourdeau et al., 2011). At the community level, a meta-analysis (Mollot et al., 2017) shows that addition of predators is significantly associated with decreases in resident species diversity in both terrestrial and aquatic ecosystems.

3.1.2 Ecological and Evolutionary Naïveté Exacerbate the Impact of Invasive Predators

A common theme in all these examples is that strong impacts occur because of "prey inexperience". The invader may be a formerly absent type of predator (i.e. tree-climbing snakes in Guam, predatory snails in Moorea) against which local species have no defence, or a large-bodied species that eats the previously top predators. Prey inexperience exists both in an ecological and in an evolutionary sense. For example, the absence of fish in a lake may represent a local, transient stage in a dynamical metacommunity with extinction-colonization dynamics (Gravel et al., 2011; Massol et al., 2017); the resulting community is "inexperienced" in an ecological sense, dominated by zooplankton transiently taking the top-predator position. On the other hand, the sensitivity of insular ecosystems to predatory invaders is an evolutionary inexperience due to isolation and lack of predators for up to millions of years; local species have not evolved—or have lost—their defences during evolution. The evolution of flightless or ground-nesting birds in remote islands, making them hypersensitive to the invasion of predators, represents an extreme case.

This situation poses a symmetry problem: why does a lack of previous coevolutionary experience seem to benefit the invasive predator more than the native prey in general? After all, the invader is as inexperienced as a predator, lacking specific foraging, or capture strategies, as the indigenous species is as a prey item, lacking specific defence. Thus, the impact of exotic predators on native prey might often be either dramatic or very slight, but not on average more negative that when predator and prey have coevolved together. This conception is illustrated in Fig. 2 showing a higher variance in interaction strength between noncoevolved pairs, but the same mean as in coevolved pairs. However, invasive species seem to exploit their prey more rapidly and efficiently than natives (e.g. Morrison and Hay, 2011) and can reduce more severely native prey than do native predators (Salo et al., 2007). Several studies have shown that the functional response coefficients (i.e. the relationship between resource density and consumer consumption rate) of invasive species were generally higher than those of comparable native species (Barrios-O'Neill et al., 2014; Dick et al., 2013). We hypothesize that this asymmetry results from the **"invasion filter"**. Invasive–native pairs are a nonrandom subset of all pairs of noncoevolved predator and prey because inefficient predators simply fail to invade. As a result, one expects an excess of asymmetrical predator–prey interactions in which invaders have a large impact on their prey (Fig. 2). For a similar reason invasive predators

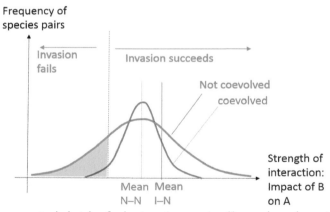

Fig. 2 An conceptual sketch of why invasive species (I) may have large impact on natives (N) through predation or parasitism. The strength of the interaction between two species can be measured by the impact of species B on the growth rate of A (predation efficiency). We expect that random noncoevolved species pairs have on average more extreme interactions (e.g. prey is totally undefended, or predator is totally unable to exploit it) than coevolved pairs. Introduced-native (I–N) pairs are not coevolved, but among introduced species only those that have a relatively high predation efficiency manage to invade (a process here represented for convenience as an invasion threshold). As a result I–N species pairs show higher average predation efficiency than native–native pairs (N–N), including a few extreme cases (right of the distribution) where the predator may drive the prey to extinction.

are often generalist (see Section 4): indeed, specialists have less chance to find exploitable prey outside their area of origin, and more often fail to invade. Importantly, unlike abiotic conditions such as temperature, this filter is expected to change as more and more invaders are incorporated as residents, with which any newcomer will have to interact. This filter is further amplified by an investigation bias, as researchers prefer to focus on invaders known to have had spectacular impact and/or proliferation.

One corollary of these asymmetrical interactions is that invasion should put a high selection pressure on resident species to evolve defences—if they are not driven to extinction first. We therefore expect rapid evolution in the resident prey while invasive predators should not experience (or not immediately) as much of an evolutionarily challenge, as they have already been filtered. Studies of rapid evolution following invasion are now accumulating (Strauss et al., 2006). They document several examples of rapid evolution of defence in native prey, such as constitutive or inducible shell thickening of marine molluscs after invasion by predatory crabs (Freeman and Byers, 2006; Seeley, 1986; Vermeij, 1982). While we know of no example of traits evolving in invasive species to increase predation efficiency on natives, it is still too early to draw general conclusions. Recent modelling suggests that it is only when invasive predator traits lie within a narrow window that evolution can help in successful invasion (Jones and Gomulkiewicz, 2012). Thus, the traits of invaders, including predatory efficiency, most likely result from filtering than from evolutionary change.

Finally, it is important to note that some resident species are themselves recent invaders. In this case, a new invader may sometimes share some coevolutionary history with recent invaders more than with native species. This is the case for example of biocontrol agents, which can be considered as intentional invasions of predators or parasites. Drawing on the lessons of known catastrophic effects of generalist invasive predators, biocontrol policies are now restricted to the release of highly specific, coevolved, enemies of a target pest species, such as specific plant-eating insects or parasitoids (see Section 3.3). However, there is no absolute certainty about the specificity of biocontrol agents. For example, parasitoid wasps originally introduced to control pests now form the vast majority of attacks on native insects in Hawaiian swamps (Henneman and Memmott, 2001).

3.1.3 Top-Down Effect of Invasions May Result in Trophic Cascades

The top-down impacts of an introduced predator may propagate several steps down a food chain in what is often called a cascading effect (White et al., 2006). The typical expectation is a "parity effect": species linked to the introduced predator through an odd number of steps, such as its prey, are negatively affected, while species at an even number of steps, such as the prey of its prey, are positively affected. Several examples have been documented in the invasion literature, with most being related to introduced top pred-ators (Gallardo et al., 2016). For example, the introduction of zooplanktivorous fish in a fishless pond can cascade down to an increase in small zooplankton. Rotifers increase as their zooplankton predators, *Daphnia*, and predatory copepods are eaten by fish, and particular groups of cyanobacteria phytoplankton become abundant as phosphorus is no lon-ger sequestered by the *Daphnia* (Reissig et al., 2006, see also Carey and Wahl, 2014). Strong cascading effects seem particularly frequent in aquatic ecosystems with algal-based primary production, perhaps because of the rel-atively high efficiency of basal links in these food webs and their relatively low diversity. Here a high proportion of the primary production is effi-ciently consumed by primary consumers and energy transfer is shared among a small set of abundant species, effectively forming simple food chains (Shurin and Borer, 2002; Shurin et al., 2006; Strong, 1992). Trophic cas-cades have even been proposed as a method for lake management, called biomanipulation (Shapiro and Wright, 1984). The idea would be to increase piscivore fish, which will decrease planktivore biomass, increase herbivore biomass, and decrease phytoplankton, resulting in higher water transpar-ency. Planktivorous fish would be removed by intensive netting and the lakes then restocked with piscivorous fish. Although strong effects of pisciv-orous fish have been observed, the results from this method have been mixed and its success requires a particular set of conditions (shallow lake, macro-phytes, etc.) and a deep understanding of local aquatic communities.

The likelihood of observing trophic-cascade effects appears lower in more complex, reticulate food webs. Evidence from terrestrial ecosystems points to frequent cascading effects of carnivores on plant damage and plant biomass, but not as strong as in aquatic ecosystems (Shurin et al., 2006), where it can lead to a rapid shift in the dominant group of primary producers (Carlsson et al., 2004). To some extent, all biocontrol strategies can be con-sidered as invasion experiments that rely on trophic cascades: for example, the aim may be to increase the yield of a cultivated plant by introducing a parasitoid of its insect enemy. However, cultivated ecosystems are artificially simplified (monocultures with low diversity of predators and few trophic levels), and in that respect are much more likely to display clear trophic cas-cades than natural, more complex terrestrial food webs.

3.2 Lateral Effects of Invaders: Exploitative Competition

3.2.1 Exploitative Competition is Expressed as a Two-Step Path in a Food Web

Exploitative competition occurs between invasive species and residents that lie at a two-step distance in the food web as they share the same resource or prey. It is often cited as a mechanism for the impacts of invasive species. However, although many studies measure the impact of invasives on resident species, the common resource is rarely identified or studied (White et al., 2006). Indeed, following the population dynamics of three interacting species in order to demonstrate exploitative competition is practically quite difficult. However, examples have been demonstrated in different organisms such as geckos (Petren and Case, 1996), snails (Byers, 2000), and fishes (Arthington, 1991; McCrary et al., 2007). The relatively low number of fully documented examples probably reflects empirical difficulty, rather than a truly low frequency; the list of resources that two species may compete for is often long, making it impossible to follow them all. Researchers may prefer to put their effort into investigating interactions that are considered more interesting, such as facilitation or apparent competition, than in illustrating a classical concept that has achieved consensus but remains difficult to fully demonstrate (or exclude). Yet competition, old-fashioned though it may be, is probably responsible for many changes in numerical dominance within a guild or trophic level, whereby an invasive species becomes quantitatively dominant and reduces resident competitors to a subordinate position (Reitz and Trumble, 2002). Accordingly, the effects of invaders tend to be negative on the diversity of species that occupy the same trophic level, while also tending to be positive on diversity of species at higher trophic levels (the **"relative trophic position hypothesis"**; Thomsen et al., 2014).

3.2.2 Extinctions by Competition Between Introduced and Native Species Are Relatively Rare

Complete extinction of native species as a consequence of competition is, however, relatively rare (Bruno et al., 2005; Davis, 2003; Gurevitch and Padilla, 2004). One of the possible explanations for this lies with the diversity of resources used by native species. Invaders are typically generalists, and the chance that the invasive is competitively superior on each and every resource is low. Resident species often have "exclusive" links with members of the resident community that are not exploited by the invader, providing them with a refuge. Residents may therefore decrease in abundance, but survive because there remain exclusive prey items. In this case, invasion by a competitor results in **ecological displacement**, with the local species restricting its realized niche to coexist with the invasives. For example, Tran et al. (2015) showed that coexistence between the invasive topmouth gudgeon, *Pseudorasbora parva*, and native fish species was possible due to niche divergence. This was shown by a trophic analysis using stable isotopes analyses, both in the wild and in the microcosms. Decreases in niche breadth have also been documented for native ants (Human and Gordon, 1996) and bees (Thomson, 2004) following invasion by competing species. All these processes tend to reduce the dietary overlap between invasive and resident species. Interestingly, this conclusion does not extend to **interference competition**, in which direct attack-and-defence mechanisms are at play and extinction is more likely, as in predator–prey interactions. For example, invasive, clonal, or quasi-clonal ants that are particularly aggressive towards other species, while not being so towards conspecific colonies, may locally exterminate other ant species (Holway et al., 2002).

3.2.3 Asymmetry in Competition Impacts, Lack of Coevolutionary History, and Invasion Filter

In the relatively few, detailed examples of exploitative competition that have been published, the invader has a competitive advantage to exploit the most abundant resource, due to better conversion efficiency of resources and/or superior harvesting ability (Byers, 2000). We believe that this advantage reflects the same process as for invasions by predators (see earlier): the lack of coevolution may result in a large variance in reciprocal competitive impacts between introduced and natives, but the overall bias in favour of invaders results from the filtering of successful invaders. Again, we expect that local species face an evolutionary challenge. Although most residents escape extinction by leaving the most abundant resource to an invader with superior competitive ability, and switching to a minor resource, this situation modifies the overall selection pressure they experience and encourages specialization. Thus, investing in traits that optimize the exploitation of the minor resource, possibly at the expense of losing the ability to exploit the resource recently monopolized by the invader, will be selected. As an example of this **evolutionary character displacement**, limb morphology has evolved in a native Caribbean lizard when it was restricted to terminal branches of trees, by an invasive species that exploited the main tree trunk (Stuart et al., 2014).

3.2.4 Exploitative Competition May Be Mixed With Other Interactions

Cases of mixed **exploitative competition** and **interference competition** between invasive and native species have been described (Crowder and Snyder, 2010). Interference may reinforce the asymmetry of competition and the impact on native species (Amarasekare, 2002). An extreme example of such interaction is intraguild predation, where the invasive species feeds on its competitor. Intraguild predation seems to be involved in many cases of invasions by insect generalist predators (Crowder and Snyder, 2010). Similarly, introduced crayfish in streams eat both invertebrates and decayed vegetation, and thus, affect invertebrates both through competition and through predation (Bobeldyk and Lamberti, 2008).

3.2.5 The Case of Invasive Fruit Flies Illustrates Asymmetric Competitive Interactions Between Invaders and Residents

The family Tephritidae (true fruit flies) is invasive worldwide (Duyck et al., 2004) and will serve us as an example of how invaders affect residents through competition. Where polyphagous tephritid species have been introduced in areas already occupied by a polyphagous tephritid, interspecific competition has repeatedly resulted in a decrease in abundance of the resident species, rather than extinction (Duyck et al., 2007). These invasions appear to follow a "hierarchy" and no reciprocal invasions have yet been observed (Fig. 3). This supports our hypothesis that invaders are filtered to competitively dominate residents and illustrates the dynamic nature of this filter, as each new invader, once incorporated into the community, adds a new constraint that introduced species have to overcome to successfully invade.

This hypothesis is further supported by the detailed study of the island of La Réunion. Four very similar fruit-infesting tephritids inhabit the island, one endemic and three introduced species that have successively invaded from 1939 to 1991. Asymmetrical, hierarchical interactions of **exploitative competition** among species, both as larvae and as adults, and adult **interference competition**, have been demonstrated (Duyck et al., 2006a,b). Recent invaders competitively dominate old ones in these lab experiments. Traits such as the production of fewer, but larger juveniles, delayed onset but longer duration of reproduction, longer lifespan, and slower senescence are associated with competitive ability (Duyck et al., 2007). While each invasion has led to a reduction in abundance and niche breath of a previously established species, climatic, and host plant niche differentiation have eventually allowed coexistence (Duyck et al., 2006a,b).

The lack of reciprocal invasions (Duyck et al., 2004) remains to be explained. If, for example, species A can coexist with a new invader, B,

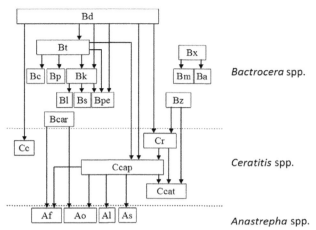

Fig. 3 A hierarchy of asymmetrical invasion links among polyphagous tephritids. Each *arrow* represents the direction of invasion and competitive **displacement** of a resident tephritid species by an introduced one. The residents can be native or previous invaders. A single link may represent two or several independent occurrences of the same invasion history (such as Hawaii, 1945; and Kenya, 2003 for *Bactrocera dorsalis* on *Ceratitis capitata*). These links follow a hierarchical structure, as they are transitive but unidirectional (no reciprocal links). *Af, Anastrepha fraterculus; Al, A. ludens; Ao, A. obliqua; As, A. suspensa; Ba, Bactrocera atra; Bc, B. curvipennis; Bcar, B. carambolae; Bd, B. dorsalis; Bk, B. kirki; Bl, B. luteola; Bm, B. melanota; Bp, B. psidii; Bpe, B. perfusca; Bs, B. setinervis; Bt, B. tryoni; Bx, B. xanthodes; Bz, B. zonata; Cc, Ceratitis cosyra; Ccap, C. capitata; Ccat, C. catoirii; Cr, C. rosa. Reproduced with permission from Duyck, P.F., David, P., Quilici, S., 2004. A review of relationships between interspecific competition and invasions in fruit flies (Diptera: Tephritidae). Ecol. Entomol. 29 (5), 511–520.*

by exploiting a particular host plant or microclimate, why is A not able to invade these specific environments when B is present? We suggest that in the case of the tephritids in La Réunion, and perhaps more generally, the opportunity to invade is concentrated on a particular habitat and resource. In La Réunion these consist of human modified, cultivated lowlands (a rich and warm habitat). This habitat may act as a gateway, where any candidate invader must be able to establish a viable population and resist competition from residents, before spreading to other habitats (Fig. 4). Here, the **invasion filter** therefore favours species that are competitively superior in the most abundant or accessible resource, irrespective of their performance elsewhere. Their performance in other habitats will be, however, important in determining their prospects of coexistence with future invaders that may outcompete them for the major resource.

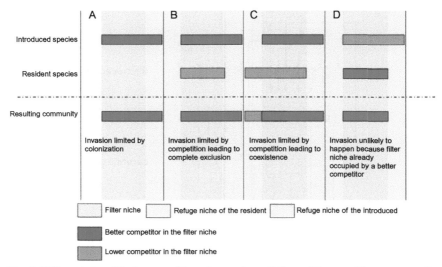

Fig. 4 Different simplified cases of invasion and prediction of the resulting community according to the presence of a resident species, to the competition ability, and to the niche breadth of this species. The "filter" niche corresponds to the habitat/resource in which the introduced species needs to establish a viable population (and thus resist to competition by the resident species) before spreading in other habitats. The "refuge" niche of a species X corresponds to habitats where it competitively dominates the other species ($X > Y$). (A) No resident species, invasion requires only the ability to colonize the filter niche. In the other three cases, in the presence of a resident, the introduced species must competitively dominate the resident in the filter niche to invade (invasion limited by competition). (B) The resident species does not have refuge: it is excluded by the introduced species. (C) The resident species is displaced to its refuge niche. (D) The introduced species could survive in its refuge niche but does not reach it because it is dominated by the resident in the filter niche.

3.3 Bottom-Up Effects of Invaders

3.3.1 Invaders Provide Direct Benefits but Indirect Costs to Local Predators

Invasive species represent a new, often abundant, resource to local predators or parasites. As such, they are expected to have direct positive consequences on resident species at higher trophic levels, and the residents may change their diet to exploit them. For example, endangered endemic water snakes from Lake Erie seem to draw large benefit from the introduction of gobies, which now make up more than 90% of their diet (King et al., 2006). However, bottom-up positive effects on local predators/parasites might be partly or completely offset, by negative indirect effects. The most obvious occurs through competition: if an invasive species competitively displaces a more profitable local prey, the net effect can be negative on the predator. Thus, the replacement of native ants by less profitable invasive Argentine ants (*Linepithema humile*) has contributed to a decline in ant-eating native lizards in California (Suarez and Case, 2002). In a meta-analysis, Pintor and Byers (2015) did not find invasive animals to be on average more or less profitable to local predators than residents, in laboratory studies, but there was a trend for positive effects on populations of local predators in the field. This suggests that the high abundance of invaders often outweighs indirect effects via competition. Reciprocally, healthy populations of resident predators are assumed to limit or prevent invasions and their impacts, according to the **"biotic resistance hypothesis"**. This is supported by experiments of predator removal leading to strong increases in invasive prey populations. For example, the predatory native blue crab, *Callinectes sapidus*, controls populations of invasive green crabs, *Carcinus maenas* (DeRivera et al., 2005), and possibly also of invasive whelks (Harding, 2003) and zebra mussels (Molloy et al., 1994) in North America. The same interaction between a native predator and a native prey can be viewed both as a positive bottom-up effect of invaders and as a **biotic resistance** exerted by the local species.

It would be incorrect, however, to be overly optimistic about positive effects of invaders on higher trophic levels. The effects shown in the meta-analysis of Pintor and Byers (2015) might be biased by an invasive species-centred view that systematically excludes predators unable to exploit a particular invasive species from the study; and by restricting the analysis to animals eating animals. Invasions by plants may not benefit herbivore invertebrate communities as much as animal invasions benefit carnivores. Indeed, unlike carnivores, invertebrate herbivores often include many specialists that may fail to exploit introduced plants; instead, the latter will be exploited by a few generalists, often themselves introduced, and the diversity of local herbivores will decrease. This has been observed for example in a suite of elegant studies of changes in herbivore insect and parasitoid communities following the spread of invasive plants (Carvalheiro et al., 2010; Heleno et al., 2009). These processes have two implications. First, we expect invasives to attract other invasives at higher trophic levels, and to see some clustering of invasives together in the resulting food webs; second, the introduction of specialist enemies of an invasive plant, taken from its area of origin, could be a way to—at least partly—offset its initial advantage on native competitors. The latter principle indeed forms the basis of modern biological control practise (McFadyen, 1998).

3.3.2 Coevolutionary History and Invasion Filter Underlie Enemy Release

The logic already applied to top-down and competition effects can also be applied to the relationship between invasives and their native enemies. The lack of coevolutionary history has often been expressed as the "**enemy release hypothesis**" (ERH), which states that invaders dominate their local competitors because they are not controlled by the enemies that used to keep them in check in their area of origin; and because the native enemies in the area of introduction have lower impact on them than on competing local species (Carpenter and Cappuccino, 2005; Parker and Gilbert, 2007). However, Colautti et al. (2004) showed that although the first premise is supported by data (as expected because the probability to introduce a species together with its enemies is not one), there is no significant statistical trend for the second statement. They argue that the empirical support for ERH has been overstated. Strikingly, one explanation made by Colautti et al. (2004) relies on the "increased susceptibility hypothesis", whereby invasives are disproportionately impacted by a few native local enemies—an argument that seems to be the exact opposite of the ERH. We believe that the solution of this paradox is that random noncoevolved species pairs should show more variance in interaction strength but no directional bias, compared to coevolved pairs. Rather, the bias again results from a filtering process: introduced species with defences against local enemies enjoy higher invasion success (see Fig. 2).

As a consequence, resident predator species should be more evolutionarily challenged than their invasive prey. Carlsson et al. (2009) provide arguments supporting this view. Indeed, invasive species often strongly and definitively alter the available prey base for local predators; they often come to constitute the main component of their diet, irrespective of their profitability as a prey. As a result, any trait change in the predator that facilitates exploitation of invasive prey comes with large fitness benefits. Of course, changes in diet may reflect the immediate consequences of the replacement of local prey by invasives, rather than changes in specific traits. However, trait changes are expected to happen through either phenotypic plasticity and learning or, more slowly, evolutionary change (Carlsson et al., 2009). For example, fruit-eating bugs (*Leptocoris tagalicus*) in Australia have rapidly evolved larger mouthparts that increase their efficiency at attacking the fruit of an invasive vine (Carroll et al., 2005). In other cases, predators evolve avoidance of toxic prey: some Australian snakes have evolved relatively smaller heads, which prevents them from eating large individuals of invasive cane toads, and dying from ingestion of high doses of toxin (Phillips and Shine, 2004). On the whole, most described cases of evolution of native species in response to invasion are cases of herbivores, and particularly phytophagous insects adapting to introduced plants (Strauss et al., 2006). Simberloff and Gibbons (2004) have proposed that adaptation of resident predators or parasites might be responsible for delayed declines of invasive populations after an initial proliferation (the "boom and bust phenomenon"). This remains to be fully tested, as other mechanisms (such as the recruitment of new invasive enemies) might also be important.

Interestingly, there are also reasons to expect evolution of invasives in response to a low impact of enemies—an evolutionary opportunity (relaxation of a constraint) rather than an evolutionary challenge (new constraint). The evolution of increased competitive ability (EICA) hypothesis (Blossey and Notzold, 1995) proposes that invaders can reallocate resources from defence into growth and development. This intuitive hypothesis could explain the existence of lags between introduction and proliferation of invasive species, and why their competitive ability seems usually higher in invaded compared to native areas. While EICA has been often mentioned to explain plant invasions (Maron et al., 2004a,b; Vilà et al., 2003), it has rarely been fully validated and therefore its true importance is currently unknown (Gurevitch et al., 2011).

3.4 Apparent Competition Between Invaders and Residents

Apparent competition is an indirect effect that occurs when one a species has a negative impact on another through the increase in abundance of a common predator or parasite (Holt, 1977). As noted earlier, invasive species often boost the population of generalist predators (resident or introduced) and can therefore be expected to contribute to the **displacement** of their resident prey. Many spectacular cases, intuitively interpreted as competitive displacement, may in fact involve apparent competition (in addition to direct competition). Apparent competition will, however, not systematically result in an asymmetrical advantage to an invasive species. For this, at least one of two conditions must be fulfilled: (i) the invasive species may have an advantage in direct competition that explains its increase in relative abundance, which is then further amplified by apparent competition and/or (ii) the invasive species must be profitable enough to sustain large populations of predators or parasites, with the latter possibly having a stronger negative impact on resident competitors. Thus, natives can face increased predation by resident predators when compared to the preinvasion state. Roemer et al. (2002) illustrated this case by modelling the indirect consequences of invasions of the California Channel islands by feral pigs. The invasion made it possible for a top predator, the golden eagle, to establish permanent populations on the islands where it exerted strong predation on terrestrial carnivores (foxes). The decrease in fox populations in turn allowed carnivore competitors (skunks) to increase in abundance, these being relatively protected against eagle predation by their nocturnal habit. Predator–mediated apparent competition can also result in "**hyperpredation**" (Courchamp et al., 2000) and is the main reason why introducing predators to control invasive species (e.g. cats to control rabbits in an island) is usually a bad idea. Invasive populations may be demographically or behaviourally more tolerant to predation than insular endemics (e.g. birds). Similar situations have been mentioned for invasive plants, which may eliminate other plants by maintaining shared herbivore populations but being less attacked than the natives. This can occur if they provide some resource other than food to the herbivores, such as shelter. For example, the introduced seaweed, *Bonnemaisonia hamifera*, provides refuge to herbivorous crustaceans against fish predation. In turn these crustaceans attack competitor seaweed (Enge et al., 2013). Parasites and pathogens can also mediate apparent competition, with potential devastating effect on local species when the invader arrives together with a pathogen from its area of origin, against which local species have no immunity. For instance, squirrels (Tompkins et al., 2003) or ladybirds (Vilcinskas et al., 2013) introduced into Europe seem to replace local competitors through the activity of viruses and microsporidiae, respectively, carried from their homelands. These cases are often referred to as pathogen "spill-over". Colautti et al. (2004) and White et al. (2006) list some more examples of apparent competition between invasives and natives.

The **biotic resistance** offered by healthy predator populations against invasion, necessarily represents a form of apparent competition exerted by resident prey. Like **exploitative competition**, apparent competition is difficult to demonstrate. When demonstrated it is often presented as an exciting, neglected process (White et al., 2006); however, both exploitative and apparent competition are ubiquitous and occur wherever prey and/or predators are shared between species. In other terms, every time a species is shown to be under top-down control by a generalist predator, it is necessarily a case of apparent competition by the other prey of the same predator. The null hypothesis would be better stated as exploitative and apparent competition exist (possibly together), rather than that they are absent.

Logically, the asymmetrical impacts of invaders through apparent competition should be explained by the same mechanisms as the ERH: a mixture of the lack of coevolutionary history and filtering. Here, the **invasion filter** tends to select introduced species that tolerate local generalist predators better than residents; as well as those that carry easily transmitted pathogens against which they have an efficient immune defence. Interestingly, this hypothesis would predict that resident species may often be challenged to evolve increased defence not only against introduced predators and parasites but also against resident generalist predators whose populations are boosted by the arrival of an invasive species. We are, however, not aware of any such example and this might be due to a lack of investigation of this kind of evolution and/or it being interpreted as the indirect consequence of invasion by a competitor.

3.5 Facilitation, Mutualisms, and Engineering: Nontrophic Indirect Interactions

Trophic links do not underlie all the indirect impacts of invaders. In the literature (White et al., 2006) indirect effects is a general term for effects conveyed by trophic interactions but requiring two or more steps (competition, cascades, see earlier) and modifications of species abundances due to providing or removing something other than a prey or a predator of the focal species. This includes agonistic interactions such as **interference competition**, but also some forms of mutualisms including habitat creation, bioturbation, mycorrhization, and pollination, as well as changes in physicochemical variables such as water clarification and capture of nutrients. Ecosystem engineers that modify or create new habitats (e.g. zebra mussels, and in general many reef-building species) have the potential to change the base of primary production in an ecosystem (e.g. by clarifying water) and facilitate the installation of many sessile or benthic species (Bruno et al., 2005). This is illustrated in the meta-analysis of Mollot et al. (2017), which shows that introduced detritivores are associated with increases in biodiversity in aquatic ecosystems; the detritivore category indeed includes some of the well-known filter-feeding invasive ecosystem engineers such as bivalves. Here, we do not try to review these interactions, given our focus on the consequences of trophic interactions of invasive species with the resident species. However, ecosystem engineering by some invaders has spectacularly altered entire communities and food webs and will be, as such, mentioned in Section 4.

4. GLOBAL EFFECTS: INVASIONS AT FOOD WEB SCALE

HIGHLIGHTS
- *Diversity is positively related to robustness to invasion, but other food web attributes need to be considered. Connectance is the main property that has been investigated so far, but its effect on food web robustness is not yet clear (Section 4.1).*
- *The relation between trophic position in food webs and invasiveness seems context-dependent and depends on where "available niches" are in the native food web (Section 4.2).*
- *Effects of invasions can propagate through food webs and strongly modify food web structure. This occurs not only through species extinctions but also through changes in relative abundance of trophic levels, interactions between resident species, or ecosystem engineering activity. However, current evidence might be biased towards observations of spectacular effects (Section 4.3).*

As seen in the discussion of local effects, in Section 3, each invader interacts directly and indirectly with its neighbours in the food web. Do these impacts propagate further? To address this, we need to move beyond pairwise, food-chain interactions towards a food web approach of trophic interactions at the community scale. Effects of invasions on food webs will depend not only on the individual effect of one invader in the food web but also on how many introduced species succeed at invasion and where they insert themselves in the network. The questions developed in the context of local effects can then be reformulated in a network context, as:

- What makes a food web resistant to invasion?
- Is invasion success related to a peculiar trophic position of invasive species in food webs or are some parts of the food web more likely to be invaded?
- When do invasions strongly modify food web structure?

4.1 Food Web Structure as a Biotic Filter

4.1.1 Species Diversity Might Increase Food Web Resistance to Invasion

Among food web characteristics, species diversity is the primary attribute that has been related to resistance to invasion. Theory predicts that species-rich communities are less invasible due to a more complete use of available resources by diverse species (niche packing), leaving less resources for potential invaders (Kennedy et al., 2002; Tilman, 2004). Experiments on plant communities generally support this prediction (Tilman et al., 2014). However, most studies have focused on the basal trophic level, and fewer on higher trophic levels (Carey and Wahl, 2014; Shurin, 2000). The latter have mostly considered the effects of diversity on invasion success within a given trophic level (but see France and Duffy, 2006; McGrady-Steed et al., 1997). Most results match the theoretical predictions. For example, invasive common carp had a weaker negative effect on the growth of native fish when fish diversity was higher in a mesocosm experiment (Carey and Wahl, 2014). France and Duffy (2006) found that the diversity of crustacean resident grazers decreased the establishment success of other introduced grazers. Additionally, crustacean diversity decreased the natural colonization of other sessile invertebrates. This result suggests that diversity at one trophic level might have consequences on invasion resistance across trophic levels. Observational studies, however, suggest that species-rich communities support more invasive species (e.g. Levine and D'Antonio, 1999). Discrepancies between experiments and observational studies may reflect correlations between diversity and other factors affecting invasions in natural systems. Stachowicz and Byrnes (2006) suggested that species diversity decreased invasibility only when resource was limiting and when there were few foundation species. Other factors might determine invasibility at larger scale or in more complex communities. Mechanisms behind the diversity effect are based on interspecific competition, implicitly considering **horizontal diversity** (Duffy et al., 2007) rather than overall food web diversity. However, vertical links might also play a role. For example, Sperfeld et al. (2010) did not find effects of phytoplankton diversity on community invisibility, invasion success being instead related to nutrient supply and herbivory. Other components of food web structure thus require consideration in addition to species diversity.

4.1.2 Effects of Food Web Structure on Invasion: Beyond Diversity

Experiments manipulating complex food webs are scarce (but see Brown et al., 2011 or Gauzens et al., 2015) as describing food webs is still highly work intensive. There have been a few investigating the consequences of the presence of herbivores or predators on invasive establishment in simplified food webs (e.g. food webs in the water-filled leaves of the plant *Sarracenia purpurea*; Gray et al., 2015; Miller et al., 2002), aquatic food webs based on phytoplankton and cladoceran consumers (Dzialowski et al., 2007; Sperfeld et al., 2010), mussel beds (Needles et al., 2015). These studies showed that consumers either decrease invasive success (Gray et al., 2015;

Miller et al., 2002; Sperfeld et al., 2010), have no effect (Dzialowski et al., 2007), or increase invasive establishment (Needles et al., 2015). As discussed in Section 3, these contrasting results can be explained by the interplay of negative effects of residents on nonnative species through direct consumption and apparent competition mediated by consumers, enhancing the dominance of the invader over residents.

Given the difficulty of assembling large food webs, the relationship between food web complexity and invasion has mainly been studied via numerical simulations. These studies have highlighted a relationship between food web **connectance** and invasibility. While Romanuk et al. (2009) predicted that invasion success decreased as connectance increased, Galiana et al. (2014) and Lurgi et al. (2014) found that more connected networks were less resistant to invasion. These contrasted results stem, in part, from the process of construction of the networks during the simulations. This process determines which food web properties covary with connectance, such as species diversity and the proportion of species at the different trophic levels, and these properties affect invasibility. Baiser et al. (2010) found, for example, that invasion success of basal species and top predators increased with connectance because highly connected networks tended to have a low proportion of herbivores and a high proportion of intermediate species. In contrast, the invasion success of herbivores was low in highly connected networks because the proportion of basal species was low and that of omnivore and intermediate species was high (Baiser et al., 2010). On the empirical side, a recent meta-analysis suggested that habitat types characterized by high-connectance food webs (e.g. marine habitats) were less invaded than habitat types with low connectance such as grasslands (Smith-Ramesh et al., in press). However, this correlation provides no direct evidence that connectance itself increases resistance to invasion, as the very general habitat categories that were compared are likely to differ by many other factors affecting invasion. Overall, theoretical and empirical results suggest that food web connectance might not be the best measure to assess food web robustness to invasion. Other food web measures, better reflecting potential niche availability for invaders, need to be investigated in this context, such as trophic complementarity. Moreover, we lack experimental and empirical studies directly relating food web structure and robustness to invasion. Recently, Wei et al. (2015) provided a first experimental evidence that fine network structure can strongly determine invasion success and establishment. They manipulated the structure of the **bipartite networks** describing the consumption links between different bacteria in the soil and carbon substrates exuded by roots to investigate the relationships between diversity, network connectance, and nestedness and invasion success of a root pathogen. They showed that in this system, competition across the network better explained invasion resistance than bacterial diversity. In addition, high connectance and low nestedness lead to reduced pathogen invasion, due to a more efficient use of carbon resources (Wei et al., 2015). With the development of molecular techniques and new methods to describe interaction networks, this study points the way to future empirical investigations of the links between network structure and invasibility.

4.1.3 Food Web Ecological and Evolutionary History Might Affect Its Biotic Resistance

Two types of models predict that food webs with long evolutionary history are more resistant to invasion. Assembly models, in which communities are built by successive species invasion events, show that invasion success decreases with time (Drake, 1990; Law and Daniel Morton, 1996; Post and Pimm, 1983). Similarly, models simulating food web evolution show that new mutants have lower probabilities of invasion as the community evolves through time (Brannstrom et al., 2011). These results are in agreement with the idea that island communities, because they have less history of species invasions than mainland communities, are more susceptible to species invasion (Yoshida, 2008a). Other models of food web evolution, however, predict that food webs can become more sensitive to invasion by primary producers (i.e. invasion leads to more extinctions), if they evolve for a long time, due to increased consumer to producer species ratios (Yoshida, 2008b). Yoshida (2008b) differs from Brännström et al. (2012) by assuming that interactions depend on the matching of 10 characters, not on a single trait, body size. In addition, candidate invaders are immigrants in Yoshida (2008b) while they are mutants, hence derived from resident types, in Brannstrom et al. (2011). The contrasted expectations of the models suggest that different mechanisms involved in community evolution, as well as different histories between invaders and resident species, might strongly determine the consequences of food web evolutionary history on robustness to invasion (see also Jones et al., 2013).

Disturbances can make food webs more invasible, by reducing the diversity and abundance of communities, thus alleviating competition and freeing niche space (Shea and Chesson, 2002; St. Clair et al., 2016). In a microcosm experiment with a simple aquatic food web of protozoans and rotifers,

Kneitel and Perrault (2006) showed that invasion success increased in the presence of drought disturbance. Observational studies provide other examples. Goudswaard et al. (2008) hypothesized that the 25-year delay between the introduction of Nile perch in Lake Victoria (in 1954) and its invasion (in 1979) was due to control by native haplochromines fishes that compete with and prey on juvenile perch; overfishing of haplochromines in the 1970s released the invasion (but see Downing et al., 2013). Nutrient enrichment can also increase invasion success because higher levels of resources might decrease resource limitation for invaders in food webs (Miller et al., 2002, but see Lennon et al., 2003). Some studies have also suggested that the presence of an invasive in a community might facilitate the success of new invaders (**invasional meltdown**) (Simberloff, 2006). For example, the invasion of a land snail on Christmas island was strongly facilitated by the presence of a highly invasive ant species (Green et al., 2011). Ants killed a land crab, which preyed upon the snail. However, the hypothesis of invasional meltdown remains a topic of some debate. Several studies have not found positive interactions between invaders (Orchan et al., 2013) and some have even shown that invaders can interact negatively (Griffen et al., 2008). The type of interactions between invaders will likely depend on their ecological position in the food web.

4.2 Position of Invasive Species in Food Webs

Whether invasive species have particular traits is important for predicting their establishment and impact in communities. Traits correlated with invasiveness have been found for different taxa, although they often differ between studies. For example, invasiveness has been related to growth rate or root allocation in plants (Van Kleunen et al., 2010), habitat generalism (Cassey et al., 2004), and tolerance of new environments (Blackburn et al., 2009) in birds, spawning habitat requirements and life history in fishes (Mandrak, 1989; Olden et al., 2006). We believe, however, that attempts to find traits characterizing efficient invaders in general, without any reference to resident species, are bound to fail, as the probability to invade will be found to be higher in species that are more fecund, long-lived, resistant, voracious, mobile, and plastic than others, i.e., Darwinian demons. The **"invasion filter"** must be relative to resident species, especially for traits involved in trophic and competitive interactions. In a food web context, the question is whether successful invaders have a particular position in food webs (e.g. trophic level, generalism) compared to natives.

4.2.1 Trophic Level May Act as an "Invasion Filter"

Invasive species can be found in all trophic levels, from basal (e.g. plants such as the N-fixing tree *Myrica faya* in Hawaii; Vitousek et al., 1987) to top levels (e.g. top predator such as the brown tree snake in Guam; Fritts and Rodda, 1998). It has been suggested that species feeding at lower trophic levels should have higher invasion success because there are more resources available at the bottom of food webs (Gido and Franssen, 2007). A few studies suggest that invasive species belong more often to the second trophic level (i.e. herbivores and detritivores) than to higher trophic levels. For mammals and birds, herbivore species had higher success than carnivore species at the introduction stage (more likely to be introduced) (Jeschke and Strayer, 2006). By analysing the distribution of marine invasive species among trophic groups in four regions, Byrnes et al. (2007) found that 70% of the invasions belonged to the second trophic level (herbivores, deposit feeders, and detritivores). Strong and Leroux (2014) also found a greater proportion of invasive species at intermediate trophic levels compared to basal and top levels in a terrestrial mammal food web. However, the relation between species trophic level and invasion success remains equivocal. Indeed, Jeschke and Strayer (2006) found that once mammal carnivores passed the introduction stage, they had a higher probability of establishment and spread than herbivores (see also Forsyth et al., 2004). Ruesink (2005) also showed for fishes that omnivorous species had higher establishment success than herbivores and zooplanktivores. In addition, comparison of the trophic level of invasive fishes between their native and introduced ranges indicates that species tend to shift to intermediate trophic positions in the food web, adopting higher trophic levels or lower positions as appropriate (Comte et al., 2016). Furthermore, theoretical models give contradictory predictions with greater invasion success being expected for either herbivores (Romanuk et al., 2009) or larger species generally occupying high trophic levels (Lurgi et al., 2014). The relation between trophic level and invasiveness thus seems context-dependent and it might depend on the degree of "available niches" at the different trophic levels in the food web.

4.2.2 Generalism, Vulnerability, and Interaction Strength of Invasive Species

In Section 3, when considering local effects, we suggested that a lack of coevolutionary history, when combined with the **invasion filter** can result in imbalanced pairwise interactions between invasive and resident species, to the benefit of the invasive. At the food web level, the number of links that an invasive species has might also determine its invasion success. Theoretical studies indeed predict that high generalism and low diversity of enemies facilitate invasion in food webs (Galiana et al., 2014; Lurgi et al., 2014; Romanuk et al., 2009). However, empirical studies do not show such significant correlations between diet breadth and invasion success in large datasets of mammals and birds (Cassey et al., 2004; Jeschke and Strayer, 2006), although invasive fishes tend to be more omnivorous in their introduced range than in their native range (Comte et al., 2016). Few studies have considered invasive generalism and vulnerability at food web scale. Strong and Leroux (2014) reconstructed the highly invaded terrestrial mammal food web of the island of Newfoundland. They found that nonnative herbivores and insectivores without predators tended to be more generalist than natives but differences between food web positions were less clear for other functional groups. While we expect that invasive species might have greater generalism and higher interaction strengths, as well as lower vulnerability, the scarcity of food web data with quantified interaction strengths has precluded the test of these hypotheses at the food web level.

4.2.3 Dissimilarity Between Invasive and Noninvasive Positions in Food Webs: Weak Evidence?

The idea that phylogenetic, functional, or ecological originality of an alien species (relative to a recipient community) favours invasion is well developed in the literature. Invaders that are more phylogenetically distant to the native community are expected to have greater invasion success (Davies et al., 2011). Similarly, in fish communities, successful invasive species tend to be positioned outside or at the border of the morphological (Azzurro et al., 2014) or life-history and trophic (Olden et al., 2006) trait space defined by the resident community, supporting the niche opportunity hypothesis. It seems, however, that trophic links are not the only niche component to consider. Penk et al. (2015) found that invasion of the mysid shrimp *Hemimysis anomala* in aquatic mesocosms was unaffected by the density of a native shrimp (*Mysis salemaai*). Although both species shared similar trophic niches, they strongly differed in their life-history traits, *M. salemaai* was univoltine, whereas *H. anomala* had several broods per year. This would suggest that trophic similarity is not always the main determinant of invasion success.

While differences between invasive and native species might provide niche opportunities to invasive species, these differences might also prevent invasion if the invasive species is unable to exploit available resources or to fit into its new environment. Thus, the relationship between phylogenetic distance to native species and establishment success is not always positive and could even be negative (Jones et al., 2013). Jones et al. (2013) model suggests that the sign of the relationship may actually depend on the modes of inter-specific interactions that prevail during community evolution (**phenotype matching** or **phenotype differences**). Future studies will be needed to assess the relative importance of trophic dissimilarity of invasive species in this context. Petchey et al. (2008) proposed a measure of trophic uniqueness, based on the overlap among species in their prey and predators, which could be useful for such studies.

Following on from our arguments for invasion at the local scale, in Section 3, we apply the same argument to provide a solution to the ambiguity of results of trait/trophic position comparisons between invasives and natives. Long-term evolutionary and ecological divergence between source and recipient communities implies that combinations of traits (and subsequently resource use) displayed by invasives should often lie outside, or in a marginal position to, the distribution of traits in native congeners. Among these possible combinations, however, the **invasion filter** retains only those that allow the introduced species to overcome impacts of competition, predation, or apparent competition exerted by the natives, or to exploit them as prey, host, or mutualists. These might be called the "available niches", in the broadest sense. The variety of impacts that invasives have on other species (see Section 3) and on the food webs as a whole (see later), would indicate that "available" does not mean "empty": invasives not only fill available gaps but also do often divert resources that were previously consumed by other species.

4.3 Impacts of Invasions on Food Web Structure

4.3.1 Additive Effects on Food Web Structure and Beyond

The invasion of a species into a community modifies food web structure through the addition of a new node and new links. These additive effects will depend both on the number of invasive species, a numerical effect that will be determined by food web resistance to invasion, discussed in Section 4.1, and on how adding invaders shifts the average characteristics of the food web, a reweighting effect determined by the difference between native and invasive trophic positions, as discussed in Section 4.2 (see Fig. 5). These additive effects can significantly affect food web structure. Byrnes et al. (2007) suggested that invasions should affect the shape of the food web of the Wadden sea with more species at intermediate levels because invaders tend to occupy the second trophic level in this marine ecosystem. Strong and Leroux (2014) studied a mammal community (see Fig. 6B) and found that link density increased with the addition of nonnative species, as well as the fraction of intermediate species and food web generality, while food web vulnerability decreased.

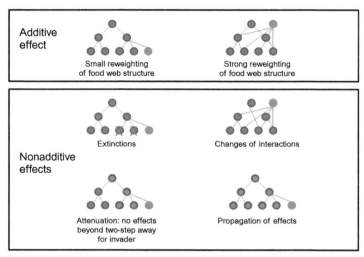

Fig. 5 Potential effects of invaders on food web structure. The native food webs are represented in *blue*, while invaders and their trophic interactions are in *red*. In the two bottom webs, native species affected by invaders are represented with *blue circles* filled in *red*.

However, invaders might also have impacts on food web structure beyond simple additive effects, for example, by modifying the abundance and interaction strengths in the community, by triggering species extinctions or changes in species diets. Investigation of these nonadditive effects requires data comparing the communities with and without invaders. Such comparison can either be achieved experimentally (Salvaterra et al., 2013) by compiling data before and after invasion (e.g. Woodward and Hildrew, 2001; Fig. 6C and D) or by comparing field sites with varying invader densities (e.g. Carvalheiro et al., 2008; Heleno et al., 2010; Fig. 6A). Additionally, well-resolved food web data are necessary to assess changes in links and interaction strengths following shifts in species diets. In the following, we report existing evidence for several types of nonadditive consequences of species invasion on food web structure: species extinction, shifts in interactions, and propagation of effects (Fig. 5).

4.3.2 Effects Beyond Additions: Is Species Loss a Common Effect?

Several well-known invasions have been related to dramatic species extinctions, mostly through direct predation. Extinctions of several hundred species of haplochromine fishes in Lake Victoria in the 1970s have been attributed to predation by adults of Nile perch (Goudswaard et al., 2008; Witte et al., 1992). The brown tree snake in Guam has also exterminated

A

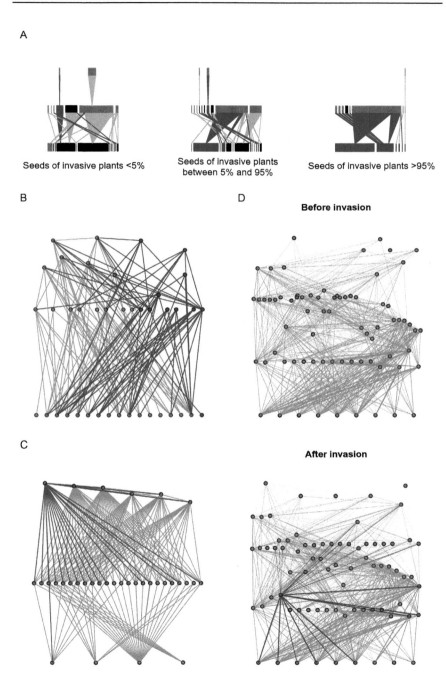

Seeds of invasive plants <5%

Seeds of invasive plants between 5% and 95%

Seeds of invasive plants >95%

B

C

D

Before invasion

After invasion

Fig. 6 See legend on next page.

many native vertebrates, by predation and apparent competition with introduced alternative prey (Fritts and Rodda, 1998). However, these studies are mainly correlative and it is difficult to precisely identify the causes of extinctions, as other environmental changes can cooccur (Gurevitch and Padilla, 2004). For example, although predation by adult Nile perch gave the final blow to many haplochromine species, their stocks had already declined due to overharvesting and pollution, prior to the boom of the Nile Perch. Populations of large haplochromines, by predation and competition with juvenile Nile Perch, previously prevented them from becoming adult and therefore may have provided resistance to the Nile Perch invasion. Their decline allowed adult Nile Perch to become abundant and haplochromines then became their prey and many went extinct (Downing et al., 2013; Goudswaard et al., 2008; Witte et al., 1992).

Few studies have considered the consequences of invasion on native species richness through an explicit food web approach (examples in Fig. 6). Carvalheiro et al. (2008) investigated how an invasive seed–herbivore altered the structure of a plant–herbivore–parasitoid network. They found a negative impact on native species richness of both seed herbivores and their specialized parasitoids, likely resulting from strong apparent competition

Fig. 6 Examples of four studies using a food web approach to analyse consequences of invaders on ecosystems. In all networks, species considered as invasive are highlighted in *red* as well as their trophic interactions. (A) Comparison between food webs from sites with different percentages of invasive plants in the Azores. The three networks represent observed interactions between plants, seed herbivores, and parasitoids, aggregated across several locations according to their percentages of seeds from invasive plants. Each *rectangle* represents a species which width is proportional to the number of interactions observed for this species. Endemic species to the Azores are represented in *black*. (B) Terrestrial mammal food web of the island of Newfoundland in Canada. Data were compiled from literature analysis of the species diets. (C) Food web of the Broadstone stream in Sussex, UK, after invasion by the dragonfly *Cordulegaster boltonii*. The food web was described thanks to gut content analyses of top predators in the web. (D) Comparison between the food web of the Oneida Lake, USA, before and after zebra mussel invasion. Carbon flows between species in the food web were estimated from data on biomass, production, respiration to biomass ratios, and diet proportions of taxa from the literature and expert knowledge on the study system. *(A) Data retrieved from Heleno, R.H., et al., 2009. Effects of alien plants on insect abundance and biomass: a food-web approach. Conserv. Biol. 23 (2), 410–419. (B) Data retrieved from Strong, J.S., Leroux, S.J., 2014. Impact of non-native terrestrial mammals on the structure of the terrestrial mammal food web of Newfoundland, Canada. PLoS One 9 (8), e106264. (C) Data retrieved from Woodward, G., Hildrew, A.G., 2001. Invasion of a stream food web by a new top predator. J. Anim. Ecol. 70 (2), 273–288. (D) Data retrieved from Jaeger Miehls, A.L., et al., 2009. Invasive species impacts on ecosystem structure and function: a comparison of Oneida Lake, New York, USA, before and after zebra mussel invasion. Ecol. Model. 220 (22), 3194–3209.*

between the invasive seed-herbivore and native ones (through an increase in generalist parasitoids). Heleno et al. (2009) studied the consequences of an invasive plant on plant–herbivore–parasitoid food webs (see Fig. 6A). Plant and herbivore diversities decreased, and the loss of herbivore richness likely resulted from a decrease in seed production during some time of the year. These examples show that local extinctions following species invasion can have multiple and cooccurring trophic causes, from direct predation or competition for resources to apparent competition. Invasions can also increase species richness. A meta-analysis on field experiments in marine systems (Thomsen et al., 2014) documented predominantly negative effects of invaders on diversity of species at the same trophic level, but positive effects on diversity at higher levels, suggesting that species at higher trophic levels come to rely on food and habitat provisioning by invaders.

Several studies report strong changes of invaded food web structure that are not associated with extinctions of resident species. Woodward and Hildrew (2001) compared streams before and after the invasion of a polyphagous top predator and found increased complexity and food chain length but did not report any species extinction. Salvaterra et al. (2013) showed that food web structure in freshwater mesocosms was significantly affected by an algal invader, with higher connectance and proportion of intermediate species; but species richness did not change significantly. Thus, changes in food web structure following invasion might, in most cases, be mainly related to changes in trophic group abundances rather than to species extinctions, as suggested by a recent meta-analysis on aquatic ecosystems (Gallardo et al., 2016).

4.3.3 Effects Beyond Additions: Changes in Interactions Might Be Frequent

Few studies have investigated how specific interactions between resident species are modified by invasion. Most studies on food webs infer links from the identity of the species present combined with literature and expert knowledge (e.g. Strong and Leroux, 2014). To our knowledge, the studies that have assessed the consequences of invaders on food web structure by direct quantification of trophic links (through gut content analysis: Woodward and Hildrew, 2001, direct observation: Carvalheiro et al., 2008, 2010; Heleno et al., 2009) did not analyse how invasion modified interaction strengths between species or induced shifts in diets. Jaeger Miehls et al. (2009) suggested that invasion by the zebra mussel strongly modified the distribution of carbon fluxes between species in the Oneida Lake food web by reinforcing the benthic pathway to the detriment of the pelagic pathway (see Fig. 6D). However, this study did not assess fully shifts in species interactions, as they were not quantified by direct observation but rather inferred from data on species biomasses and a literature search on species production and diet.

Those studies that have investigated trophic shifts in resident species following invasion have not directly observed interactions between species but have rather assessed trophic position through isotope analyses (Tran et al., 2015; Vander Zanden et al., 2010). Vander Zanden et al. (1999) found that following the invasion of two species of bass in Canadian lakes. The resident trout shifted from having a predominantly littoral fish diet to a predominantly pelagic zooplankton diet and thus dropped in trophic level. Tran et al. (2015), in investigating the trophic shifts of competing native and invasive fishes, found that trophic niches of invasive and native species were more divergent when they were in competition than when the fish species were growing in isolation. These two studies would suggest that invaders induce phenotypic shifts in trophic interactions for native species either under direct competition or under predation by the invader. As noted in Section 3, evolution of resident species in response to invasion might also play a role in such shifts.

4.3.4 Effects Beyond Additions: Propagation Through the Food Web or Attenuation?

We might expect an attenuation of impacts of an invasion in a food web as we move away from the invader position in the network, with species close to the invader being expected to be more affected than species further away. Indeed, models on the consequences of extinction predict decreasing effects on species abundances as the degree of separation from the extinct species in the food web increases (Berlow et al., 2009). However, in reality, most species might be within two or three links to each other in food webs (Montoya et al., 2006). There is now abundant evidence that the effects of the introduction of invasive species are not always restricted to the abundance of species one or two links away from the invader. Rather, invasion can trigger trophic cascades, affecting the abundance of the main trophic groups in the food web (e.g. Penk et al., 2015, reviewed in Gallardo et al., 2016). Large impacts are also likely to occur when the invader is an ecosystem engineer that strongly modifies the physical environment (Sanders et al., 2014). For example, the bioturbation of the invasive common carp and red swamp crayfish increases sediment resuspension which can lead to major shifts in food webs in lakes, from macrophyte dominated ecosystems with clear waters to phytoplankton dominated systems with turbid waters (Matsuzaki et al., 2009). Large-scale food web impacts are also expected when the invader affects the abundance of a keystone species or an ecosystem engineer. The consequences of the invasion of the crazy ant *Anoplolepis gracilipes* on Christmas island are a good example of such case. The ant has a strong negative impact on an endemic land crab that burrows and indirectly modifies the plant composition and diversity in forests, as well as associated insect communities (O'Dowd et al., 2003).

It is, however, difficult to assess whether these large effects of invasion are general or not. Effects of invaders might not always spread through the food web. Current evidence on the consequences of invasions on food webs is probably biased towards observations of spectacular effects since they are effects noticeable directly. It is also unclear whether the food web consequences of invasive species depend on the ecosystems concerned. Aquatic ecosystems often show stronger trophic cascades than terrestrial ecosystems (Shurin et al., 2006; Strong, 1992), which might make them more sensitive to invasions. A recent meta-analysis also suggests that invasive plants have different effects on green food webs, based on primary producers, and brown food webs, based on decomposers, which depend on the type of ecosystem (McCary et al., 2016). Invasive plants appear to decrease the abundance of decomposers and herbivores in woodlands but not in grasslands, and brown food webs are more affected by invasive plants in woodlands than in wetlands. Future studies will need to collect more data on food webs before and after invasion to assess the generality of the effects of invaders at food web scale and how they depend on ecosystem types.

5. HOW DOES A NETWORK PERSPECTIVE INFLUENCE THE MANAGEMENT OF INVASIVE SPECIES?

HIGHLIGHTS
- *Prevention of invasion should incorporate information on local interaction networks. Quantifying network diversity is a necessary first step, but trait distributions give powerful additional information to identify empty niches (Section 5.1).*
- *Knowing the relative importance of bottom-up and top-down controls at different trophic levels in communities can help one to find the best ways to control an invader's growth rate, depending on its trophic position (Section 5.2).*
- *Eradication programs may have unexpected impacts, especially indirect effects on nontarget species mediated by network structure. The efficiency of eradication is also limited by evolutionary processes such as the cost of adaptation of residents to the presence of invaders, and evolutionary rescue of invaders through the acquisition of traits allowing to survive eradication methods (Section 5.3).*

5.1 Preventing Invasions From a Network Perspective

Our increasing knowledge of the structure and functioning of ecological networks is opening new doors to research that may eventually help in the management of invasive species. Given that invasive species are notoriously difficult to remove (Gardener et al., 2010; Kettenring and Adams, 2011), it may be interesting to try to maintain ecological networks that are inherently difficult to invade. Resistance to invasion is one of the many measures of stability in community ecology (McCann, 2000) that has been amply studied from a theoretical point of view and that is not too difficult to assess, so that we could rely on a few mechanisms based on theory or experiments to help the management of biological invasions.

Let us focus first on the concept of species packing along niche axes (MacArthur, 1970) and community closure (Brännström et al., 2012). As mentioned in Section 4, diversity protects against invasions. Thus, the maintenance of diversity per se could be an important management target. Low-diversity webs inherently attract invaders, and invasive removal itself is unlikely to solve the problem, as long as niche opportunities persist. Diversity-oriented management is particularly important in the case of well-defined, isolated systems, such as lakes or islands. In such instances, empty niches may exist, for example, abundant prey without predator, due to dispersal limitation. One may want to preserve the situation as it is, if some of the prey are endemic or patrimonial, but the vacant niches make the network inherently vulnerable to invasions (Gravel et al., 2011). It may well be that protecting the network by removing invasive predators in such instances simply delays the inevitable. Management may then have to trade-off the benefits and costs of current invaders with the ones of other possible other invaders. "Vaccinating" such a system with an invader with mild effect that fills an empty niche, thus preventing subsequent invasions by higher impact invaders might be more suitable in the long run. On a related note, any success at removing an invasive may indicate that the removed invader was not very high on the fast reproduction/high competitive ability scale. Its removal may open the door for faster or more competitive invaders that may not be so easily removed.

Next to diversity, several aspects of web structure or functioning may also be managed to produce a network that is resistant against invasions. From the observed importance of top-down controls and indirect competitive effects, it can be inferred that maintaining healthy populations of top predators as well as omnivore predators is an important management target as they reduce the probability of invasion (Hawkes, 2007; Shea and Chesson,

2002). More generally, in order to avoid or limit the impacts of invasions, functional diversity may matter more than species diversity. Indeed, if an invader creates a new functional group within the receiving network, this invasion will by definition modify the overall functioning of the web. Consequences can then be far reaching (Ehrenfeld, 2010), as such invasive species can redistribute biomass constraints and nutrient fluxes within the network. As an example, Yelenik and Antonio (2013) showed that invasive C4 species in Hawaiian woodlands have favoured the subsequent invasions by nitrogen fixers through modifications of plant–soil nutrient feedbacks. One corollary of this is that, if it were possible to maintain ecological networks that contain a large set of functions (i.e. a high level of functional diversity), any incomer would likely be redundant with species already present. This would reduce the probability that the impact of the invasive species is large.

Aspects of network functioning are increasingly linked to species traits within the network (Cadotte et al., 2015; Deraison et al., 2015). As a consequence, networks that contain complete sets of traits may be less vulnerable to biological invasions or suffer less impact when invaded due to trait and function redundancy. Trait distributions are increasingly measured and can be compared among networks and ecosystems (Ackerly and Cornwell, 2007; Stegen et al., 2009). By doing such comparisons, it is possible to assess how complete the network is in terms of trait (e.g. in terms of body size distribution), compared to other equivalent networks. For instance, it may be possible to compare the body size distribution in different lakes of similar sizes within a region to infer whether networks differ in trait distribution, outliers being especially interesting in terms of vulnerability to invasions. Such an analysis, however, requires that comparable ecosystems exist, and also that the key traits can be identified for comparison. Such traits exist for the plant compartments in the food webs, in which a slow/fast structuring continuum of traits has been clearly identified (Wright et al., 2004). Among the animals in food webs, there is ample evidence that body size determines interactions and global functioning (Brose et al., 2006; Cohen et al., 2003; Loeuille and Loreau, 2009). Body size also has the advantage of often being easier to measure, in comparison to other traits. Therefore, an analysis of size distribution of different ecosystems in a region may allow the identification of some gaps in local distributions. These gaps may be likely candidates for future invasions.

From the consideration of all these points, a general pattern emerges. Communities that have a high species diversity, healthy top predator populations, and complete sets of traits and functions are less likely to be invaded. Therefore, the best way to deal with biological invasions is to act preventively and to keep ecosystems in good shape (Chapin et al., 2000). Biological invasions should not be dissociated from other aspects of the current biodiversity crisis, as they are very much a consequence of it. For example, if an ecosystem is inherently vulnerable to invasions by predators because it lacks a trophic level given the amount of energy available (Oksanen et al., 1981; Persson et al., 1992), removing the invader will not provide a long-term solution, unless the surplus of energy (e.g. due to nutrient enrichment) is treated as well.

While these principles are supported by scientific knowledge and theory, putting them into practise is not easy. Indeed, most ecosystems are affected simultaneously by multiple stressors (Crain et al., 2008; Tylianakis et al., 2008). An integrated vision of ecosystem functioning has inspired European directives to monitor and maintain the quality of inland water ecosystems. However, there is still room to extend this vision to legal and technical tools for invasive species management, such as "Invasive Species Predictive Schemes" (Whitney and Gabler, 2008). These schemes are, for the most part, species–centred, and they try to determine and prioritize potential invaders that should be put under surveillance. Such schemes may benefit from moving to a species-and-network information framework. Our review suggests that invasions are context-dependent, so that an invasive species in a given area may not be a likely invader in another. For example, the large meta-analysis by Van Kleunen et al. (2010) showed that predictions based on traits of invasives are greatly improved when comparing them to those of resident species from the local network, rather to noninvasive aliens (a purely species-centred view). An example of this is the Tephritidae case study in Section 3. In a given territory, traits of successive invaders follow a directional trend (recent invaders have slower life-histories, and produce fewer but larger juveniles), indicating that each newly established species in a community modifies the trait distribution from which future successful invaders might be drawn. For a given location, such as an island, the analysis should start by identifying gaps (traits, functional groups, trophic levels), then cross-referencing this information with the characteristics of potential invaders.

We anticipate two difficulties in achieving this, linked to the interaction types involved and to evolutionary dynamics. The food web approaches, that are the subject of this review, only consider trophic interactions (Cohen, 1977; Pimm, 1979; Pimm and Lawton, 1978; Williams and

Martinez, 2000). Other network approaches include mutualistic interactions (Bascompte, 2006; Jordano et al., 2003; Rohr et al., 2014). The rules of system assembly or stability (including resilience to invasion) change from networks of one interaction type to another. For instance, it has been proposed that network nestedness contributes to the stability of mutualistic networks, but weakens the stability of antagonistic networks, while the reverse seems to be true for network **compartmentalization** (Thébault and Fontaine, 2010). Because invasive species impact local networks through negative (Dorcas et al., 2012) or positive (Yang et al., 2011) interactions, this separation of interactions is an important shortcoming for our understanding of biological invasions. The structure and stability of networks containing different interaction types is a recent topic (Fontaine et al., 2011; Georgelin and Loeuille, 2014; Kéfi et al., 2012) and we have little data on such complete networks (but see Melian et al., 2009; Pocock et al., 2012). Another related issue is that ecosystem engineering (Jones et al., 1994) is seldom considered in network approaches. However, some of the most spectacular effects of invasive species happen through ecosystem engineering or niche construction (reviewed in Ehrenfeld, 2010). It is therefore important to develop conceptual models that merge classic trophic or mutualistic networks with ecosystem engineering effects.

A second limit concerns the role of evolution. We proposed earlier the use of the species traits and traits already present in the network to predict which species are likely to be invaders. While this is surely an adequate first step, a potential problem is that evolution can quickly modify the traits of resident and invasive species (Lambrinos, 2004; Moran and Alexander, 2014; Stockwell et al., 2003). Invasive species create large selective pressures on species at the same trophic level, as well as a few steps up and down. Many instances of evolution of the recipient communities have now been observed, from morphological adaptation of local herbivores to new plants (Carroll et al., 2005) to adaptation of predators to the defences of invasive prey (Phillips and Shine, 2004). As a consequence of these multiple observations, it has been suggested that such evolutionary or plastic trait change should also be considered in "Invasive Species Predictive Schemes" (Whitney and Gabler, 2008).

5.2 Using Ecological Networks to Limit Impacts or Exclude Invaders

A network perspective may help in reducing the impacts of invasive species or excluding them. Indirect effects due to bottom–up controls (species limitation by resources) and top-down controls (limitation by enemies) have been studied extensively in the past 50 years and that this work may give some insight into the management of biological invasions.

Species competition for resources is the dominant force determining the distribution of biomass within networks with bottom-up control. Increasing resources allows for a general increase in biomass and food chain length. As a consequence, increased nutrient or energy availability facilitates the maintenance of invasive predators; and, for management, it may be important to focus on the interaction between biological invasions and nutrient imbalance (e.g. eutrophication) rather than on invasion per se. This idea has received lots of theoretical (Loeuille and Loreau, 2006; Oksanen et al., 1981) and empirical support (Arim et al., 2007; Dickman et al., 2008; Townsend et al., 1998, but see Post et al., 2000). In such instances, the long-term solution may be to reduce the primary production in the system. Another option might be to allow another invasive predator with mild effect to occupy the niche. However, such controlled invasion touches on many issues of ethics and societal norms that go beyond the scope of this review. Reduction of available energy and primary production may also have some system-dependent effects. It may be efficient in eutrophicated waters, where reducing nitrogen or phosphorus can only make the ecosystem healthier. For instance in several marine ecosystems large amounts of nutrient (and overfishing) have fostered the invasion by different species of jellyfish (Richardson et al., 2011). In coastal ecosystems, a global correlation between eutrophication and species invasion has been found (Lotze et al., 2006). When the local ecosystem is not obviously eutrophicated, however, reduction of local energy may have many different and potentially hard to predict side effects. In food webs that are already weakened by the recent invasion of a species, for example, reducing the available energy may be the final straw for some of its competitors.

While such ideas of fighting invasions by bottom-up control are rarely applied, currently, top-down controls to limit the impact of invasive species are much more widely used. Much of biological control uses local or introduced predators to fight the invasion of a species in a given ecosystem. Much of this literature is devoted to the management of pests in agrosystems (Bohan et al., 2013; McFadyen, 1998), but in several instances, biological control has been used in natural ecosystems to limit the growth of an invasive species (Hoddle, 2004; Shea et al., 2010). As biological control requires the introduction of a predator to limit the focal species, decisions for which predator to use have to be made. Particularly, the traits of the predator have to be chosen carefully to limit potential side effects. While a specialist predator may seem ideal, as it will likely focus on the invasive species, this strategy has two drawbacks. First, a specialist able to eliminate the invasive species will, by definition, not exist in the native community, so that such species often have to be taken from the invasive range, technically opening the door for a new invasion (Hoddle, 2004). Second, theoretical work predicts that very specialized consumers will likely lead to boom and crash cycles (Hanski et al., 1991, 2001), so that the maintenance of the control agent is not guaranteed in the long term. One proposed solution is to introduce and association of two specialist enemies (Ong and Vandermeer, 2015). The alternative, of choosing a generalist consumer, will likely create many side effects on the community. Most particularly, generalist control agents may switch to concentrate on native species. They also have a much larger chance to become pervasive in the ecosystem, becoming an invasive species themselves. As an example, the cane toad (a generalist consumer) was introduced as control agent in many different sugar cane ecosystems, and became invasive on a number of islands, most notably having large effects on Australian ecosystems (Shine, 2010). While the recommendation for choosing specialists over generalists might seem simple and intuitive, it may not be so easy in practise. Often, species that seem specialized may reveal themselves to be generalist, either because they were constrained to a realized (specialized) niche or because the breadth of the feeding niche changed when used as a control agent, due to plastic or evolutionary changes (Jones and Gomulkiewicz, 2012; Loeuille et al., 2013). The consumption (and extinction) of local endemic snails by the invasive predatory snail, *E. rosea* (a biocontrol agent), in the Pacific islands clearly demonstrates the possible speed of adaptation to the consumption of nontarget, native species.

Bottom–up and top–down controls often occur simultaneously in nature. In general, it has been pointed out that bottom–up control likely dominates at lower trophic levels (i.e. the plant compartment, and sometimes the herbivores), while the top–down controls are dominant at higher trophic levels. This view is consistent with empirical observations both in freshwater (McQueen et al., 1989) and terrestrial systems (Halaj and Wise, 2001). Trophic cascades have indeed been observed at higher trophic levels (Halaj and Wise, 2001), while at the plant trophic level, variations in edibility is thought to largely reduce the amplitude of top–down effects (Hunter and Price, 1992; Loeuille and Loreau, 2004; Strong, 1992; White, 2005). These large amounts of plant defences may in turn severely limit plant consumption by herbivores, so that bottom–up control may also apply at this trophic level

(White, 2005). If this scheme is correct, then our invasion management recipes based on bottom-up control should mainly be used to control invasive species at lower trophic levels (e.g. plants and herbivores), while top-down control solutions should be targeted at invaders into the upper trophic levels.

5.3 Implications of Invader Removal From a Network Perspective

The most commonly applied policy against invasion remains the direct removal of the invasive species itself. Invasive rats or cats on islands are often the direct targets of eradication programmes (Courchamp et al., 2003; Nogales et al., 2004). While some of these removals act by hunting individuals, in most cases the eradication methods rely on poisonings (Courchamp et al., 2003; Nogales et al., 2004). Chemical warfare against an invasive species may have large inadvertent impacts on other species (Courchamp et al., 2003). For instance, agricultural pesticides have important effects on nontarget species (Bourguet and Guillemaud, 2016; Theiling and Croft, 1988). While the effects likely vary depending on the network structure, the accumulation of pesticide up food chains has been widely documented (Vo et al., 2011), so that native top predators can be impacted by poison-based eradication programmes that target lower trophic levels.

Whatever eradication method is used, indirect effects following eradication may propagate within the network. Once an invader is eliminated, local species may take time, or even never manage, to restore the functioning in its preinvasion state. In some instances, the invasive species created new habitats that were then used by native species or modified the ecosystem through engineering effects thereby acting positively on some species of the network. Eradication will therefore involve clear trade-offs. Zavaleta et al. (2001) details how invasive shrubs of the genus *Tamarix*, despite their detrimental effects on many riparian communities in North America, provide important nesting sites for the endangered willow flycatcher, *Empidonax traillii extimus*. Similarly, removing an invasive species that had a large biomass or productivity will likely create important bottom-up effects (see Section 3.3). In a Portuguese marsh, the invasive crayfish *Procambarus clarkii* is an important resource for nine endemic mammal and bird species. Removing such invasive species may undermine the resource basis of many native species, at least during the period that native prey populations may take to recover. Unfortunately, recovery is not systematic, making the whole process risky. The removal of alien species may thus lead to further disequilibrium within the community (Courchamp et al., 2003). When the invasive species is a predator, indirect effects due to top-down controls are similarly expected, especially through trophic cascades and relaxation of apparent competition (see Section 3). Again, such indirect effects are now well documented (Ehrenfeld, 2010; Zavaleta et al., 2001).

While accounting for ecological indirect effects is certainly important before planning an eradication, it is also fruitful to consider the evolutionary perspective. Eradication policies may affect both the evolution of the invasive species and that of residents. The extra mortality exerted on invasive species through hunting, poisoning, etc., coupled to the high abundance and/or rapid growth rate, make them prime candidates for **evolutionary rescue** (Gomulkiewicz and Holt, 1995). Some striking examples of evolutionary rescue come from eradication programme of invasive species (Carlson et al., 2014; Stockwell et al., 2003; Vander Wal et al., 2013). These include multiple observations of evolved resistance to pesticides by agricultural pests and to antibiotics by bacteria (Carlson et al., 2014) or to poisoning (Vander Wal et al., 2013). It seems that whatever the aptness of the eradication plan, invasive species have traits that make it possible for them to adapt. Indeed, most eradication programmes fail. Even on islands, and for invasive species that are quite large (e.g. cats), only a small minority of the eradication programmes attempted have succeed (Nogales et al., 2004). Confronted to increased extrinsic mortality, invasives are selected to evolve faster growth rates, earlier maturation and smaller body sizes (Coltman et al., 2003; Grift et al., 2003; Law, 2000; Olsen et al., 2004), which make eradication increasingly unlikely.

Eradication programmes may also interfere with the evolution of resident species. Resident species, where they do not go extinct, often adapt to the invasion (see Section 3). However, such evolutionary responses come with associated trade-offs or costs, either at the expense of the reproductive or at the growth rate of the species (Herms and Mattson, 1992; Reznick et al., 2000; Steiner and Pfeiffer, 2007) or by changing the cost/benefits of other ecological interactions (Müller-Schärer et al., 2004; Strauss et al., 2002). These costs are only offset because they confer an advantage in the presence of the invasive species. When the invasive species is removed, such species will be subject to the costs without any benefit, thereby experiencing reduced growth rates and being at a disadvantage in the invader-free environment. In such conditions, we do not expect the system to return immediately to its preinvasion state. Indeed, further invasions may be facilitated by the investment of local species to defence against a past invader at the expense of their competitive ability.

Finally, although we have tried to illustrate how network thinking can be used in management, we acknowledge that management is often based on practical, correlative approaches rather than conceptual understanding. A species observed to be invasive somewhere (or closely related to a known invader), will be treated as a potential invader elsewhere (Bellard et al., 2013; Bertelsmeier et al., 2014; Vander Zanden et al., 2010), and the recipe for its management will be largely drawn from previous experience (Courchamp et al., 2003). Although intuitive, the correlative approach leaves out important ecology necessary for prediction. First, it will not help the management of a new problem (i.e. a new invasive species or an old one in a new context), for which no or limited data exist. That is, it is only a good cure if one has already been sick. Second, correlation may not give optimal guidance from an allocation of resource viewpoint. Correlative policies to either prevent introduction or extirpate a known invader entail considerable costs. Knowledge of the species traits and the composition of local networks (in terms of trait, functions, etc.) would reduce such costs by identifying where and when this particular species might be dangerous, neutral, or beneficial, thus tuning the management approach. While previous work on this question indicates that past history may be used to forecast impacts (Kulhanek et al., 2011; Ricciardi, 2003), these studies incorporate the biomass of the invader, which is tightly linked to the network context, and the previous state of the invaded community (network information per se). Purely correlative approaches may also fail to predict the consequences of indirect ecological effects and evolutionary dynamics. To take just one example, ignoring the possibility of **evolutionary rescue** will likely yield strong underestimates of the effort and money to be devoted to invasive species management or eradication. Evolution of resistance to pesticides and antibiotics are now major constraints affecting economies worldwide (Carlson et al., 2014).

6. CONCLUSION

Invasion biology needs a network perspective. We found that many empirical studies have adopted the species-centred viewpoint while few adopt the food-web perspective, possibly as a consequence of intrinsic costs and difficulties of their construction. This situation poses practical problems to our effort to understand invasion: (i) the species-centred approach runs the risk of large biases as researchers will tend to focus on species with particularly important effects or that are particularly impacted;

(ii) by focussing on a single pathway, the species-centred approach will tend to ignore interaction between different causes of invasion and (iii) invasions are a global problem and providing solution by focusing on each individual invader in turn will prove to be an endless job unless generalities emerge from the accumulation of case studies—a likelihood we believe is too optimistic given the biases in (i) and (ii). It is necessary to address, therefore, the question of impacts at the ecosystem level, potentially at the risk of losing species-specific details, and ecological networks such as food webs are one of the most intuitive conceptualizations we can make of an ecosystem. Unfortunately, to date, the empirical data lag well behind theory.

The structure of resident food webs determines the extent of potential impacts of invasive species. The complexity of most food webs leads to a simple expectation that we call the **attenuation principle**. Impacts are usually attenuated by dilution when they propagate along more steps in the food web because each species has many interactions. For the same reason, we expect negative impacts to be more evident at lower trophic levels below the invasive species. Evidence from species-centred studies support this view, as strong negative impacts, especially extinctions of resident species, result more frequently from direct predation than from indirect interactions such as exploitative or apparent competition. However, not all food webs are equally prone to this attenuation. Low diversity of the resident food web, with few but strong trophic links and efficient consumption of primary production, such as in some lake ecosystems, can ensure higher propagation of impacts along several steps, sometimes down to primary producers with the potential for impacts on the whole ecosystem. The impact is also enhanced when the invader is an ecosystem engineer or belongs to a new functional group or trophic level, such as a new top predator or type of herbivore. Such invaders may produce new ways to exploit resources and divert lots of the energy and matter flow away from previously dominant pathways.

Incidence of invasions depends on trait complementarity of invaders relative to residents and reflect insaturation of the resident community. Community ecology theory, in addition to several lines of empirical evidence, suggests that invaders should mostly position themselves in "available niches". In a food web, these "niches" are defined by sets of potential ascending or descending trophic links, such as the ability to exploit species A and B efficiently but not C and D, or to be an easy prey for X but defended against Y, which together allow a positive rate of population growth. Introduced species will tend to establish themselves if their traits result in a unique combination of links dissimilar to that of any resident, either within a functional group already present (e.g. new body size in lake fish) or by being from a formerly absent functional group. At first glance, however, this seems at odds with the expectation that invaders are often generalists and may have strong negative impacts on resident species. If invasives usually fill local trait gaps in the resident community, should they not rather be new local specialists than ubiquitous generalists?

There are several solutions to this apparent paradox. First, even when some niches are vacant, theory predicts little competitive impact but does not preclude strong **top-down impacts**. Second, **anthropization may create a "global niche"** with a variety of resources available to species that are able to tolerate the conditions of anthropized systems. This tends to favour species that are "generalists" in the sense that they exploit many different prey, but could otherwise be treated as "specialists" of anthropogenic situations. The consequence is that invaders may often be limited by competition with other invaders (cf. the Tephritid fly case study in Section 3). In addition, invasion can often be considered a symptom of previous disturbance or anthropization. Thus, trying to remove the invader is not a good long-term strategy if the root cause of the disturbance is not also dealt with (cf. Section 4). Third, **long-term evolutionary isolation** (islands) also creates ecosystems with less functional groups than usual, such as those that do not have certain types of predator, and that are sensitive to invasions. In both cases, whether of anthropized systems or islands, invaders will tend to appear as generalists in the resident community because the latter is "immature". A newcomer will then be able to exploit lots of resource because the system is evolutionarily and ecologically young and trait combinations that make an invader more efficient at exploitation than resident species are readily found.

The lack of common ecological and evolutionary history between invaders and residents is involved in, but not entirely responsible for, negative impacts of invaders on recipient food webs. While the role of evolutionary naïveté is frequently mentioned in the invasion literature, this role has been shrouded in ambiguity and contradiction. We believe that this stems from a biased view of the lack of coevolutionary history, which is often considered to provide a systematic advantage to outsiders. The lack of coevolution between two (or more) potentially interacting species would predict more variance in interaction strength, such as a very high or very low predation efficiency. Indeed extreme interactions are unlikely to persist over long ecological or evolutionary time in resident communities, as one species evolves defences or goes extinct. However, lack of coevolution by itself does not explain an asymmetrical advantage to the outsider. This asymmetry results from the condition that the growth rate of the alien species must be positive in the resident community, and possibly even very high. This bias is reinforced by the propensity of researchers to study only spectacular invasions.

Finally, these considerations pose questions for the long-term readjustment of food webs and communities after invasion. It has long been observed that some invaders decline after a period of extremely high abundance (Simberloff and Gibbons, 2004). This is to be expected if they exploit some resource better than residents, as they will reduce the equilibrium abundance of this resource. Invaders may, for example, grow fast initially by consuming lots of available prey and as a result the abundance of this prey community decreases, and so subsequently does the invader. However, evolutionary responses of resident species may also contribute to the modification of the long-term equilibria of invaded food webs. Because invaders have already passed the "**invasion filter**" the evolutionary challenge posed by invasions is initially towards the resident species, and many examples of relatively rapid evolution among native species have been described that tend to mitigate the negative impacts of invaders. While there is certainly room to debate the optimism of this "Community-strikes-back" scenario, the opposing, apocalyptic vision of impacts of invasion has no better grounding but still pervades the literature. It may not, therefore, always be a good idea to try to remove an invader if resident species are either extinct, or largely engaged in a process of adaptation to interactions with the newcomer. Practical invasion management policies should accept that sometimes there is simply no way back to the preinvasion state. Rather, it would be better to manage ecosystems to render them resistant to invasion in the first place, by limiting nutrient input for example, or to influence the new network, with its invasive species, to best fit societal demands for biodiversity and ecosystem service.

6.1 Future Directions

More data are needed on invasions at the food web scale. While evidence is growing for a role of diversity in protecting against invasions or buffering their impact, we still do not have enough data to construct general patterns about the kind of position invaders take in a food web, and the associated changes in overall food web structure. The data, ideally of food webs before and after invasion(s), or of more or less invaded food webs, are exceedingly rare. Measuring species abundances is not enough, because invasive and resident species may shift their diet in ways that are not straightforward following invasion. Therefore, trophic links need to be evaluated before and after invasion. To date, direct observation (as in plant–herbivore–parasitoid network) and isotopes were the methods of choice for determining trophic links. The development of high-throughput metabarcoding of gut contents will certainly offer a rich source of data in the future, although not devoid of their own limitations (Kamenova et al., 2017). In addition, when the data allow it, invasion into food webs should ideally be considered in a spatial context that accounts for processes limiting dispersal (Massol et al., 2017).

There is a clear need to systematically incorporate evolutionary thinking into invasion and network analysis. There are more and more examples of rapid evolution of local species in response to trophic or competitive interactions with invaders, yet each is treated as an "isolated" case study. Invaders often change the interactions context of many species in a food web. For example, we would predict that indirect competition with invaders would challenge local species to adapt to increased predation by their usual, local predators—even though there is as yet no such example in the literature. Life-history traits are also relatively understudied, although they are very good candidates for rapid evolutionary responses given that they usually display high genetic variance, and can modulate the demographic impact of new interspecific interactions, from predation to competition, or reflect growth or fecundity costs of the acquisition of other traits, such as defences. As such, screening changes in life-history traits in local species following invasion would be a worthwhile goal.

Finally, we would point out that there is a great deal more to ecological network analysis than food webs. Here, we have focused on networks of predators and prey, but network thinking can be applied to all ecological interactions, including mutualisms (e.g. pollinator plant), metabolic interactions in microbial communities, and host–pathogen interactions (see Médoc et al., 2017), with the potential of great insight into related aspects of invasion.

ACKNOWLEDGEMENTS

This work is part of the COREIDS project supported by the CESAB (Centre d'Analyse et Synthèse sur la Biodiversité) and the FRB (Fondation pour la Recherche sur la BIodiversité). P.D. and E.C. were supported by a grant from Agence Nationale de la Recherche (project AFFAIRS, ANR-12SV005).

GLOSSARY

Biotic–resistance hypothesis the idea that healthy and/or diverse resident populations or communities of predators limit the success and the impact of invasions in an ecosystem.

Bipartite network a network of interaction composed of two groups of nodes, interactions being restricted to pairs of nodes that belong to different groups (e.g. plant pollinator networks or plant–herbivore networks).

Competition *interspecific competition* is defined as a reduction in individual fecundity, survival, or growth as a result either of exploitation of resources or of interference with individuals of another species. *Exploitative competition* occurs through the consumption of a common limiting resource, which might be another species, nutrients, water, sunshine, etc. *Interference competition* requires direct negative interactions (e.g. fights, poisoning).

Compartmentalization (of a network) the tendency for species within a network to form separate clusters based on interaction frequency.

Connectance (of a network) the proportion of realized over all potential interactions in the network.

Ecological displacement change in the range of diet, resource, or habitat used by a species as a consequence of competition with another species.

Enemy release (hypothesis) the idea that invasive species have a high population growth rate in their area of introduction because their populations are not or little controlled by specialist, coevolved enemies.

Evolutionary rescue the process by which a population confronted with a deteriorating environment evolves traits that allow it to escape extinction.

Filter (invasion filter) a biotic or abiotic constraint that makes the distribution of a trait in successful invaders different from that of introduced species that did not manage to invade. Biotic filters are due to interspecific interactions between resident and introduced species, such as competition or predation, while abiotic filters are due to the suitability or insuitability of physicochemical conditions such as soil acidity and temperature.

Food web a set of taxa linked to one another by trophic interactions (is eaten by).

Horizontal diversity (in a network) diversity of species within a given trophic level.

Hyperpredation a special case of apparent competition, that occurs when an introduced prey boosts resident populations of a generalist predator, therefore increasing predation pressure on a resident prey.

Invasional meltdown a hypothesis stating that the presence of invaders increases the probability of further invasion, inducing a positive feedback loop.

Invasive species (or invader) in this chapter, we adopt the following definition: an alien, introduced species that is able to grow in numbers, spread in space, and maintain itself in its area of introduction.

Native species a species present in a given area since prehuman times.

Phenotype matching and phenotype differences two ways to model how phenotypes may determine predation efficiency. Under the phenotype matching paradigm, efficiency is highest when predator traits match prey traits as exactly as possible; while under the phenotype difference paradigm, predation efficiency depends on the signed phenotypic distance between the two species (e.g. predation efficiency increases as predator phenotype gets increasingly larger than prey phenotype).

Relative trophic position hypothesis the idea that invaders have positive effects on resident species at higher trophic levels, while they have negative effects on species at the same trophic level.

Resident species (relative to a particular invasive species) a species that was present and maintaining itself in a given area before the arrival of a particular invasive species.

REFERENCES

Ackerly, D.D., Cornwell, W.K., 2007. A trait-based approach to community assembly: partitioning of species trait values into within- and among-community components. Ecol. Lett. 10 (2), 135–145.

Amarasekare, P., 2002. Interference competition and species coexistence. Proc. R. Soc. Lond. B Biol. Sci. 269 (1509), 2541–2550.

Arim, M., Marquet, P.A., Jaksic, F.M., 2007. On the relationship between productivity and food chain length at different ecological levels. Am. Nat. 169 (1), 66–72.

Arthington, A.H., 1991. Ecological and genetic impacts of introduced and translocated freshwater fishes in Australia. Can. J. Fish. Aquat. Sci. 48 (Suppl. 1), 33–43.

Azzurro, E., et al., 2014. External morphology explains the success of biological invasions. Ecol. Lett. 17 (11), 1455–1463.

Baiser, B., Russell, G.J., Lockwood, J.L., 2010. Connectance determines invasion success via trophic interactions in model food webs. Oikos 119 (12), 1970–1976.

Barrios-O'Neill, D., et al., 2014. Fortune favours the bold: a higher predator reduces the impact of a native but not an invasive intermediate predator. J. Anim. Ecol. 83 (3), 693–701.

Bascompte, J., 2006. Asymmetric coevolutionary networks facilitate biodiversity maintenance. Science 312 (5772), 431–433.

Bellard, C., et al., 2013. Will climate change promote future invasions? Glob. Chang. Biol. 19 (12), 3740–3748.

Berlow, E.L., et al., 2009. Simple prediction of interaction strengths in complex food webs. Proc. Natl. Acad. Sci. U.S.A. 106 (1), 187–191.

Bertelsmeier, C., et al., 2014. Worldwide ant invasions under climate change. Biodivers. Conserv. 24 (1), 117–128.

Blackburn, T.M., Cassey, P., Lockwood, J.L., 2009. The role of species traits in the establishment success of exotic birds. Glob. Chang. Biol. 15 (12), 2852–2860.

Blackburn, T.M., et al., 2011. A proposed unified framework for biological invasions. Trends Ecol. Evol. 26 (7), 333–339.

Blossey, B., Notzold, R., 1995. Evolution of increased competitive ability in invasive nonindigenous plants: a hypothesis. J. Ecol. 83 (5), 887–889.

Bobeldyk, A.M., Lamberti, G.A., 2008. A Decade after invasion: evaluating the continuing effects of rusty crayfish on a Michigan river. J. Great Lakes Res. 34 (2), 265–275.

Bohan, D.A., et al., 2013. Networking agroecology: integrating the diversity of agroecosystem interactions. Adv. Ecol. Res. 49, 1–67.

Bourdeau, P.E., Pangle, K.L., Peacor, S.D., 2011. The invasive predator Bythotrephes induces changes in the vertical distribution of native copepods in Lake Michigan. Biol. Invasions 13 (11), 2533–2545.

Bourguet, D., Guillemaud, T., 2016. The hidden and external costs of pesticide use. Sustainable Agricultural Review. vol. 19. pp. 35–120.

Brannstrom, A., et al., 2011. Emergence and maintenance of biodiversity in an evolutionary food-web model. Theor. Ecol. 4 (4), 467–478.

Brännström, Å., et al., 2012. Modeling the ecology and evolution of communities: a review of past achievements, current efforts, and future promises. Evol. Ecol. Res. 14, 601–625.

Brose, U., Williams, R.J., Martinez, N.D., 2006. Allometric scaling enhances stability in complex food webs. Ecol. Lett. 9 (11), 1228–1236.

Brown, L.E., et al., 2011. Food web complexity and allometric scaling relationships in stream mesocosms: implications for experimentation. J. Anim. Ecol. 80 (4), 884–895.

Bruno, J.F., et al., 2005. Insights into biotic interactions from studies of species invasions. In: Species Invasions Insights Into Ecology, Evolution, and Biogeography. Sinauer Associates, Inc., Sunderland, MA, pp. 13–40.

Byers, J.E., 2000. Competition between two estuarine snails: implications for invasions of exotic species. Ecology 81 (5), 1225–1239.

Byrnes, J.E., Reynolds, P.L., Stachowicz, J.J., 2007. Invasions and extinctions reshape coastal marine food webs. PLoS One 2 (3), 1–7.

Cadotte, M.W., et al., 2015. Predicting communities from functional traits. Trends Ecol. Evol. 30 (9), 510–511(Box 1).

Carey, M.P., Wahl, D.H., 2014. Native fish diversity alters the effects of an invasive species on food webs. Ecology 91 (10), 2965–2974.

Carey, C.C., et al., 2012. Eco-physiological adaptations that favour freshwater cyanobacteria in a changing climate. Water Res. 46 (5), 1394–1407.

Carlson, S.M., Cunningham, C.J., Westley, P.A.H., 2014. Evolutionary rescue in a changing world. Trends Ecol. Evol. 29 (9), 521–530.

Carlsson, N.O.L., Brönmark, C., Hansson, L.A., 2004. Invading herbivory: the golden apple snail alters ecosystem functioning in Asian wetlands. Ecology 85 (6), 1575–1580.

Carlsson, N.O.L., Sarnelle, O., Strayer, D.L., 2009. Native predators and exotic prey—an acquired taste? Front. Ecol. Environ. 7 (10), 525–532.

Carpenter, D., Cappuccino, N., 2005. Herbivory, time since introduction and the invasiveness of exotic plants. J. Ecol. 93 (2), 315–321.

Carroll, S.P., et al., 2005. And the beak shall inherit—evolution in response to invasion. Ecol. Lett. 8 (9), 944–951.

Carvalheiro, L.G., et al., 2008. Apparent competition can compromise the safety of highly specific biocontrol agents. Ecol. Lett. 11 (7), 690–700.

Carvalheiro, L.G., Buckley, Y.M., Memmott, J., 2010. Diet breadth influences how the impact of invasive plants is propagated through food webs breadth influences how the impact of invasive plants is propagated through food webs. Ecology 91 (4), 1063–1074.

Cassey, P., et al., 2004. Global patterns of introduction effort and establishment success in birds. Proc. R. Soc. Lond. Ser. B Biol. Sci. 271 (Suppl. 6), S405–S408.

Chapin, F.S., et al., 2000. Consequences of changing biodiversity. Nature 405 (6783), 234–242.

Charnov, E.L., 1976. Optimal foraging, the marginal value theorem. Theor. Popul. Biol. 9 (2), 129–136.

Clarke, B., Murray, J., Johnson, M., 1984. The extinction of endemic species by a program of biological control. Pac. Sci. 38 (2), 97–104.

Cohen, J.E., 1977. Food webs and the dimensionality of trophic niche space. Proc. Natl. Acad. Sci. U.S.A. 74 (10), 4533–4536.

Cohen, J.E., Jonsson, T., Carpenter, S.R., 2003. Ecological community description using the food web, species abundance and body size. Proc. Natl. Acad. Sci. U.S.A. 100 (4), 1781–1786.

Colautti, R.I., et al., 2004. Is invasion success explained by the enemy release hypothesis? Ecol. Lett. 7 (8), 721–733.

Coltman, D.W., et al., 2003. Undesirable evolutionary consequences of trophy hunting. Nature 426, 655–658.

Comte, L., Cucherousset, J., Olden, J.D., 2016. Global test of Eltonian niche conservatism of nonnative freshwater fish species between their native and introduced ranges. Ecography 39, 1–9.

Courchamp, F., Langlais, M., Sugihara, G., 2000. Rabbits killing birds: modelling the hyperpredation process. J. Anim. Ecol. 69 (1), 154–164.

Courchamp, F., Chapuis, J.-L., Pascal, M., 2003. Mammal invaders on islands: impact, control and control impact. Biol. Rev. 78 (3), 347–383.

Crain, C., Kroeker, K., Halpern, B., 2008. Interactive and cumulative effects of multiple human stressors in marine systems. Ecol. Lett. 11, 1304–1315.

Crowder, D.W., Snyder, W.E., 2010. Eating their way to the top? Mechanisms underlying the success of invasive insect generalist predators. Biol. Invasions 12 (9), 2857–2876.

Davies, K.F., Cavender-Bares, J., Deacon, N., 2011. Native communities determine the identity of exotic invaders even at scales at which communities are unsaturated. Divers. Distrib. 17 (1), 35–42.

Davis, M.A., 2003. Biotic globalization: does competition from introduced species threaten biodiversity? Bioscience 53 (5), 481–489.

Deraison, H., et al., 2015. Functional trait diversity across trophic levels determines herbivore impact on plant community biomass. Ecol. Lett. 18, 1346–1355.

DeRivera, C.E., et al., 2005. Biotic resistance to invasion: native predator limits abundance and distribution of an introduced crab. Ecology 86 (12), 3364–3376.

Deudero, S., et al., 2014. Benthic community responses to macroalgae invasions in seagrass beds: diversity, isotopic niche and food web structure at community level. Estuar. Coast. Shelf Sci. 142, 12–22.

Dick, J.T.A., et al., 2013. Ecological impacts of an invasive predator explained and predicted by comparative functional responses. Biol. Invasions 15 (4), 837–846.

Dickman, E.M., et al., 2008. Light, nutrient, and food chain length constrain planktonic energy transfer efficiency across multiple trophic levels. Proc. Natl. Acad. Sci. U.S.A. 105, 18408–18412.

Dorcas, M.E., et al., 2012. Severe mammal declines coincide with proliferation of invasive Burmese pythons in Everglades National Park. Proc. Natl. Acad. Sci. U.S.A. 109 (7), 2418–2422.

Downing, A.S., et al., 2013. Was lates late? A null model for the Nile perch boom in Lake Victoria. PLoS One 8 (10). e76847.

Drake, J.A., 1990. The mechanics of community assembly and succession. J. Theor. Biol. 147 (2), 213–233.

Duffy, J.E., et al., 2007. The functional role of biodiversity in ecosystems: incorporating trophic complexity. Ecol. Lett. 10 (6), 522–538.

Duyck, P.F., David, P., Quilici, S., 2004. A review of relationships between interspecific competition and invasions in fruit flies (Diptera: Tephritidae). Ecol. Entomol. 29 (5), 511–520.

Duyck, P.-F., et al., 2006a. Importance of competition mechanisms in successive invasions by polyphagous tephritids in La Réunion. Ecology 87 (7), 1770–1780.

Duyck, P.F., David, P., Quilici, S., 2006b. Climatic niche partitioning following successive invasions by fruit flies in La Réunion. J. Anim. Ecol. 75 (2), 518–526.

Duyck, P.F., David, P., Quilici, S., 2007. Can more K-selected species be better invaders? A case study of fruit flies in La Réunion. Divers. Distrib. 13 (5), 535–543.

Dzialowski, A.R., Lennon, J.T., Smith, V.H., 2007. Food web structure provides biotic resistance against plankton invasion attempts. Biol. Invasions 9 (3), 257–267.

Ehrenfeld, J.G., 2010. Ecosystem consequences of biological invasions. Annu. Rev. Ecol. Evol. Syst. 41, 59–80.

Enge, S., Nylund, G.M., Pavia, H., 2013. Native generalist herbivores promote invasion of a chemically defended seaweed via refuge-mediated apparent competition. Ecol. Lett. 16 (4), 487–492.

Fontaine, C., et al., 2011. The ecological and evolutionary implications of merging different types of networks. Ecol. Lett. 14 (11), 1170–1181.

Forsyth, D.M., et al., 2004. Climatic suitability, life-history traits, introduction effort, and the establishment and spread of introduced mammals in Australia. Conserv. Biol. 18 (2), 557–569.

France, K.E., Duffy, J.E., 2006. Consumer diversity mediates invasion dynamics at multiple trophic levels. Oikos 113 (3), 515–529.

Freeman, A.S., Byers, J.E., 2006. Divergent induced responses to an mussel populations. Science 58, 831–833.

Fritts, T.H., Rodda, G.H., 1998. The role of introduced species in the degradation of island ecosystems: a case history of Guam. Annu. Rev. Ecol. Evol. Syst. 29 (1998), 113–140.

Galiana, N., et al., 2014. Invasions cause biodiversity loss and community simplification in vertebrate food webs. Oikos 123 (6), 721–728.

Gallardo, B., et al., 2016. Global ecological impacts of invasive species in aquatic ecosystems. Glob. Chang. Biol. 22, 151–163.

Gardener, M.R., Atkinson, R., Rentería, J.L., 2010. Eradications and people: lessons from the plant eradication program in Galapagos. Restor. Ecol. 18 (1), 20–29.

Gauzens, B., et al., 2015. Trophic groups and modules: two levels of group detection in food webs. J. R. Soc. Interface 12 (106). 20141176.

Georgelin, E., Loeuille, N., 2014. Dynamics of coupled mutualistic and antagonistic interactions, and their implications for ecosystem management. J. Theor. Biol. 346, 67–74.

Gido, K.B., Franssen, N.R., 2007. Invasion of stream fishes into low trophic positions. Ecol. Freshw. Fish 16 (3), 457–464.

Gomulkiewicz, R., Holt, R.D., 1995. When does evolution by natural selection prevent extinction? Evolution 49 (1), 201–207.

Goudswaard, K., Witte, F., Katunzi, E.F.B., 2008. The invasion of an introduced predator, Nile perch (Lates niloticus, L.) in Lake Victoria (East Africa): chronology and causes. Environ. Biol. Fish. 81 (2), 127–139.

Gravel, D., et al., 2011. Trophic theory of island biogeography. Ecol. Lett. 14 (10), 1010–1016.

Gray, S.M., Dykhuizen, D.E., Padilla, D.K., 2015. The effects of species properties and community context on establishment success. Oikos 124 (3), 355–363.

Green, P.T., et al., 2011. Invasional meltdown: invader-invader mutualism facilitates a secondary invasion. Ecology 92 (9), 1758–1768.

Griffen, B.D., Guy, T., Buck, J.C., 2008. Inhibition between invasives: a newly introduced predator moderates the impacts of a previously established invasive predator. J. Anim. Ecol. 77 (1), 32–40.

Grift, R., et al., 2003. Fisheries-induced trends in reaction norms for maturation in North Sea plaice. Mar. Ecol. Prog. Ser. 257, 247–257.

Gunderson, A.R., Leal, M., 2016. A conceptual framework for understanding thermal constraints on ectotherm activity with implications for predicting responses to global change. Ecol. Lett. 19 (2), 111–120.

Gurevitch, J., Padilla, D.K., 2004. Are invasive species a major cause of extinctions? Trends Ecol. Evol. 19 (9), 470–474.

Gurevitch, J., et al., 2011. Emergent insights from the synthesis of conceptual frameworks for biological invasions. Ecol. Lett. 14 (4), 407–418.

Halaj, J., Wise, D.H., 2001. Terrestrial trophic cascades: how much do they trickle? Am. Nat. 157 (3), 262–281.

Hanski, I., 1999. Metapopulation Ecology. Oxford University Press, Oxford.

Hanski, I., Hansson, L., Henttonen, H., 1991. Specialist predators, generalist predators, and the microtine rodent cycle. J. Anim. Ecol. 60 (1), 353–367.

Hanski, I., et al., 2001. Small-rodent dynamics and predation. Ecology 82 (6), 1505–1520.

Harding, J.M., 2003. Predation by blue crabs, Callinectes sapidus, on rapa whelks, Rapana venosa: possible natural controls for an invasive species? J. Exp. Mar. Biol. Ecol. 297 (2), 161–177.

Hawkes, C.V., 2007. Are invaders moving targets? The generality and persistence of advantages in size, reproduction, and enemy release in invasive plant species with time since introduction. Am. Nat. 170 (6), 832–843.

Heleno, R.H., et al., 2009. Effects of alien plants on insect abundance and biomass: a food-web approach. Conserv. Biol. 23 (2), 410–419.

Heleno, R., et al., 2010. Evaluation of restoration effectiveness: community response to the removal of alien plants. Ecol. Appl. 20 (5), 1191–1203.

Henneman, M.L., Memmott, J., 2001. Infiltration of a Hawaiian community by introduced biological control agents. Science 293 (5533), 1314–1316.

Herms, D.A., Mattson, W.J., 1992. The dilemma of plants: to grow or to defend. Q. Rev. Biol. 67, 283–335.

Higgins, S.N., Vander Zanden, M.J., 2009. What a difference a species makes: a meta-analysis of Dreissena mussel impacts on freshwater ecosystems. Ecol. Monogr. 79 (1), 3–24.

Hoddle, M.S., 2004. Restoring balance: using exotic species to control invasive exotic species. Conserv. Biol. 18 (1), 38–49.

Holt, R.D., 1977. Predation, apparent competition, and the structure of prey communities. Theor. Popul. Biol. 12 (2), 197–229.

Holway, D.A., et al., 2002. The causes and consequences of ant invasions. Annu. Rev. Ecol. Syst. 33, 181–233.

Human, K.G., Gordon, D.M., 1996. Exploitation and interference competition between the invasive Argentine ant, Linepithema humile, and native ant species. Oecologia 105 (3), 405–412.

Hunter, M.D., Price, P.W., 1992. Playing chutes and ladders: heterogeneity and the relative roles of bottom-up and top-down forces in natural communities. Ecology 73 (3), 724–732.

Ings, T.C., et al., 2009. Ecological networks—beyond food webs. J. Anim. Ecol. 78 (1), 253–269.

Jaeger Miehls, A.L., et al., 2009. Invasive species impacts on ecosystem structure and function: a comparison of Oneida Lake, New York, USA, before and after zebra mussel invasion. Ecol. Model. 220 (22), 3194–3209.

Jeschke, J.M., Strayer, D.L., 2006. Determinants of vertebrate invasion success in Europe and North America. Glob. Chang. Biol. 12 (9), 1608–1619.

Jeschke, J.M., et al., 2014. Defining the impact of non-native species. Conserv. Biol. 28 (5), 1188–1194.

Jones, E.I., Gomulkiewicz, R., 2012. Biotic interactions, rapid evolution, and the establishment of introduced species. Am. Nat. 179 (2), E28–E36.

Jones, C.G., Lawton, J.H., Shachak, M., 1994. Organisms as ecosystem engineers organisms as ecosystem engineers. Oikos 69 (3), 373–386.

Jones, E.I., Nuismer, S.L., Gomulkiewicz, R., 2013. Revisiting Darwin's conundrum reveals a twist on the relationship between phylogenetic distance and invasibility. Proc. Natl. Acad. Sci. U.S.A. 110 (51), 20627–20632.

Jordano, P., Bascompte, J., Olesen, J.M., 2003. Invariant properties in coevolutionary networks of plant-animal interactions. Ecol. Lett. 6, 69–81.

Kamenova, S., Bartley, T.J., Bohan, D.A., Boutain, J.R., Colautti, R.I., Domaizon, I., Fontaine, C., Lemainque, A., Le Viol, I., Mollot, G., Perga, M.-E., Ravigné, V., Massol, F., 2017. Invasions toolkit: current methods for tracking the spread and impact of invasive species. Adv. Ecol. Res. 56, 85–182.

Kéfi, S., et al., 2012. More than a meal…integrating non-feeding interactions into food webs. Ecol. Lett. 15, 291–300.

Kennedy, T.A., et al., 2002. Biodiversity as a barrier to ecological invasion. Nature 417 (6889), 636–638.

Kettenring, K.M., Adams, C.R., 2011. Lessons learned from invasive plant control experiments: a systematic review and meta-analysis. J. Appl. Ecol. 48 (4), 970–979.

King, R.B., Ray, J.M., Stanford, K.M., 2006. Gorging on gobies: beneficial effects of alien prey on a threatened vertebrate. Can. J. Zool. 84 (1), 108–115.

Kneitel, J.M., Perrault, D., 2006. Disturbance-induced changes in community composition increase species invasion success. Community Ecol. 7 (2), 245–252.

Kulhanek, S.A., Ricciardi, A., Leung, B., 2011. Is invasion history a useful tool for predicting the impacts of the world's worst aquatic invasive species? Ecol. Appl. 21 (1), 189–202.

Lambrinos, J.G., 2004. How interactions between ecology and evolution influence contemporary invasion dynamics. Ecology 85 (8), 2061–2070.

Law, R., 2000. Fishing, selection, and phenotypic evolution. ICES J. Mar. Sci. 57 (3), 659–668.

Law, R., Daniel Morton, R., 1996. Permanence and the assembly of ecological communities. Ecology 77 (3), 762–775.

Leibold, M.A., et al., 2004. The metacommunity concept: a framework for multi-scale community ecology. Ecol. Lett. 7 (7), 601–613.

Lennon, J.T., Smith, V.H., Dzialowski, A.R., 2003. Invasibility of plankton food webs along a trophic state gradient. Oikos 103, 191–203.

Leung, B., et al., 2002. An ounce of prevention or a pound of cure: bioeconomic risk analysis of invasive species. Proc. Biol. Sci. 269 (1508), 2407–2413.

Levine, J.M., D'Antonio, C.M., 1999. Elton revisited: a review of evidence linking diversity and invasibility. Oikos 87, 15–26.

Levins, R., 1969. Some demographic and genetic consequences of environmental heterogeneity for biological control. Bull. Entomol. Soc. Am. 15 (3), 237–240.

Loeuille, N., Loreau, M., 2004. Nutrient enrichment and food chains: can evolution buffer top-down control? Theor. Popul. Biol. 65, 285–298.

Loeuille, N., Loreau, M., 2006. Evolution of body size in food webs: does the energetic equivalence rule hold? Ecol. Lett. 9, 171–178.

Loeuille, N., Loreau, M., 2009. Emergence of complex food web structure in community evolution models. In: Morin, P.J., Verhoef, H. (Eds.), Community Ecology. Oxford University Press, Oxford, pp. 163–178.

Loeuille, N., et al., 2013. Eco-evolutionary dynamics of agricultural networks: implications for sustainable management. Adv. Ecol. Res. 49, 339–435.

Lotze, H.K., et al., 2006. Depletion, degradation, and recovery potential of estuaries and coastal seas. Science 312, 1806–1809.

Lurgi, M., et al., 2014. Network complexity and species traits mediate the effects of biological invasions on dynamic food webs. Front. Ecol. Evol. 2, 36.

MacArthur, R.H., 1970. Species packing and competitive equilibrium for many species. Theor. Popul. Biol. 1, 1–11.

Mandrak, N.E., 1989. Potential invasion of the Great Lakes by fish species associated with climatic warming. J. Great Lakes Res. 15 (2), 306–316.

Maron, J.L., Vilà, M., Bommarco, R., et al., 2004a. Rapid evolution of an invasive plant. Ecol. Monogr. 74 (2), 261–280.

Maron, J.L., Vilà, M., Arnason, J., 2004b. Loss of enemy resistance among introduced populations of St. John's wort (Hypericum perforatum). Ecology 85 (12), 3243–3253.

Massol, F., Dubart, M., Calcagno, V., Cazelles, K., Jacquet, C., Kéfi, S., Gravel, D., 2017. Island biogeography of food webs. Adv. Ecol. Res. 56, 183–262.

Matsuzaki, S.I.S., et al., 2009. Contrasting impacts of invasive engineers on freshwater ecosystems: an experiment and meta-analysis. Oecologia 158 (4), 673–686.

McCann, K., 2000. The diversity-stability debate. Nature 405, 228–233.

McCary, M.A., et al., 2016. Invasive plants have different effects on trophic structure of green and brown food webs in terrestrial ecosystems: a meta-analysis. Ecol. Lett. 19, 328–335.

McCrary, J.K., et al., 2007. Tilapia (Teleostei: Cichlidae) status in Nicaraguan natural waters. Environ. Biol. Fish. 78 (2), 107–114.

McFadyen, R.E., 1998. Biological control of weeds. Annu. Rev. Entomol. 43, 369–393.

McGrady-Steed, J., Harris, P.M., Morin, P.J., 1997. Biodiversity regulates ecosystem predictability. Nature 390 (6656), 162–165.

McQueen, D.J., et al., 1989. Bottom-up and top-down impacts on freshwater pelagic community structure. Ecol. Monogr. 59 (3), 289–309.

Médoc, V., Firmat, C., Sheath, D.J., Pegg, J., Andreou, D., Britton, J.R., 2017. Parasites and Biological Invasions: Predicting ecological alterations at levels from individual hosts to whole networks. Adv. Ecol. Res. 57, 1–54.

Melian, C.J., et al., 2009. Diversity in a complex ecological network with two interaction types. Oikos 118, 122–130.

Miller, T.E., Kneitel, J.M., Burns, J.H., 2002. Effect of community structure on invasion success and rate. Ecology 83 (4), 898–905.

Mollot, G., et al., 2014. Cover cropping alters the diet of arthropods in a banana plantation: a metabarcoding approach. PLoS One 9 (4). e93740.

Mollot, G., Pantel, J.H., Romanuk, T.N., 2017. The effects of invasive species on the decline in species richness: a global meta-analysis. Adv. Ecol. Res. 56, 61–83.

Molloy, D.P., Powell, J., Ambrose, P., 1994. Short-term reduction of adult zebra mussels (Dreissena-Polymorpha) in the Hudson River near Catskill, New York—an effect of juvenile blue-crab (Callinectes-Sapidus) predation. J. Shellfish Res. 13 (2), 367–371.

Montoya, J.M., Pimm, S.L., Solé, R.V., 2006. Ecological networks and their fragility. Nature 442 (7100), 259–264.

Moran, E.V., Alexander, J.M., 2014. Evolutionary responses to global change: lessons from invasive species. Ecol. Lett. 17, 637–649.

Morrison, W.E., Hay, M.E., 2011. Herbivore preference for native vs. exotic plants: generalist herbivores from multiple continents prefer exotic plants that are evolutionarily naive. PLoS One 6 (3). e17227.

Müller-Schärer, H., Schaffner, U., Steinger, T., 2004. Evolution in invasive plants: implications for biological control. Trends Ecol. Evol. 19, 417–422.

Murphy, G.E.P., Romanuk, T.N., 2014. A meta-analysis of declines in local species richness from human disturbances. Ecol. Evol. 4 (1), 91–103.

Needles, L.A., et al., 2015. Trophic cascades in an invaded ecosystem: native keystone predators facilitate a dominant invader in an estuarine community. Oikos 124 (10), 1282–1292.

Nogales, M., et al., 2004. A review of feral cat eradication on islands. Conserv. Biol. 18 (2), 310–319.

O'Dowd, D.J., Green, P.T., Lake, P.S., 2003. Invasional "meltdown" on an oceanic island. Ecol. Lett. 6 (9), 812–817.

Oksanen, L., et al., 1981. Exploitation ecosystems in gradients of primary productivity. Am. Nat. 118, 240–261.

Olden, J.D., LeRoy Poff, N., Bestgen, K.R., 2006. Life-history strategies predict fish invasions and extirpations in the Colorado River Basin. Ecol. Monogr. 76 (1), 25–40.

Olsen, E.M., et al., 2004. Maturation trends indicative of rapid evolution preceded the collapse of northern cod. Nature 428, 932–935.

Ong, T.W.Y., Vandermeer, J.H., 2015. Coupling unstable agents in biological control. Nat. Commun. 6, 5991.

Orchan, Y., et al., 2013. The complex interaction network among multiple invasive bird species in a cavity-nesting community. Biol. Invasions 15 (2), 429–445.

Parker, I.M., Gilbert, G.S., 2007. When there is no escape: the effects of natural enemies on native, invasive, and noninvasive plants. Ecology 88 (5), 1210–1224.

Penk, M., Irvine, K., Donohue, I., 2015. Ecosystem-level effects of a globally spreading invertebrate invader are not moderated by a functionally similar native. J. Anim. Ecol. 84 (6), 1628–1636.

Persson, L., et al., 1992. Trophic interactions in temperate lake ecosystems: a test of food chain theory. Am. Nat. 140 (1), 59–84.

Petchey, O.L., et al., 2008. Trophically unique species are vulnerable to cascading extinction. Am. Nat. 171 (5), 568–579.

Petren, K., Case, T.J., 1996. An experimental demonstration of exploitation competition in an ongoing invasion. Ecology 77, 118–132.

Phillips, B.L., Shine, R., 2004. Adapting to an invasive species: toxic cane toads induce morphological change in Australian snakes. Proc. Natl. Acad. Sci. U.S.A. 101 (49), 17150–17155.

Pimm, S.L., 1979. The structure of food webs. Theor. Popul. Biol. 16, 144–158.

Pimm, S.L., Lawton, J.H., 1978. On feeding on more than one trophic level. Nature 275, 542–544.

Pintor, L.M., Byers, J.E., 2015. Do native predators benefit from non-native prey? Ecol. Lett. 18, 1174–1180.

Pocock, M.J.O., Evans, D.M., Memmott, J., 2012. The robustness and restoration of a network of ecological networks. Science 335 (6071), 973–977.

Pompanon, F., et al., 2012. Who is eating what: diet assessment using next generation sequencing. Mol. Ecol. 21 (8), 1931–1950.

Post, W.M., Pimm, S.L., 1983. Community assembly and food web stability. Math. Biosci. 64, 169–192.

Post, D.M., Pace, M.L., Hairston, N.G., 2000. Ecosystem size determines food-chain length in lakes. Nature 405 (6790), 1047–1049.

Reissig, M., et al., 2006. Impact of fish introduction on planktonic food webs in lakes of the Patagonian Plateau. Biol. Conserv. 132 (4), 437–447.

Reitz, S.R., Trumble, J.T., 2002. Competitive displacement among insects and arachnids 1. Annu. Rev. Entomol. 47 (1), 435–465.

Reznick, D.N., Nunney, L., Tessier, A.J., 2000. Big houses, big cars, superfleas and the costs of reproduction. Trends Ecol. Evol. 15 (10), 421–425.

Ricciardi, A., 2003. Predicting the impacts of an introduced species from its invasion history: an empirical approach applied to zebra mussel invasions. Freshw. Biol. 48 (6), 972–981.

Ricciardi, A., MacIsaac, H.J., 2000. Recent mass invasion of the North American Great Lakes by Ponto-Caspian species. Trends Ecol. Evol. 15 (2), 62–65.

Richardson, A.J., et al., 2011. The jellyfish joyride: causes, consequences and management responses to a more gelatinous future. Trends Ecol. Evol. 24 (6), 312–322.

Roemer, G.W., Donlan, C.J., Courchamp, F., 2002. Golden eagles, feral pigs, and insular carnivores: how exotic species turn native predators into prey. Proc. Natl. Acad. Sci. U.S.A. 99 (2), 791–796.

Rohr, R.P., Saavedra, S., Bascompte, J., 2014. On the structural stability of mutualistic systems. Science 345 (6195), 1253497.

Romanuk, T.N., et al., 2009. Predicting invasion success in complex ecological networks. Philos. Trans. R. Soc. Lond. B Biol. Sci. 364 (1524), 1743–1754.

Ruesink, J.L., 2005. Global analysis of factors affecting the outcome of freshwater fish introductions. Conserv. Biol. 19 (6), 1883–1893.

Salo, P., et al., 2007. Alien predators are more dangerous than native predators to prey populations. Proc. R. Soc. Lond. B Biol. Sci. 274 (1615), 1237–1243.

Salvaterra, T., et al., 2013. Impacts of the invasive alga Sargassum muticum on ecosystem functioning and food web structure. Biol. Invasions 15 (11), 2563–2576.

Sanders, D., et al., 2014. Integrating ecosystem engineering and food webs. Oikos 123 (5), 513–524.

Savidge, J.A., 1987. Extinction of an island forest avifauna by an introduced snake. Ecology 68 (3), 660–668.

Seeley, R.H., 1986. Intense natural selection caused a rapid morphological transition in a living marine snail. Proc. Natl. Acad. Sci. U.S.A. 83 (18), 6897–6901.

Shapiro, J., Wright, D.I., 1984. Lake restoration by biomanipulation: round Lake, Minnesota, the first two years. Freshw. Biol. 14 (4), 371–383.

Shea, K., Chesson, P., 2002. Community ecology theory as a framework for biological invasions. Trends Ecol. Evol. 17 (4), 170–176.

Shea, K., et al., 2010. Optimal management strategies to control local population growth or population spread may not be the same. Ecol. Appl. 20 (4), 1148–1161.

Shine, R., 2010. The ecological impact of invasive cane toads (Bufo marinus) in Australia. Q. Rev. Biol. 85 (3), 253–291.

Shurin, J.B., 2000. Dispersal limitation, invasion resistance, and the structure of pond zooplankton communities. Ecology 81 (11), 3074–3086.

Shurin, J., Borer, E., 2002. A cross ecosystem comparison of the strength of trophic cascades. Ecol. Lett. 5, 785–791.

Shurin, J.B., Gruner, D.S., Hillebrand, H., 2006. All wet or dried up? Real differences between aquatic and terrestrial food webs. Proc. R. Soc. B 273 (1582), 1–9.

Simberloff, D., 2006. Invasional meltdown 6 years later: important phenomenon, unfortunate metaphor, or both? Ecol. Lett. 9 (8), 912–919.

Simberloff, D., Gibbons, L., 2004. Now you see them, now you don't!—population crashes of established introduced species. Biol. Invasions 6, 161–172.

Smith-Ramesh L.M., Moore A.C. and Schmitz O.J., Global synthesis suggests that food web connectance correlates to invasion resistance, Glob. Chang. Biol., http://dx.doi.org/10.1111/gcb.13460, in press.

Sperfeld, E., et al., 2010. Productivity, herbivory, and species traits rather than diversity influence invasibility of experimental phytoplankton communities. Oecologia 163 (4), 997–1010.

St. Clair, S.B., et al., 2016. Biotic resistance and disturbance: rodent consumers regulate postfire plant invasions and increase plant community diversity. Ecology 97, 1700–1711.

Stachowicz, J.J., Byrnes, J.E., 2006. Species diversity, invasion success, and ecosystem functioning: disentangling the influence of resource competition, facilitation, and extrinsic factors. Mar. Ecol. Prog. Ser. 311, 251–262.

Stegen, J.C., Enquist, B.J., Ferriere, R., 2009. Advancing the metabolic theory of biodiversity. Ecol. Lett. 12 (10), 1001–1015.

Steiner, U.K., Pfeiffer, T., 2007. Optimizing time and resource allocation trade-offs for investment into morphological and behavioral defense. Am. Nat. 169 (1), 118–129.

Stockwell, C.A., Hendry, A.P., Kinnison, M.T., 2003. Contemporary evolution meets conservation biology. Trends Ecol. Evol. 18 (2), 94–101.

Strauss, S.Y., et al., 2002. Direct and ecological costs of resistance to herbivory. Trends Ecol. Evol. 17 (6), 278–285.

Strauss, S.Y., Lau, J.A., Carroll, S.P., 2006. Evolutionary responses of natives to introduced species: what do introductions tell us about natural communities? Ecol. Lett. 9 (3), 357–374.

Strong, D.R., 1992. Are trophic cascades all wet? Differentiation and donor-control in speciose ecosystems. Ecology 73 (3), 747–754.

Strong, J.S., Leroux, S.J., 2014. Impact of non-native terrestrial mammals on the structure of the terrestrial mammal food web of Newfoundland, Canada. PLoS One 9 (8), e106264.

Stuart, Y.E., et al., 2014. Rapid evolution of a native species following invasion by a congener. Science 346 (6208), 463–466.

Suarez, A.V., Case, T.J., 2002. Bottom-up effects on persistence of a specialist predator: ant invasions and horned lizards. Ecol. Appl. 12 (1), 291–298.

Thébault, E., Fontaine, C., 2010. Stability of ecological communities and the architecture of mutualistic and trophic networks. Science 329, 853–856.

Theiling, K.M., Croft, B.A., 1988. Pesticide side-effects on arthropod natural enemies: a database summary. Agric. Ecosyst. Environ. 21 (3–4), 191–218.

Thomsen, M.S., et al., 2014. Impacts of marine invaders on biodiversity depend on trophic position and functional similarity. Mar. Ecol. Prog. Ser. 495, 39–47.

Thomson, D., 2004. Competitive interactions between the invasive European honey bee and native bumble bees. Ecology 85 (2), 458–470.

Tilman, D., 2004. Niche tradeoffs, neutrality, and community structure: a stochastic theory of resource competition, invasion, and community assembly. Proc. Natl. Acad. Sci. U.S.A. 101 (30), 10854–10861.

Tilman, D., Isbell, F., Cowles, J.M., 2014. Biodiversity and ecosystem functioning. Annu. Rev. Ecol. Evol. Syst. 45 (1), 471.

Tompkins, D.M., White, A.R., Boots, M., 2003. Ecological replacement of native red squirrels by invasive greys driven by disease. Ecol. Lett. 6 (3), 189–196.

Townsend, C.R., et al., 1998. Disturbance, resource supply, and food-web architecture in streams. Ecol. Lett. 1, 200–209.

Tran, T.N.Q., et al., 2015. Patterns of trophic niche divergence between invasive and native fishes in wild communities are predictable from mesocosm studies. J. Anim. Ecol. 84, 1071–1080.

Tylianakis, J.M., et al., 2008. Global change and species interactions in terrestrial ecosystems. Ecol. Lett. 11 (12), 1351–1363.

Vacher, C., et al., 2016. Learning ecological networks from next-generation sequencing data. Adv. Ecol. Res. 54, 1–39.

Van Kleunen, M., Weber, E., Fischer, M., 2010. A meta-analysis of trait differences between invasive and non-invasive plant species. Ecol. Lett. 13 (2), 235–245.

Vander Wal, E., et al., 2013. Evolutionary rescue in vertebrates: evidence, applications and uncertainty. Philos. Trans. R. Soc. Lond. B Biol. Sci. 368 (1610), 20120090.

Vander Zanden, M.J., Casselman, J.M., Rasmussen, J.B., 1999. Stable isotope evidence for the food web consequences of species invasions in lakes. Nature 401 (6752), 464–467.

Vander Zanden, M.J., et al., 2010. A pound of prevention, plus a pound of cure: early detection and eradication of invasive species in the Laurentian Great Lakes. J. Great Lakes Res. 36 (1), 199–205.

Vermeij, G.J., 1982. Phenotypic evolution in a poorly dispersing snail after arrival of a predator. Nature 299, 349–350.

Vilà, M., Gómez, A., Maron, J.L., 2003. Are alien plants more competitive than their native conspecifics? A test using Hypericum perforatum L. Oecologia 137 (2), 211–215.

Vilcinskas, A., et al., 2013. Invasive harlequin ladybird carries biological weapons against native competitors. Science 340 (6134), 862–863.

Vitousek, P.M., et al., 1987. Biological invasion by Myrica faya alters ecosystem development in Hawaii. Science 238, 802–804.

Vo, A.E., et al., 2011. Temporal increase in organic mercury in an endangered pelagic seabird assessed by century-old museum specimens. Proc. Natl. Acad. Sci. U.S.A. 108, 7466–7471.

Wei, Z., Yang, T., Friman, V.P., Xu, Y., Shen, Q., Jousset, A., 2015. Trophic network architecture of root-associated bacterial communities determines pathogen invasion and plant health. Nat. Commun. 6:8413.

White, T.C.R., 2005. Why Does the World Stay Green? Nutrition and Survival of Plant Eaters. CSIRO Publishing, Collingwood, VIC, Australia.

White, E.M., Wilson, J.C., Clarke, A.R., 2006. Biotic indirect effects: a neglected concept in invasion biology. Divers. Distrib. 12 (4), 443–455.

Whitney, K.D., Gabler, C.A., 2008. Rapid evolution in introduced species, "invasive traits" and recipient communities: challenges for predicting invasive potential. Divers. Distrib. 14 (4), 569–580.

Williams, R.J., Martinez, N.D., 2000. Simple rules yield complex food webs. Nature 404, 180–183.

Witte, F., et al., 1992. The distribution of an endemic species flock: quantitative date on the decline of the haplochromine cichlids of Lake Victoria. Environ. Biol. Fish 23, 1–28.

Woodward, G., Hildrew, A.G., 2001. Invasion of a stream food web by a new top predator. J. Anim. Ecol. 70 (2), 273–288.

Wright, I.J., et al., 2004. The worldwide leaf economics spectrum. Nature 428 (6985), 821–827.

Yang, S., Ferrari, M.J., Shea, K., 2011. Pollinator behavior mediates negative interactions between two congeneric invasive plant species. Am. Nat. 177 (1), 110–118.

Yelenik, S.G., Antonio, C.M.D., 2013. Self-reinforcing impacts of plant invasions change over time. Nature 503 (7477), 517–520.

Yoshida, K., 2008a. Evolutionary cause of the vulnerability of insular communities. Ecol. Model. 210 (4), 403–413.

Yoshida, K., 2008b. The relationship between the duration of food web evolution and the vulnerability to biological invasion. Ecol. Complex. 5 (2), 86–98.

Zavaleta, E.S., Hobbs, R.J., Mooney, A.H., 2001. Viewing invasive species removal in a whole ecosystem-context. Trends Ecol. Evol. 16 (8), 454–459.

CHAPTER TWO

The Effects of Invasive Species on the Decline in Species Richness: A Global Meta-Analysis

G. Mollot[*,†,1], J.H. Pantel[‡,§], T.N. Romanuk[¶]

[*]SupAgro, UMR CBGP (INRA/IRD/CIRAD/Montpellier SupAgro), Montferrier-sur-Lez, France
[†]CESAB-FRB, Immeuble Henri Poincaré, Aix en Provence, France
[‡]Centre d'Ecologie Fonctionnelle et Evolutive, UMR 5175, CNRS-Université de Montpellier-UMIII-EPHE, Montpellier, France
[§]Centre for Ecological Analysis and Synthesis, Foundation for Research on Biodiversity, Bâtiment Henri Poincaré, Rue Louis-Philibert, 13100 Aix-en-Provence, France
[¶]Dalhousie University, Halifax, NS, Canada
[1]Corresponding author: e-mail address: gregory.mollot@supagro.inra.fr

Contents

Abstract

Biological invasions are one of the most important ecological disturbances that threaten native biodiversity. An expected increase in the rate of species extinction will have major effects on the structure and function of ecosystems worldwide. The goal of our study is

Advances in Ecological Research, Volume 56
ISSN 0065-2504
http://dx.doi.org/10.1016/bs.aecr.2016.10.002

to determine which ecological properties mediate the impact of invasive species on biodiversity loss on a global scale using a meta-analysis. We considered the role of properties such as the trophic and taxonomic position of invaders, taxonomic groups of invaded systems, the type of habitats invaded and whether the invasive species is included in a list of the most harmful invasive species for biodiversity loss. We compiled 185 studies that included 253 numerical values of changes of species richness due to species invasion. We investigated the role of trophic and taxonomic parameters of invaders, as well as the role of abiotic parameters of habitat on changes in species richness due to biological invasions. Our results show that plant invaders are highly represented (85% of all invaders studied), especially those belonging to the Poaceae family. For animals, predation seems to be the feeding behaviour associated with the greatest decrease in species richness and this relationship is independent of habitat type, with a 21% decline observed in aquatic habitats and a 27% decline in terrestrial habitats. In invaded communities, birds suffer the greatest decline in species richness (41% decline). Finally, we found that species richness declines in Europe are spatially autocorrelated, suggesting that the consequences of invasive species cannot be understood through local-scale analysis alone.

1. INTRODUCTION

While biological invasions have long been considered as a disturbance that can alter ecosystem function and services (Powell et al., 2010; Vila et al., 2006), the effect of the trophic and taxonomic position of invaders on native species loss has not been previously addressed at a global scale. Food web structure can be viewed as an ecological network, with nodes (species or group of species) and directed links (consumer/resource interactions), which are dynamically changing over time and space (Cohen et al., 2003; Dunne et al., 2002; Krause et al., 2003; Paine, 1980; Pimm et al., 1991). Biological invasions alter the structure of food webs initially by increasing the species richness of the ecosystem through the addition of a new species. The new species changes network structure by adding new sets of interactions, both as a resource (unless it is a top predator) and as a consumer. The effects of invasive species have been shown to propagate diffusely in food web networks (Gallardo et al., 2016). For instance, the invasion of the crazy ant, *Anoplolepis gracilipes*, in rain forests of Christmas Island led to rapid changes affecting three trophic levels, including the displacement of the red land crab in the forest above-ground level, new interactions with honeydew-producers in the trees and an increase in generalist arthropods in the canopy (O'Dowd et al., 2003). Similarly, the unintentional introduction of the brown tree snake *Boiga irregularis* in Guam caused severe changes to the food web structure of the recipient communities with multiple extinctions of bats, birds and reptiles (Fritts and Rodda, 1998).

Confounding effects of invasive species make it difficult to determine to what extent food web structure is altered, and whether the alteration can be predicted on the basis of the trophic parameters of invaders (primary resources, herbivores, predators, omnivores and detritivores). Both simulations (Romanuk et al., 2009) and reviews (Howeth et al., 2015; Hui et al., 2016) have shown that trophic role is a strong predictor of the invasibility of a species. In addition, model results suggest that invader properties are better indicators for predicting successful invasion than the recipient food web structure itself (Romanuk et al., 2009).

One of the major consequences of biological invasions is reflected by the irreversible loss of species (Newbold, 2015). The increase in species loss rate challenges biological conservation programs, and negative effects of species loss are associated with the loss of genetic diversity in both natural and agricultural communities, which threatens human well-being (Stearns, 2009). For instance, in cropping systems, the availability of different genotypes of a given crop enables to switch from sensitive to resistant cultivars in case of a significant change in temperature, ensuring the sustainability of agricultural outputs in future global warming scenarios. Indeed, healthy ecosystems with unthreatened biodiversity can be highly productive and have a greater ecological resilience, i.e. they are better able to withstand and recover from disturbance. For example, in aquatic ecosystems, the positive effects of biodiversity include more efficient recycling of organic matter, increased chelation of toxins, increased food production for many species (including humans) and enhanced transformation of carbon dioxide into food and oxygen (Munguía-Vega et al., 2015). Biodiversity is declining worldwide, and at the current rate, ecosystems may shift to alternative stable states from which they do not return to the original condition (Scheffer et al., 2001). As human activities are also dependant on the well-being of ecosystems (Pejchar et al., 2009), it is increasingly important to understand and identify those properties that allow a species to become invasive.

The ecological properties of invaders themselves, such as their trophic position and taxonomic classification, might provide an important framework by which to disentangle the mechanisms that lead to species loss with biological invasions. As comprehensive datasets of food web structure are scarce in the literature, we used species richness (S, a structural metric used to characterize network properties; Martinez, 1993) as one possible proxy or indicator of food web structure that could be used to evaluate changes between invaded and uninvaded ecosystems. It should be noted that while invasive species can disturb the entire network of interactions (Gallardo et al., 2016), the effects on changes in species richness in relation to ecological properties of food webs and invaders are not fully understood.

In this study, we compiled articles that report the effect of the presence of an invasive species on changes in species richness into a dataset. The data includes 253 empirical values of changes in species richness from 185 previously published experimental and observational invasion studies. We first determined whether the fraction of change in species richness caused by the invasive species differed based on the trophic position of the invader (for a summary of the factors and their respective variable, see Table 1). The trophic position of invaders may indicate the way they are connected with other species in a trophic network and it may be that invasive species with more connections (primary resources, omnivores) are more likely to influence other species and reduce native species richness. Second, we investigated the potential for the taxonomic group of the invader and the invaded resident species to be considered as a factor explaining species loss. The taxonomic group could be a valuable predictor of the effect of the invasion on species richness because it is often correlated with the trophic position, especially when considered with habitat type. Third, beside the trophic and taxonomic consideration of species themselves, habitat types are characterized by a set of abiotic factors that could influence the fraction of change in species loss and we analysed this property's effect on biodiversity loss. We expected that biodiversity loss due to invasive species would increase in habitats with more variable abiotic conditions because of the emergence of empty ecological niche due to the succession of species with the seasons. Harmful impact for biodiversity is one of the criteria used to include species in the ISSG Top 100 of the worst invasive species list. While we did not expect the studies included in the meta-analysis to be particularly restricted to species on this list, we did expect that invasive species included in the list to have a stronger reduction for species richness than invasive species not included on the list. Finally, we also investigated the geographical distribution of species loss and the existence of a spatial autocorrelation between the distribution of species loss. This analysis could identify potential "hotspots" of biodiversity loss due to invasive species and can help researchers determine whether richness loss effects occur in isolation or have a spatial association.

Table 1 Summary of Factors With Their Respective Categorical Variable Used in the Meta-Analysis

Factor	Categorical Variable
Trophic position of the invader (5)	Primary producers Herbivores Predators Omnivores Detritivores
Taxonomic group of the invader (5)	Viridiplantae Arthropoda Mollusca Stramenopile Chordata
Taxon of the invaded communities (6)	Birds Algae Plants Arthropods Fishes Microorganisms
Habitat type (9)	Aquatic (3) Wetland Freshwater Marine Terrestrial (6) Temperate forest Desert Mediterranean Grassland Boreal Tropical
Occurrence in the ISSG list (2)	Included Not Included

For each factor analysed, the number under brackets is the number of categorical variable that is included.

2. METHODS

2.1 Selection Criteria

We used an existing dataset built from literature search results of studies reporting the effect of invasive species on species richness (Murphy and Romanuk, 2016) to which we added new references using a modified literature search method.

We performed the literature search using the ISI Web of Science database of the following research areas: "Agriculture", "Biodiversity & Conservation", "Marine & Freshwater Ecology", "Entomology", "Environmental Sciences & Ecology" and "Zoology". We used the following search expressions: biodiversity loss OR species loss OR species richness OR community change OR species removed AND invasi* species OR invasi* OR alien species OR invad*. The search of literature was completed on 15 April 2016. We then searched for studies that experimentally manipulated species invasions ($n = 31$) or observational studies comparing invaded ecosystem with a control (without invader). The literature search yielded 9149 citations, of which 185 studies presented 253 numerical values of change in species richness due to species invasion were used in the final dataset. All the selected papers reported a mean measure of species richness and a corresponding error measure in both invaded and uninvaded ecosystems. Values were given in the text for 94 studies, and graphical results were extracted and converted to numerical values using GetData Graph Digitizer software version 2.26 (Fedorov, 2013). Studies that presented multiple responses, such as those measured in different geographical regions or for the same community with different invader species, were treated as separate results. Multiple responses that did not differ from one another were averaged. Values from studies that gave species richness measures at different densities of invader species were also averaged. For the controls, we excluded studies that had artificially removed invader from the invaded habitat and studies that used lower densities of the invader species than in the treatment. Geographic coordinates were directly extracted from the text when given, or estimated using Google Earth. We determined habitat types by entering locations of all responses into ArcGIS 10.1 (ESRI, 2011) with an implemented layer of the global terrestrial ecoregions identified by Olson et al. (2001). Studies that reported the species richness changes under a range of human densities were not included.

The invader species analysed in the current study fell into five trophic position categories (see Table 1). Primary producers are organisms that convert sunlight into chemical components which can be processed along the food chain of organisms, i.e. plants and algal species. Invader species that feed only on primary producers were ranked as herbivores. In our dataset, we called a species feeding strictly on herbivores a "predator"; we found no study reporting predation of other predators and detritivores. Invader species feeding on more than one trophic position category were ranked as omnivores; in this category, species are able to feed on primary producers and on animal resources. Filter feeding species were ranked in the detritivore category.

Particular attention is often given to the invader species itself, and the trophic properties of invaded communities are often not mentioned. Rather, many invaded community descriptions rely on the taxonomic groups of species that are present. For each record in the final dataset, we therefore ranked the invaded communities by their respective taxonomic group. Invaded communities were found to fall into six broad categories (Table 1).

In addition, we compared invader species from our dataset with the ISSG Top 100 worst invader species in the world list to determine the effects of highly invasive species on species richness changes. This was done to better understand whether human perspectives on invasion and invasive species match objective measures of invasion and its realised consequence on biodiversity.

2.2 Data Analysis

The meta-analysis was done to compare species richness between experimental and control habitats with invasive vs no invasive species. Since the studies included in our dataset differed greatly from each other in both methodology and biological factors, we used a random effects meta-analysis, which considers the effect sizes to exhibit random variation among studies (after Murphy and Romanuk, 2014). Inverse variance weighting was used to pool richness estimates (R core team, 2013; Schwarzer, 2007). The response variable of the meta-analysis was the response ratio (RR) of species richness in experimental vs. control treatments, calculated using the standard equation of Murphy and Romanuk (2014), $RR = \ln (S_{\text{experimental}}/S_{\text{control}})$, which is frequently used to measure effect sizes in ecological meta-analysis (Hedges et al., 1999). The log-transformation of the ratio was used to approximate a normal distribution and give a scaling such that values significantly differ from zero indicate a change in species richness, and the sign of the RR indicates the direction of the effect with respect to the control. The responses were weighted to give greater weight to experiments with a smaller standard error. Variance for each response was calculated as:

$$v = \left(S_e^2 / n_e X_e^2 \right) + \left(S_c^2 / n_c X_c^2 \right),$$

where v is the variance, (e) is experimental, (c) is control, S is the species richness in that response, n is sample size and X is the averaged measure of species richness.

We performed categorical meta-analyses (in contrast to continuous meta-analyses that use continuous variables) to assess the effect of five different factors on the magnitude of change in species richness between the control and invaded treatments (Table 1).

We used the between-class heterogeneity statistic (Qb) to test whether the observed differences are due to sampling error and not due to the effect of the category. This means that a significant between-group heterogeneity statistic suggests effect sizes between the different levels of a factor are heterogenous, i.e. the differences are not due to sampling error alone.

We tested for the presence of spatial autocorrelation between the RR calculated for each individual study. Moran I coefficients were calculated using the R package "ape" (Paradis et al., 2004) from a binary distance matrix with all pairwise geographic distances between study sites. Moran's I is a measure of how connected the values of RR are, based on the locations where they were measured. A map of the distribution of species loss was computed with Google Earth (Mountain View, CA).

Significant differences in an effect size from zero were assessed using 95% CI of the RR, indicating a decrease or increase in the species richness in the invaded ecosystems, compared to the control treatment (without invader). In the text, the percentage of change in the responses is also presented. This was calculated as the percentage of change in species richness in the experimental treatment (e) compared to the control (c) with the following expression:

$$\% = ((X_e/X_c) - 1) \times 100$$

where % is the percentage of change in species richness, X_e is the averaged measure of the species richness in the experimental treatment and X_c is the averaged measure of the species richness in the control.

2.3 Publication Bias

Published studies of invasion are skewed towards those that find highly significant results. The results of invasion into communities that then show little or no change in species richness are much less likely to appear in the literature (Møller and Jennions, 2001). We assessed the publication bias with a funnel plot of the sample size of each study as a function of the respective effect size measure. Visual inspection of the distribution of the data will suggest a publication bias when results are skewed in favour of studies that have small sample sizes compared to those with large sample sizes, partly due to different sampling effort. The test for publication bias, based upon funnel plot asymmetry, used a weighted linear regression of the treatment effect on the effect standard error (Egger and Smith, 1997), tested as Student's t distribution.

3. RESULTS

3.1 Trophic Position of Invaders

Our results show that on average, invasions by a single species lead to a 16.6% decrease in total species richness ($n=253$). For habitat type, between-class heterogeneity was significant (Qb$=35.71$, $P<0.0001$, $n=219$), suggesting that the magnitude of species loss differed between terrestrial and aquatic habitats. This difference is likely explained by the particular effect of detritivores in aquatic habitats. In aquatic habitats, we observed a significant increase in species richness when the invader was a detritivore (29.91% increase, $n=10$, Fig. 1). In terrestrial habitats, we observed a significant decrease in species richness when the invader was a primary producer (16.46% decrease, $n=177$). In both terrestrial and aquatic habitats, we observed a significant decrease in species richness when the invader was a predator (21.06% decrease, $n=4$ for aquatic; and 27.69% decrease, $n=3$ for terrestrial). Within the terrestrial habitats, between-class heterogeneity was not significant (Qb$=5.18$, $P=0.2695$, $n=167$), suggesting that the magnitude of species loss did not depend on invader trophic role. Within the aquatic habitat, between-class heterogeneity was marginally nonsignificant (Qb$=8.82$, $P=0.0657$, $n=52$), suggesting that the magnitude of species loss might differ with invader trophic role.

3.2 Taxonomic Classification of the Invaders

When considering the taxonomic classification of the invaders (Fig. 2), we observed a significant decrease in species richness when the invader belonged to the Arthropoda, Chordata and Viridiplantae (Table 2). We found no significant change in species richness when the invader belonged to Mollusca and Stramenopiles. Between-class heterogeneity was significant (Qb$=14.06$, $P=0.0071$), suggesting that the magnitude of species loss differed across taxonomic classification of the invaders.

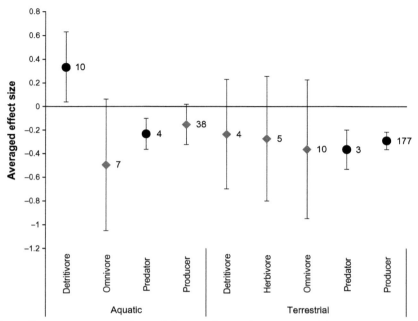

Fig. 1 Change in species richness following the trophic position of the invader. Average response ratios (averaged effect size) and 95% confidence intervals of species richness for each trophic position of the invasive species. The graph is splitted in two categories corresponding to habitat type. The values beside each point represent the number of response included in the analysis. Values that significantly differ from zero according to the 95% confidence intervals are plotted with a *black circle*.

3.3 Effects of Invasion Within Recipient Communities

Within the terrestrial habitat, we observed a significant decrease in species richness when the taxonomic group of the invaded resident community (Fig. 3) were birds (41.31% decrease, $n=8$), plants (28.42% decrease, $n=119$) and arthropods (19.12% decrease, $n=50$). Within the aquatic habitat, we observed a significant decrease in species richness when the taxonomic group of the invaded resident community (Fig. 3) was algae (31.29% decrease, $n=10$). Between-class heterogeneity was significant (Qb$=35.71$, $P<0.0001$, $n=219$), suggesting that the magnitude of species loss differed according to habitat type. We found that between-class heterogeneity was significant within the terrestrial habitat (Qb$=19.54$, $P=0.0002$, $n=164$) and within the aquatic habitat (Qb$=9.00$, $P=0.0293$, $n=49$), suggesting that the magnitude of species loss differed according to the taxonomic group of the invaded resident species.

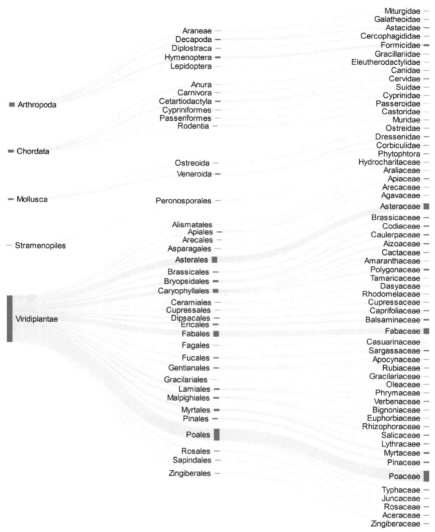

Fig. 2 Taxonomic classification of the invaders. Diagram of the taxonomic groups of the invaders ranked by phylum, order and family. Each taxonomic rank was assigned to each species on the basis of the taxonomy of NCBI. The width of each link represents the sample size of each taxon included in the current meta-analysis (phylum with $n < 3$ were discarded).

3.4 Effects of Invasion Depending on Habitat Types

Within the terrestrial habitat, our results showed a decrease in species richness when the biological invasion occurred in temperate forest (31.62% decrease, $n = 61$, Fig. 4), Mediterranean (23.84% decrease, $n = 45$) and grassland

Table 2 Percentage Change Depending on the Taxonomic Classification of the Invaders

Taxon	Sample Size	Mean RR	CI (95%)	Change (%)
Chordata	15	−0.2623	0.2314	−26.60*
Arthropoda	16	−0.4112	0.3596	−20.47*
Viridiplantae	198	−0.2694	0.0714	−14.69*

For each taxonomic group, the table displays the sample size in the current meta-analysis (group with $n < 3$ were discarded), the mean response ratio (Mean RR, which is the averaged effect size), the 95% confidence interval and the percentage of change. Values of percentage of change that not overlap with the x-axis are considered significantly different from zero and are embedded with an *asterisk*.

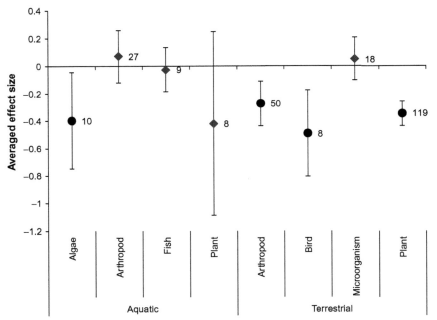

Fig. 3 Change in species richness following invasion when the taxonomic position of the invaded communities varied. Average response ratio and 95% confidence intervals of species richness for invaded communities from different taxonomic positions (taxa in which the species richness changes was measured). The values beside each point represent the number of response included in the analysis. Values that significantly differ from zero according to the 95% confidence intervals are plotted with a *black circle*.

(16.30% decrease, $n = 26$). Between-class heterogeneity was significant (Qb $= 31.66$, $P < 0.0001$, $n = 190$), suggesting that the magnitude of species loss differed according to habitat type. Within the aquatic habitat, between-class heterogeneity was not significant (Qb $= 9.67$, $P = 0.0852$, $n = 142$).

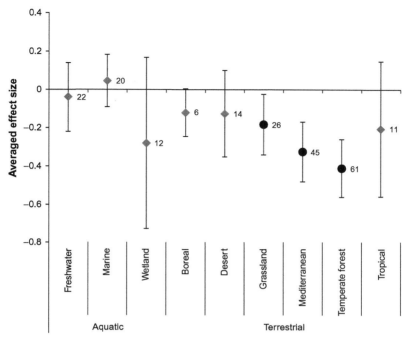

Fig. 4 Change in species richness depending on the type of habitat that is invaded. Average response ratio and 95% confidence intervals of species richness for each type of habitat that was invaded. The values beside each point represent the number of response included in the analysis. Values that significantly differ from zero according to the 95% confidence intervals are plotted with a *black circle*.

3.5 Spatial Distribution of Species Loss

When considering the spatial distribution of effect size over the entire globe, we found no significant spatial autocorrelation (Moran's I autocorrelation coefficient $= -0.0087$, $P=0.07$; Fig. 5). Given that the majority of studies were carried out in North America and Europe, we estimated the occurrence of a spatial autocorrelation in these restricted regions separately. Our results revealed a significant spatial autocorrelation between the distribution of effect size in Europe ($n=86$, Moran's I autocorrelation coefficient $=0.1307$, $P=0.0004$), while we found no significant spatial autocorrelation in the distribution of effect size in North America ($n=95$, Moran's I autocorrelation coefficient $= -0.0227$, $P=0.80$).

Fig. 5 Map of the spatial distribution of species loss. The figure is a planisphere of the distribution of the effect size of the studies included in the meta-analysis. The colour of each point displays the intensity of the effect size (averaged response ratio). Positive effect refers to values of mean ln RR > 0 (*white circles*); negative effect refers to values of mean ln RR between [−0.15;0]; Decline refers to values of mean ln RR between [−0.35;−0.15]; high decline refers to the values of mean ln RR between [−0.60;−0.35] and very high decline refers to the values of mean ln RR < − 0.60 (*black circles*). The map was created with Google Earth (Mountain View, CA).

3.6 Comparison to the ISSG Top 100 Worst Invasive Species List

Our results showed that invader species that are not included in the ISSG list had a significant negative effect size on species richness (14.85% decrease, $n=212$; Fig. 6). For invader species in our dataset that are also listed in the ISSG database, the observed change in species richness did not significantly differ from zero ($n=34$). Between-class heterogeneity was not significant ($Qb=1.09$, $P=0.2964$, $n=210$), suggesting that the magnitude of species loss did not accord well with the ISSG list.

3.7 Publication Bias

The funnel plot of sample size against effect size displayed a funnel shape (Fig. 7). Studies with small sample sizes made up the majority of the dataset. Variance was found to increase with decreasing sample size increase. Statistical testing showed low asymmetry in the funnel plot ($t=-4.39$, df$=217$, $P<0.0001$, slope$=0.28$), suggesting a low publication bias in the data (Fig. 7). The funnel shape we observed corresponds to that expected when manipulating large number of studies with similar research methods (Møller and Jennions, 2001).

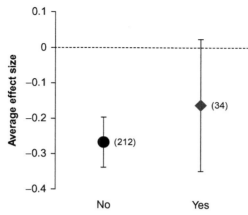

Fig. 6 Change in species richness depending on whether the invader species was listed in the ISSG Top 100 list of the worst invasive species. The average response ratio and 95% confidence intervals of species richness for invaders included and not included in the ISSG Top 100 worst invasive species are shown. The values in parentheses represent the number of responses included in the analysis. Values that significantly differed from zero according to the 95% confidence intervals are plotted with a *black circle*.

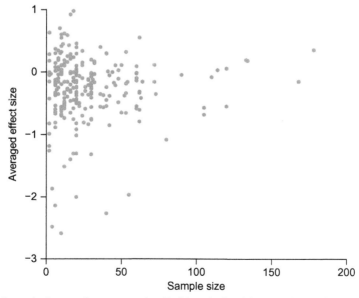

Fig. 7 Funnel plot used to assess the likelihood of publication bias. The effect sizes (ln RR) are plotted as a function of the corresponding sample sizes for each response in the meta-analysis to assess for a possible asymmetry in the distribution of responses.

4. DISCUSSION

4.1 Species Richness Changes

Species invasions have been considered second only to land-use change as a major cause of declines in biodiversity (Vitousek, 1997). Invasions by a single species are associated with substantial species loss, potentially even rivalling land-use change as the most serious disturbance when it comes to loss of native species richness (Murphy and Romanuk, 2014). The case of the invasive Nile perch *Lates niloticus* into Lake Victoria is a great illustration of the severe multitrophic effects that can follow an invasion. The invader perch not only caused the extinction of two-thirds of the endemic fish fauna but also reset the entire structure of the resident food web by reducing phytoplanktivore grazing (Bulleri et al., 2008). Thus, while invasions by nonnative species are in general recognized as a major threat for biodiversity decline, and it is likely the number of invasions will increase in the future (Kolar and Lodge, 2001), the specific patterns of extinction resulting from biological invasions seem to be case- dependent.

Effects of invading primary producers appear disproportionately represented in the literature (Lowry et al., 2012), and this high representation was observed in the current meta-analysis, with some 85% of all invasive species being either plants or algae. Within primary producers, Viridiplantae are the most studied invaders, especially those belonging to the Poaceae family which represent almost one fifth (18.6%) of all invaders analysed in the current study. The Poaceae family includes grasses and bamboos and are one of the most important plant families, with a worldwide distribution and a large representation in cultivated crops and ornamentals as well as in biofuel production. Bottom-up effects of plant diversity are of particular importance to food web structure and stability, and these effects can be altered after species invasions (Scherber et al., 2010). Decrease in plant diversity following species invasion is expected to have a strong effect on multitrophic interaction networks, and these effects increase with the degree of omnivory (Scherber et al., 2010).

Predator invasions lead to a significant decrease in species richness, independent of the habitat type, but the literature suffers from an insufficient number of studies. Seven studies reporting a change in species richness following predator invasion were included in the present meta-analysis. Arctic fox (*Vulpes lagopus*), northern yellow sac spider (*Cheiracanthium mildei*), common coqui (*Eleutherodactylus coqui*), bullfrog (*Lithobates catesbeianus*) and

Centrarchids are the predator invaders that accounted for the greatest declines in species richness in invaded ecosystems. Among them, arctic fox was the predator species with the greatest effect on species richness (36% decline) in the plant community (Maron et al., 2006). The mechanisms by which predator invaders cause a significant decline in species richness have typically been studied within the framework of trophic cascades, where feeding at one level influences feeding at the next trophic level (Paine, 1980). However, indirect effects on species richness can also occur through pathways that are not exclusively trophic. For instance, the presence of arctic foxes in the Aleutian Islands reduced abundant seabird populations, disrupting the nutrient subsidies provided by the seabirds from the sea to the land (Maron et al., 2006). Consequently, soils on fox-invaded islands had low phosphorus and nitrogen concentrations when compared to fox-free islands, challenging the development and richness of the plant community (Maron et al., 2006). Therefore, both trophic and nontrophic top-down and bottom-up effects of predator invasions can significantly decrease species richness.

Top-down effects of invasive predators might, however, be dampened by their degree of omnivory. While our results did not show omnivorous invaders to have a substantial effect on the decline of species richness in recipient ecosystems, the role of omnivory in biological invasions is believed to be important but remains unclear (David et al., 2017). On the one hand, omnivory can have a strong impact on the structural dynamics of food webs since it can break down trophic cascades and induce an alteration of energy flow by modifying energy channels (Jackson et al., 2014; Lodge et al., 1994; Nystrom et al., 1999). Many of the best known invaders are omnivores, such as rats, ants and humans. An omnivorous diet allows prey to feed on a variety of resources (Pimm, 1982) when invasion occurs in the nonnative range of the invader, where resource availability does not necessarily match the diet preference of the invader. Ability of the invader to switch from its preferred prey to another one in the recipient food web would be the key to becoming established and possibly having a strong impact in resident species richness. On the other hand, the fact that omnivores can find resources easily may lead to successful invasion, but it will not necessarily exclude native species. For instance, the successful invasion of the omnivorous gizzard shad (*Dorosoma cepedianum*) in Lake Powell (Utah, USA) occurred over just 4 years, but the invasion had limited effect on the species richness of recipient ecosystem because of the spatial segregation of predators (Vatland et al., 2007).

By considering the trophic position of invaders, we found the interesting effect that detritivorous invaders have a positive effect on the magnitude of species richness changes in aquatic ecosystems. This finding is discussed further in David et al. (2017) and agrees with the findings of the meta-analysis of Gallardo et al. (2016), which showed a very large increase in the abundance and diversity of benthic invertebrates and in the abundance of macrophytes following invasion. Filter-feeders have an important role in aquatic ecosystems, since they can process a large amount of organic matter from their habitat and enhance water quality. Invasive dreissenid filter feeders have been shown to have such large effects on water quality that they act as ecosystem engineers (Jones et al., 1994), restructuring energy from pelagic to benthic pathways (Karatayev et al., 2002). A meta-analysis focusing on the effects of dreissenid mussels for biomass change in freshwater ecosystems (Higgins and Vander Zanden, 2010) showed dramatic increases in algal and macrophytic biomass, especially for sediment-associated bacteria which showed a 2000% biomass increase. Another study showed increases in invertebrate abundance following zebra mussel invasion (Reed et al., 2004).

Finally, the analysis of the effect of invasion within recipient communities revealed that birds suffered from a strong decrease in species richness in terrestrial habitat when invaded (41% decrease). However, this finding must be considered in light of effects of habitat fragmentation that accompany invasion, which is known to have a great impact in bird communities (Andrén, 1994; Mortelliti et al., 2010).

4.2 Geographical Distribution of Species Loss

Many studies of biological invasions and richness decline focus on invasive plants (Lowry et al., 2012) occurring in North America and Europe, and those typically in temperate terrestrial environments (Cardinal et al., 2012). Temperate forests are the invaded habitat that is most studied worldwide (one-fourth of all studies) and appear to experience the greatest decrease in species richness (32% decline). It is striking, therefore, that research into invasion in the forests of South America and Africa is less represented. This may partly be explained by the prevalence of English as the primary language of articles in the ISI web of knowledge database. Compilation of Spanish, Portuguese, Indian and Chinese articles might go some way towards bridging this geographic bias in the literature.

Invasive species appear more likely to establish in disturbed, fragmented habitats, where there are available ecological niches (Murphy and

Romanuk, 2014). This implies that there is a confounding effect of habitat fragmentation in the process of biological invasion that needs to be considered when explaining the factors driving species loss. Arguably, the more the recipient habitat is disturbed, the more it will experience biological invasions and further species loss. Interestingly, biological invasions of detritivore species that act as ecosystem engineers in marine ecosystems have been shown to increase species richness via facilitation (see David et al., 2017). In these systems, increase of algal and macrophytic densities led to dramatic altered basal resource availability (Gallardo et al., 2016). This could mean, in turn, that an increase of species richness can have a negative effect on food web structure. The occurrence of a biological invasion, leading to disturbed food web, then makes the resulting food web more susceptible to further species loss through yet more biological invasions. We found, for example, a significant spatial autocorrelation between declines in species richness in Europe following a biological invasion, which were independent of the invader species considered. This result would suggest that the rate and extent of species extinction could increase in the future if global warming results in increased occurrence of biological invasions (Sala et al., 2000).

4.3 Reconsidering Invasive Species Management Policies

Biodiversity has positive effects on ecosystem functions and subsequently on the services they provide (Balvanera et al., 2006), and the increase in trade linked to human activities and global warming alter the reliability of these services (McCary et al., 2016; Powell et al., 2010; Tylianakis et al., 2008). As human activities are also dependent on the well-being of ecosystems (Pejchar et al., 2009), it is important to understand and identify the ecological conditions that facilitate species invasion into the recipient food web. In the present study, we found it interesting that the effects of species that are listed into the "worst invasive species in the worlds" appear smaller than the effects of the species that are not listed. This result suggests that our definition of what is an invasive species, and particularly those species that are problematic, should be discussed and redefined to help establish suitable management policies.

While species richness increases and decreases are not directly related to the stability of food webs, food web complexity and stability are related (Pimm, 1984) and structural changes in the food web may therefore influence stability in some scenarios. Whereas we found a significant decrease in species richness following invasions by primary producers and predators, each scenario of invasion differs depending on many parameters, which suggests that there are different classes of response for which it is necessary to apply different appropriate management policies. In the case of dreissenid mussel invasion, benthic biomass increases, while pelagic biomass decreases (Higgins et al., 2010); deer invasion in North America leads to impoverishment of plant diversity (Cardinal et al., 2012; Higgins and Vander Zanden, 2010), while ground-dwelling invertebrates diversity response does not necessarily follow a negative trend (Allombert et al., 2005; Suominen et al., 1999). Consequently, management and decision-making should be tailored to a given invader and to a given place, as effects of invasions on diversity and composition of native species in the recipient food web are substantially different (Hejda et al., 2009).

4.4 Perspectives

General trends highlighted in the present study are consistent with other studies that reported a negative effect of biological invasion for species richness of recipient ecosystems (Levine, 2000; Scherber et al., 2010; Simberloff and Daniel, 2003). The increasing rate of species loss, the economic consequences of this on productivity and yield, and the cost of invasive species management and human well-being highlight that invasion could have marked impact on our everyday existence.

We need to enhance our knowledge on the effect of invasive species into the structure of food webs to develop better theory and more data is required. In particular, it would be interesting to describe the structure of food webs, in detail, before the occurrence of a biological invasion, in order to identify local structural elements that mediate the success or failure of a given invader. Comparisons of the entire ecological network would allow better predictions of the likely occurrence of a biological invasion.

ACKNOWLEDGEMENTS

This research was funded by FRB-CESAB and Total for the use of existing datasets to explore issues linked to biodiversity. Authors thank Grace Murphy, Patrice David and three anonymous referees for their valuable contribution to improve the manuscript.

REFERENCES

Allombert, S., Steve, S., Jean-Louis, M., 2005. A Natural experiment on the impact of overabundant deer on forest invertebrates. Conserv. Biol. 19, 1917–1929.
Andrén, H., 1994. Effects of habitat fragmentation on birds and mammals in landscapes with different proportions of suitable habitat: a review. Oikos 71, 355.
Balvanera, P., Pfisterer, A.B., Buchmann, N., He, J.-S., Nakashizuka, T., Raffaelli, D., Schmid, B., 2006. Quantifying the evidence for biodiversity effects on ecosystem functioning and services. Ecol. Lett. 9, 1146–1156.

Bulleri, F., Bruno, J.F., Benedetti-Cecchi, L., 2008. Beyond competition: incorporating positive interactions between species to predict ecosystem invasibility. PLoS Biol. 6, e162.

Cardinal, E., Etienne, C., Jean-Louis, M., Côté, S.D., 2012. Large herbivore effects on songbirds in boreal forests: lessons from deer introduction on Anticosti Island. Ecoscience 19, 38–47.

Cohen, J.E., Jonsson, T., Carpenter, S.R., 2003. Ecological community description using the food web, species abundance, and body size. Proc. Natl. Acad. Sci. U.S.A. 100, 1781–1786.

David, P., Thébault, E., Anneville, O., Duyck, P.-F., Chapuis, E., Loeuille, N., 2017. Impacts of invasive species on food webs: a review of empirical data. Adv. Ecol. Res. 56, 1–60.

Dunne, J.A., Williams, R.J., Martinez, N.D., 2002. Network structure and biodiversity loss in food webs: robustness increases with connectance. Ecol. Lett. 5, 558–567.

Egger, M., Smith, G.D., 1997. Meta-analysis: potentials and promise. BMJ 315, 1371–1374.

ESRI, 2011. ArcGIS Desktop: 10.1. Environmental Systems Research Institute, Redlands, CA.

Fedorov, S., 2013. GetData Graph Digitizer. Russia.

Fritts, T.H., Rodda, G.H., 1998. The role of introduced species in the degradation of Island ecosystems: a case history of Guam 1. Annu. Rev. Ecol. Syst. 29, 113–140.

Gallardo, B., Clavero, M., Sánchez, M.I., Vilà, M., 2016. Global ecological impacts of invasive species in aquatic ecosystems. Glob. Chang. Biol. 22, 151–163.

Hedges, L.V., Jessica, G., Curtis, P.S., 1999. The meta-analysis of response ratios in experimental ecology. Ecology 80, 1150.

Hejda, M., Martin, H., Petr, P., Vojtěch, J., 2009. Impact of invasive plants on the species richness, diversity and composition of invaded communities. J. Ecol. 97, 393–403.

Higgins, S.N., Vander Zanden, M.J., 2010. What a difference a species makes: a meta-analysis of dreissenid mussel impacts on freshwater ecosystems. Ecol. Monogr. 80, 179–196.

Howeth, J.G., Gantz, C.A., Angermeier, P.L., Frimpong, E.A., Hoff, M.H., Keller, R.P., Mandrak, N.E., Marchetti, M.P., Olden, J.D., Romagosa, C.M., Lodge, D.M., 2015. Predicting invasiveness of species in trade: climate match, trophic guild and fecundity influence establishment and impact of non-native freshwater fishes. Divers. Distrib. 22, 148–160.

Hui, C., Richardson, D.M., Landi, P., Minoarivelo, H.O., Garnas, J., Roy, H.E., 2016. Defining invasiveness and invasibility in ecological networks. Biol. Invasions 18, 971–983.

Jackson, M.C., Tabitha, J., Maaike, M., Danny, S., Jeff, T., Adam, E., Judy, E., Jonathan, G., 2014. Niche differentiation among invasive crayfish and their impacts on ecosystem structure and functioning. Freshw. Biol. 59, 1123–1135.

Jones, C.G., Lawton, J.H., Moshe, S., 1994. Organisms as ecosystem engineers. Oikos 69, 373.

Karatayev, A.Y., Burlakova, L.E., Padilla, D.K., 2002. Impacts of zebra mussels on aquatic communities and their role as ecosystem engineers. In: Invasive Aquatic Species of Europe. Distribution, Impacts and Management. Springer, Netherlands, pp. 433–446.

Kolar, C.S., Lodge, D.M., 2001. Progress in invasion biology: predicting invaders. Trends Ecol. Evol. 16, 199–204.

Krause, A.E., Frank, K.A., Mason, D.M., Ulanowicz, R.E., Taylor, W.W., 2003. Compartments revealed in food-web structure. Nature 426, 282–285.

Levine, J.M., 2000. Species diversity and biological invasions: relating local process to community pattern. Science 288, 852–854.

Lodge, D.M., Kershner, M.W., Aloi, J.E., Covich, A.P., 1994. Effects of an omnivorous crayfish (Orconectes rusticus) on a freshwater littoral food web. Ecology 75, 1265–1281.

Lowry, E., Rollinson, E.J., Laybourn, A.J., Scott, T.E., Aiello-Lammens, M.E., Gray, S.M., Mickley, J., Gurevitch, J., 2012. Biological invasions: a field synopsis, systematic review, and database of the literature. Ecol. Evol. 3, 182–196.

Newbold, T., Hudson, L.N., Hill, S.L., Contu, S., Lysenko, I., Senior, R.A., Börger, L., Bennett, D.J., Choimes, A., Collen, B., Day, J., De Palma, A., Díaz, S., Echeverria-Londoño, S., Edgar, M.J., Feldman, A., Garon, M., Harrison, M.L., Alhusseini, T., Ingram, D.J., Itescu, Y., Kattge, J., Kemp, V., Kirkpatrick, L., Kleyer, M., Correia, D.L., Martin, C.D., Meiri, S., Novosolov, M., Pan, Y., Phillips, H.R., Purves, D.W., Robinson, A., Simpson, J., Tuck, S.L., Weiher, E., White, H.J., Ewers, R.M., Mace, G.M., Scharlemann, J.P., Purvis, A., 2015. Global effects of land use on local terrestrial biodiversity. Nature 520 (7545), 45–50.

Maron, J.L., Estes, J.A., Croll, D.A., Danner, E.M., Elmendorf, S.C., Buckelew, S.L., 2006. An introduced predator alters Aleutian island plant community by thwarting nutrient subsidies. Ecol. Monogr. 76, 3–24.

Martinez, N.D., 1993. Effect of scale on food web structure. Science 260, 242–243.

McCary, M.A., Mores, R., Farfan, M.A., Wise, D.H., 2016. Invasive plants have different effects on trophic structure of green and brown food webs in terrestrial ecosystems: a meta-analysis. Ecol. Lett. 19, 328–335.

Møller, A.P., Jennions, M.D., 2001. Testing and adjusting for publication bias. Trends Ecol. Evol. 16, 580–586.

Mortelliti, A., Alessio, M., Stefano, F., Corrado, B., Dario, C., Luigi, B., 2010. Independent effects of habitat loss, habitat fragmentation and structural connectivity on forest-dependent birds. Divers. Distrib. 16, 941–951.

Munguía-Vega, A., Adrián, M.-V., Andrea, S.-A., Greenley, A.P., Espinoza-Montes, J.A., Palumbi, S.R., Marisa, R., Fiorenza, M., 2015. Marine reserves help preserve genetic diversity after impacts derived from climate variability: lessons from the pink abalone in Baja California. Glob. Ecol. Conserv. 4, 264–276.

Murphy, G.E.P., Romanuk, T.N., 2014. A meta-analysis of declines in local species richness from human disturbances. Ecol. Evol. 4, 91–103.

Murphy, G.E.P., Romanuk, T.N., 2016. Data gaps in anthropogenically driven local-scale species richness change studies across the Earth's terrestrial biomes. Ecol. Evol. 6, 2938–2947.

Nystrom, P., Per, N., Christer, B., Wilhelm, G., 1999. Influence of an exotic and a native crayfish species on a littoral benthic community. Oikos 85, 545.

O'Dowd, D.J., Green, P.T., Lake, P.S., 2003. Invasional "meltdown" on an oceanic island. Ecol. Lett. 6, 812–817.

Olson, D.M., Eric, D., Wikramanayake, E.D., Burgess, N.D., Powell, G.V.N., Underwood, E.C., D'amico, J.A., Illanga, I., Strand, H.E., Morrison, J.C., Loucks, C.J., Allnutt, T.F., Ricketts, T.H., Yumiko, K., Lamoreux, J.F., Wettengel, W.W., Prashant, H., Kassem, K.R., 2001. Terrestrial ecoregions of the world: a new map of life on earth. Bioscience 51, 933.

Paine, R.T., 1980. Food webs: linkage, interaction strength and community infrastructure. J. Anim. Ecol. 49, 666.

Paradis, E., Claude, J., Strimmer, K., 2004. APE: analyses of phylogenetics and evolution in R language. Bioinformatics 20, 289–290.

Pejchar, L., Liba, P., Mooney, H.A., 2009. Invasive species, ecosystem services and human well-being. Trends Ecol. Evol. 24, 497–504.

Pimm, S.L., 1982. Food webs, In: Food webs. Springer, Netherlands, pp. 1–11.

Pimm, S.L., 1984. The complexity and stability of ecosystems. Nature 307, 321–326.

Pimm, S.L., Lawton, J.H., Cohen, J.E., 1991. Food web patterns and their consequences. Nature 350, 669–674.

Powell, K.I., Krakos, K.N., Knight, T.M., 2010. Comparing the reproductive success and pollination biology of an invasive plant to its rare and common native congeners: a case study in the genus Cirsium (Asteraceae). Biol. Invasions 13, 905–917.

R core team, 2013. R: A language and environment for statistical computing. R Foundation for Statistical Computing, Vienna, Austria.

Reed, T., Tara, R., Wielgus, S.J., Barnes, A.K., Schiefelbein, J.J., Fettes, A.L., 2004. Refugia and local controls: benthic invertebrate dynamics in lower green bay, Lake Michigan following Zebra Mussel invasion. J. Great Lakes Res. 30, 390–396.

Romanuk, T.N., Zhou, Y., Brose, U., Berlow, E.L., Williams, R.J., Martinez, N.D., 2009. Predicting invasion success in complex ecological networks. Philos. Trans. R. Soc. Lond. B Biol. Sci. 364, 1743–1754.

Sala, O.E., Chapin 3rd, F.S., Armesto, J.J., Berlow, E., Bloomfield, J., Dirzo, R., Huber-Sanwald, E., Huenneke, L.F., Jackson, R.B., Kinzig, A., Leemans, R., Lodge, D.M., Mooney, H.A., Oesterheld, M., Poff, N.L., Sykes, M.T., Walker, B.H., Walker, M., Wall, D.H., 2000. Global biodiversity scenarios for the year 2100. Science 287, 1770–1774.

Scheffer, M., Marten, S., Steve, C., Foley, J.A., Carl, F., Brian, W., 2001. Catastrophic shifts in ecosystems. Nature 413, 591–596.

Scherber, C., Christoph, S., Nico, E., Weisser, W.W., Bernhard, S., Winfried, V., Markus, F., Ernst-Detlef, S., Christiane, R., Alexandra, W., Eric, A., Holger, B., Michael, B., Nina, B., François, B., Clement, L.W., Anne, E., Christof, E., Stefan, H., Ilona, K., Alexandra-Maria, K., Robert, K., Stephan, K., Esther, K., Volker, K., Annely, K., Markus, L., Dirk, L., Cornelius, M., Migunova, V.D., Alexandru, M., Ramona, M., Stephan, P., Petermann, J.S., Carsten, R., Tanja, R., Alexander, S., Stefan, S., Jens, S., Temperton, V.M., Teja, T., 2010. Bottom-up effects of plant diversity on multitrophic interactions in a biodiversity experiment. Nature 468, 553–556.

Schwarzer, G., 2007. Meta: an R package for meta-analysis. R News 7 (3), 40–45.

Simberloff, D., Daniel, S., 2003. How much information on population biology is needed to manage introduced species? Conserv. Biol. 17, 83–92.

Stearns, S.C., 2009. Book review: sustaining life: how human health depends on biodiversity. Environ. Health Perspect. 117, A266.

Suominen, O., Otso, S., Kjell, D., Roger, B., 1999. Moose, trees, and ground-living invertebrates: indirect interactions in Swedish pine forests. Oikos 84, 215.

Tylianakis, J.M., Didham, R.K., Bascompte, J., Wardle, D.A., 2008. Global change and species interactions in terrestrial ecosystems. Ecol. Lett. 11, 1351–1363.

Vatland, S., Shane, V., Phaedra, B., 2007. Predicting the invasion success of an introduced omnivore in a large, heterogeneous reservoir. Can. J. Fish. Aquat. Sci. 64, 1329–1345.

Vila, M., Montserrat, V., Marc, T., Suehs, C.M., Giuseppe, B., Luisa, C., Alexandros, G., Philip, L., Manuela, M., Frederic, M., Eva, M., Anna, T., Troumbis, A.Y., Hulme, P.E., 2006. Local and regional assessments of the impacts of plant invaders on vegetation structure and soil properties of Mediterranean islands. J. Biogeogr. 33, 853–861.

Vitousek, P.M., 1997. Human domination of Earth's ecosystems. Science 277, 494–499.

CHAPTER THREE

Invasions Toolkit: Current Methods for Tracking the Spread and Impact of Invasive Species

S. Kamenova[*,1], **T.J. Bartley**[*], **D.A. Bohan**[†], **J.R. Boutain**[‡], **R.I. Colautti**[§], **I. Domaizon**[¶], **C. Fontaine**[‖], **A. Lemainque**[#], **I. Le Viol**[‖], **G. Mollot**[**], **M.-E. Perga**[¶], **V. Ravigné**[††], **F. Massol**[‡‡]

[*]University of Guelph, Guelph, ON, Canada
[†]UMR1347 Agroécologie, AgroSup/UB/INRA, Pôle Gestion des Adventices, Dijon Cedex, France
[‡]Botanical Research Institute of Texas, Fort Worth, TX, United States
[§]Queen's University, Kingston, ON Canada
[¶]INRA, Université de Savoie Mont Blanc, UMR CARRTEL, Thonon les Bains, France
[‖]Muséum National d'Histoire Naturelle—CESCO, UMR 7204 MNHN-CNRS-UPMC, Paris, France
[#]Commissariat à l'Energie Atomique et aux Energies Alternatives (CEA), Institut de Génomique (IG), Genoscope, Evry, France
[**]SupAgro, UMR CBGP (INRA/IRD/CIRAD/Montpellier SupAgro), Montferrier-sur-Lez, France
[††]CIRAD, UMR PVBMT Pôle de Protection des Plantes, Saint-Pierre, Réunion, France
[‡‡]CNRS, Université de Lille, UMR 8198 Evo-Eco-Paleo, SPICI Group, Lille, France
[1]Corresponding author: e-mail address: stefaniya.kamenova@gmail.com

Contents

Advances in Ecological Research, Volume 56
ISSN 0065-2504
http://dx.doi.org/10.1016/bs.aecr.2016.10.009

Abstract

Biological invasions exert multiple pervasive effects on ecosystems, potentially disrupting species interactions and global ecological processes. Our ability to successfully predict and manage the ecosystem-level impacts of biological invasions is strongly dependent on our capacity to empirically characterize complex biological interactions and their spatiotemporal dynamics. In this chapter, we argue that the comprehensive integration of multiple complementary tools within the explicit context of ecological networks is essential for providing mechanistic insight into invasion processes and their impact across organizational levels. We provide an overview of traditional (stable isotopes, populations genetics) and emerging (metabarcoding, citizen science) techniques and methods, and their practical implementation in the context of biological invasions. We also present several currently available models and machine-learning approaches that could be used for predicting novel or undocumented interactions, thus allowing a more robust and cost-effective forecast of network and ecosystem stability. Finally, we discuss the importance of methodological advancements on the emergence of scientific and societal challenges for investigating local and global species histories with several skill sets.

1. INTRODUCTION

Through species extinction, immigration, displacement, and speciation, biodiversity is intrinsically dynamic. Species immigrate, shift the limits of their distribution, or go extinct, at various rates and spatial scales (Aguirre-Gutiérrez et al., 2016; Hoffmann and Courchamp, 2016; Jackson and Overpeck, 2000; Jackson and Sax, 2010; Savage and Vellend, 2015; Sax and Gaines, 2003). Over the last decade, unprecedented globalization and intensification of human activities have significantly accelerated the frequency and magnitude of biodiversity turnover (Barnosky et al., 2012; Vitousek et al., 1997). One of the most documented expressions of global anthropogenic forcing is the human–induced movement of nonnative species (Hulme et al., 2008; Levine and D'Antonio, 2003). This phenomenon usually refers to the voluntary or accidental introduction of taxa or genotypes far from their historical distributional areas as a result of trade, tourism, agriculture, or biological control programmes (e.g. Anderson et al., 2004;

Fisher et al., 2012; Geslin et al., 2017; Hulme et al., 2008; Levine and D'Antonio, 2003; Roy and Wajnberg, 2008). This introduction of non-native species has increased by orders of magnitude since the 18th century (Grosholz, 2005).

Today, biological invasions appear as an important disruptor of biotic processes at a global scale, significantly contributing to the ongoing planetary-scale transition to a starkly different, unanticipated state (Barnosky et al., 2012; Vitousek et al., 1997). Major effects of nonnative species are changes in species diversity that could lead to impoverishment and homogenization of communities through the loss of phylogenetically or functionally unique species (Zavaleta et al., 2001). Growing evidence points out that effects of invasive species could also propagate across multiple organizational levels (e.g. Desprez-Loustau et al., 2007), thus affecting evolutionary trajectories and complex interactions within entire assemblages (Grosholz, 2002; Schlaepfer et al., 2005). All these observations have raised considerable concern about the impact of invasive species on global ecosystem functioning (e.g. Butchart et al., 2010; Hooper et al., 2005).

In order to prevent any adverse consequences of biological invasions, an efficient forecasting of their impacts is required in order to draw appropriate management actions. However, successful forecasts directly depend on our capacity to quantify such impacts in terms of biodiversity dynamics (Barnosky et al., 2012; Sala et al., 2000), evolutionary history (Sakai et al., 2001; Tayeh et al., 2015), or ecosystem functioning (Stachowicz et al., 2002) at relevant spatial and temporal scales (Fig. 1). Such an endeavour currently constitutes one of the biggest challenges in invasion ecology, still hindering our ability to prompt any general, evidence-based conclusions or management recommendations. Not only is the acquisition of large-scale empirical data methodologically demanding, but there is also a general lack of conceptual agreement upon how to define (Jeschke et al., 2014; Pyšek et al., 2012) or quantify (Hulme et al., 2013) the impact of invasive species.

First, studies explicitly measuring invaders' impact on components of ecosystem functioning are rare. Traditionally, studies tend to consider invader's establishment success as an approximation of ecological impact—in most cases, successful establishment is deemed equal to a necessarily adverse ecological impact. But in the rare cases in which functional consequences of biological invasions have been considered, nondesired ecological consequences are not necessarily the rule. Examples show that despite significant changes in the community composition and/or species richness they induce, biological invasions do not always lead to the loss of native species

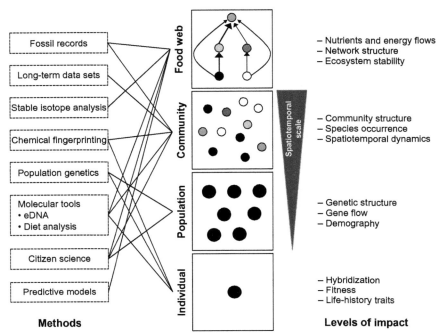

Fig. 1 Schematic overview of the utility of combining multiple methods for studying the distribution, the interactions, and the impact of invasive species at various organizational levels and spatio-temporal scales.

(Henneman and Memmott, 2001) or the disruption of local ecological processes (Carvalheiro et al., 2008). On the contrary, invasive species can actually be beneficial to local human populations (e.g. Pointier and Jourdane, 2000; Thomas et al., 2016b), suggesting that invasives' impact on ecosystem functions as a whole could be unexpected and complex to predict.

Second, the majority of studies chiefly focused on single cases (Hulme et al., 2013; Pyšek et al., 2012), usually considering only a single species within a very limited spatial and/or temporal range, preventing the incorporation of direct and indirect nonadditive effects, inherent in complex biological systems (e.g. Zavaleta et al., 2001). Moreover, they are inappropriate for establishing an explicit link or for predicting impacts on ecosystem-level processes and functions.

Network-based approaches recently emerged as a promising tool for predicting the impact of anthropogenic stressors on ecosystem function and stability (Bohan et al., 2016; Hines et al., 2015; McCann, 2007; Thompson et al., 2012). Patterns of species interactions within and among trophic levels (i.e. network structure) are likely informative about mechanisms governing ecosystem-level processes. Such mechanisms comprise bottom-up or top-down trophic cascades (Pace et al., 1999; Polis et al., 2000), community assembly and species co-occurrence (Massol et al., 2017), competition (David et al., 2017), parasitism (Lafferty et al., 2008), or functional complementarity between species (Pantel et al., 2017; Poisot et al., 2013; Yachi and Loreau, 2007). All these mechanisms have been shown to control at least partly the rate of key ecological functions such as productivity (Jenkins et al., 1992; Paine, 2002) or nutrient recycling (Handa et al., 2014; Ngai and Srivastava, 2006), and prey attack rates (Peralta et al., 2014). Consequently, interaction networks provide a powerful framework incorporating both biological complexity and stocks and flows of ecosystem functions (Memmott, 2009; Thompson et al., 2012). Building upon network theory in invasion ecology thus opens an unprecedented opportunity to improve our current predictive capacity. From a management perspective, it is also a promising approach for the implementation of rules for management and restoration strategies (Memmott, 2009) with minimal management risks (e.g. Caut et al., 2009; Courchamp et al., 2003).

Growing evidence helps reinforce these assertions in the context of biological invasions. For instance, recent theoretical advances point out the importance of network-level attributes (e.g. trophic position and diet breadth) for predicting invasion success (Romanuk et al., 2009, 2017). Empirical evidence from plant–pollinator and plant–herbivore–parasitoid networks brings additional evidence regarding the importance of a network approach when dealing with community-level impacts of biological invasions (e.g. Albrecht et al., 2014; Carvalheiro et al., 2010; Geslin et al., 2017; Lopezaraiza-Mikel et al., 2007).

Finally, species interactions are subjected to constant change and evolution through time (Eklöf et al., 2012; Hairston et al., 2005; Montoya, 2007), with biological invasions being one important driver of such dynamic changes. Consequently, integrating temporal perspective into biological invasions studies is another important prerequisite for understanding the alteration of ecosystem functioning in response to these stressors (Brose and Hillebrand, 2016; Loreau, 2010). Indeed, when looking at ecological networks as a whole, a delayed response (i.e. a time lag) to perturbations can be expected, due to the ecological inertia. Only a temporal perspective could inform about the detection of tipping points—these critical states indicative of major changes in the system. However, integrating both biological complexity at multiple trophic levels and its dynamics in a realistic ecosystem-level context could be challenging (e.g. Schindler, 1998). Especially, scalling up ecological observations could be resource-demanding resulting in a general lack of replication (e.g. Pocock et al., 2012) or in a general omission of many cryptic but important interactions (e.g. parasite or microbial communities; Amsellem et al., 2017; Médoc et al., 2017). Another major challenge relates to the general lack of "initial conditions" data, preventing a true analysis of impact.

Fortunately, during the last decade, ecologists experienced an unprecedented improvement of their methodological toolbox (summarized in Fig. 1). Thanks to methodological achievements, it is now possible to comprehensively estimate present and past biodiversity dynamics, while accounting for the numerous environmental and biological processes cited earlier. The most spectacular advance concerns the development of molecular techniques, which now allow sampling entire ecosystems or interaction networks in the form of minute amounts of ancient or modern DNA, present in the environment (eDNA; see Fig. 1). Indeed, DNA techniques offer an accessible, noninvasive, and cost-efficient tool for species surveillance, biodiversity assessment, or ecological network reconstruction. When combined with time-integrative environmental tracers, e.g., stable isotopes, molecular methods offer a powerful tool for revealing mechanisms behind energy transfer and ecosystem stability at various scales (Fig. 1). Other approaches, such as citizen sciences, are gaining important momentum and could provide valuable large-scale data about species distributions and interactions. If combined with new technologies (e.g. real-time video and GPS recording, real-time sequencing), citizen science promises to change the face of data collection and traditional ecological research as a whole. Palaeoecology and population genetics are tools that can essentially contribute to reconstructing past biodiversity dynamics, even in the absence of high-quality temporal data (Fig. 1). Finally, new modelling and statistical approaches, such as machine learning, can help optimize replicability and increase the current predictive power (Fig. 1).

In this chapter, we speculate that the inherent difficulties of collecting replicated system-level data could be significantly alleviated by mobilizing a full range of available techniques, tools, and models. The combination of multiple approaches should ensure replicable, high-resolution quantitative data, leading to a better understanding of how biological invasions affect ecosystem functioning and stability. Below we review current advances in the most popular methods in ecology and provide relevant examples of applications in the context of biological invasions.

2. DETECTING AND MONITORING SPATIOTEMPORAL CHANGES OF INVADERS AND INVADED COMMUNITIES AT DIFFERENT SCALES

One first prerequisite to any study on biological invasions is to disentangle native species from invasive species and determine when invasives were introduced. According to Webb (1985), clarifying whether a species is alien or native requires fossil and historical evidence and knowledge of its habitat, geographical distribution, cases of naturalization elsewhere, genetic diversity, reproductive pattern, and supposed means of introduction and therefore calls for large spatiotemporal perspectives. Since long-term, extensive monitoring data covering at the same time pre- and postsettlement stages have been rare, direct observation has been used in only a few studies. Current monitoring data series could be efficiently extended using retrospective approaches based on palaeorecords, museum specimens, and human historical records, while chronosequences (a series of study sites that differ primarily in the time since an event occurred, such as clearcutting, deglaciation, or species invasion) are another short-term approach to reconstructing long-term dynamics (Strayer et al., 2006).

2.1 Direct Methods for Reconstructing Past and Current Invasion History

2.1.1 Palaeogenetics and Fossil Records

Strayer et al. (2006) regretted that most studies on invasive species have been brief and lack a temporal context, with about 40% of recent studies that did not even state the amount of time that had passed since the invasion. Temporal records are crucial to detect, qualify, and explain invasion episodes. However, although palaeoecological approaches appear as the most straightforward way to complete observational data series, they have long been considered as too descriptive with little practical and conceptual applications. It is only very recently that several palaeoecological examples provided direct guidelines for the identification and management of invasives (Willis and Birks, 2006).

Lake and marine sediments preserve traces from the organisms living within the aquatic ecosystems and also act as accumulation basins for remains of organisms living in the surrounding terrestrial environments. Therefore, provided that sediment chronological accumulation is well preserved, the stratigraphical analyses of archived records can date how long a given species has been present within a given geographical area (thus invalidating or confirming its native nature), and eventually providing a timing for the first species appearance (Willis and Birks, 2006). For instance, the status of the clubmoss (*Selaginella kraussiana*, Kunze) on the oceanic Azores archipelago has long been uncertain, although its introduction was hypothesized as consequent to the discovery and settlement of the Azores (van Leeuwen et al., 2005). Yet, high numbers of spores of this taxon were already present in the sediment cores from the two island lakes several thousand years before the Portuguese discovery and the Flemish settlement in the 15th century, invalidating the introduced status for *S. kraussiana* (van Leeuwen et al., 2005). In contrast, the nonnative nature of the toxic phytoplankton species *Gymnodinium catenatum* in the Northeast Atlantic (a species responsible for major worldwide losses in aquaculture as it induces risks of toxicity in shellfish feeding upon it) could have been confirmed in cyst records from dated sediment cores originating from the West Iberian shelf (Ribeiro et al., 2012).

For long time, the study of sediment records has been restricted to macroscopic remains, providing taxonomic information only for organisms, which produce fossilizing parts (spores and pollen from terrestrial plants, zooplankton carapaces or resting eggs, diatom frustules, phytoplankton cysts). Recent advances in the extraction and analyses of free DNA preserved in sediment archives have now almost infinitely extended the field of taxa for which the reconstruction of species dynamics over time is possible (Coolen et al., 2011; Domaizon et al., 2013; Giguet-Covex et al., 2014).

Information provided by palaeorecords goes far beyond the status of the suspected nonnative species or timing of introduction. It can efficiently document the colonization mechanisms, such as founding events, source of invasions, and invasion routes, and whether observed invasions result from single or multiple introduction events (as reviewed in Cristescu, 2015). It can also address other long-standing questions on whether invasive species are the triggering mechanism for ecosystem change, or merely opportunists taking advantage of environmental change caused by other biotic or abiotic factors (Lodge, 1993).

Most breakthroughs in the area of past invasion dynamics have been provided by emblematic examples of invasions by exotic freshwater zooplankton. Indeed, crustaceans from the order Cladocera possess traits such as long-term diapausing eggs and rapid parthenogenetic reproduction that make them efficient invaders of new habitats. Because diapausing eggs contain genetic as well as morphological information, zooplankton sediment egg banks also provide opportunities for studying the phylogenetic origin and population size of founders (Duffy et al., 2000; Hairston et al., 1999; Mergeay et al., 2006), as well as population-level genetic consequences of natural colonization events (Alric et al., 2016; Brede et al., 2009).

A notorious example of the dispersive and invasive potentials of the cladoceran genus *Daphnia* comes from the "failed invasion" of Onondaga Lake, the United States (Duffy et al., 2000; Hairston et al., 1999). Palaeoecological studies of Onondaga Lake's egg bank revealed that it got simultaneously but transiently invaded (from the 1930s up to the 1980s) by two species of water flea (*Daphnia* spp.). *D. exilis* that exclusively occurs in temporary saline ponds in southwestern North America and therefore presented a range extension of 1000 km (Hairston et al., 1999), and *D. curvirostris*, an Eurasian species that has been only reported once before from North America, in extreme northwestern Canada (4500 km range extension for this species, Duffy et al., 2000). For both species, the low genetic diversity of the diapausing egg banks (allozyme diversity for Hairston et al., 1999, and genetic divergence on the 12S rRNA gene for Duffy et al., 2000) supported that dispersal resulted from an isolated event, most likely related to transport by industrial equipment. The transient settlement of both nonnative species pointed to invasibility being mediated by the lower water quality and increased lake salinity as a result of industry in the 1950s. Both invasive species disappeared as lake water quality improved ever since the 1980s (Duffy et al., 2000; Hairston et al., 1999). Most recent studies, however, highlighted that even transient invasions, resulting from time–restricted alterations of water quality (as for eutrophication and subsequent re-oligotrophication of deep European lakes), can result in irreversible changes in *Daphnia* genetic architecture via interspecific hybridization and introgression (Alric et al., 2016; Brede et al., 2009).

Mergeay et al. (2006) took palaeogenetic analysis one step further by reconstructing the invasion history of a single asexual American water flea clone (a hybrid between *Daphnia pulex* and *Daphnia pulicaria*) in Africa. They attributed the introduction of the water flea in Lake Naivasha to a single, accidental event sometime between 1927 and 1929. This period corresponded to the stocking of largemouth bass (*Micropterus salmoides*) from the United States. As a result of this concomitant introduction, authors showed that within 60 years the introduced water flea clone became the only occurring genotype, which displaced the genetically diverse, sexual population of native *D. pulex*.

The use of fossil data to address questions relative to invasions has been criticized because they do not necessarily provide accurate taxonomic information (i.e. at the species level). But recent improvements in molecular analyses of preserved resting stages promise to overcome such limits (Hofreiter et al., 2015; Leonardi et al., 2016). There are also still-persisting doubts about long-range dynamics of fossil remains (Webb, 1985) due to time and space heterogeneities in remains production, transport, and archiving to the sediment. Consequently, the taxonomic composition of the palaeorecord does not necessarily reflect the actual taxonomic composition of past communities because mechanisms by which remains are produced, transported, and archived in lake sediment may vary between ecosystems but also with time for a single ecosystem. As a result, the detection of a species in a palaeorecord constitutes more robust information as the absence of detection could be also attributed to differential production, transport, or preservation of remains. Therefore, palaeorecords might be more adequate to invalidate rather than confirm the status of a nonnative species. Nevertheless, considering these limits, when combined with contemporary records, palaeodata offer a great but so far underexploited potential to document the status, the timing, and the mechanisms of successful invasions.

2.1.2 Historical Observations and Museum Records

Fossil records and palaeoecological samples provide important historical data for analyzing the long-term ecological impacts of biological invasions. However, geological records are usually not sensitive to short-term changes, and thus other methods are needed for documenting impacts over timescales of years to decades. Financial constraints typically limit the spatial scale and temporal scope of sampling in ecological studies, and this poses a challenge for collecting consistent ecological data over periods longer than the average duration of a funded research project (Dumbrell et al., 2016). Periodic natural history surveys and environmental monitoring efforts offer important exceptions to this rule (cf. Chauvet et al., 2016; Nedwell et al., 2016; Storkey et al., 2016).

Many long-term ecological data sets have been motivated by economic interests and were only later recognized for their ecological value. For example, the Continuous Plankton Recorder survey began as a project for understanding fish stocks throughout the world's oceans and now includes more than 200,000 samples spanning eight decades (Richardson et al., 2006). Inventories of Mediterranean fish biodiversity date back from 1800, but they have only recently been used to investigate how different environmental variables and fish functional and life-history traits could predict invasion success (Ben Rais Lasram et al., 2008).

Most data sets are more regional as they are motivated and funded by local economic interests, and these can be combined to examine larger-scale patterns of invasions. The Great Lakes of North America, for example, have a history of local monitoring projects funded by nearby state governments in the United States and nearby provincial governments in Canada. As a result, aquatic invasion in the Great Lakes basin has been relatively well characterized (Grigorovich et al., 2003). More recently, the RivFunction project funded by the European Commission provided a unique opportunity to demonstrate regional-specific impact on leaf litter decomposition following the establishment of various exotic woody species in freshwater streams across France and Great Britain (Chauvet et al., 2016). These records, often collected by government organizations, remain important data sources to track the timing and spatial extent of biological invasions.

In addition to environmental monitoring, natural history surveys have a long history in biology and are probably an underutilized resource in invasion biology. For example, there are approximately 3300 herbaria worldwide that contain an estimated 350 million specimens (and associated meta-data) extending back as over 400 years (http://sweetgum.nybg.org/science/ih/). Herbarium records have become an important data source for reconstructing the spatiotemporal spread of invasive plants as well as their pathogens and microbial associations (e.g. Lavoie et al., 2012; Ristaino, 2002; Yoshida et al., 2014, 2015). However, many of these invaluable resources are themselves threatened by financial cutbacks even as interest is building in natural history collections as an important link to the ecological conditions of the past.

In contrast to reconstructing the spread of invading species over space and time, identifying the ecological consequences of these invasions is much more complicated for at least two reasons. First, ecological impacts can be complex, involving multiple trophic levels (e.g. Pantel et al., 2017), but surveys tend to focus on specific taxonomic groups (e.g. birds, insects, plankton) rather than whole ecosystems. Second, any observable changes in community structure may appear to be driven by an invasion when both are in fact driven by another factor (e.g. human disturbance). In other words, it may be unclear whether invasive species are the "drivers" of ecological changes observed over time or merely "passengers" riding along a wave of other global change factors (Didham et al., 2005; MacDougall and Turkington, 2005).

Long-term surveys again provide a potential solution to the problems of ecological complexity and distinguishing "passenger" colonizing species from invasive species driving ecological change. Ideally, such surveys would be conducted at multiple locations at time points preceding invasion by a species of interest. For example, long-term monitoring of plankton communities in Ontario lakes has proved fortuitous in providing baseline data for examining the ecological impact of *Bythotrephes longimanus*, an invasive zooplankton in both invaded and reference lakes (e.g. Palmer and Yan, 2013; Yan et al., 2002). Given that these lakes vary in chemistry, food web structure, and exposure to anthropogenic stresses including exposure to *B. longimanus*, it has been possible to show that this invader is indeed a major driver of changes to aquatic food webs.

2.1.3 Large-Scale Monitoring Through Citizen Science

Citizen sciences, where citizens take part in the data collection effort, have proven to be very efficient to monitor biodiversity over large scale and long term (Conrad and Hilchey, 2011; Couvet et al., 2008; Dickinson et al., 2010). With the development of such monitoring programmes, large data sets about species' spatial and temporal distributions, abundances, and traits are becoming available for a variety of taxa and regions. Such data sets offer promising opportunities to detect and monitor spatiotemporal changes in invasive species (Crall et al., 2010; Dickinson et al., 2010; Gallo and Waitt, 2011).

Data on species are usually collected by direct observations (vision, hearing) and, to a lesser extent, by indirect observations (traces, etc.). As for all sampling methods, they vary from unprotocoled observation reports such as single-species observations or species lists at a given times and places (e.g. http://ebird.org), to repeated standardized sampling of species abundance or species interactions within communities (e.g. http://vigienature.mnhn.fr). This variety usually goes with a trade-off between the amount of data collected and the level of constraints or skills required to collect data, and with data ranging from low to high precision (Purvis and Hector, 2000).

Citizen science monitoring programmes also collect data using digital technologies such as photographs (Stafford et al., 2010), camera traps (O'Brien et al., 2010), or sound recordings (Blumstein et al., 2011). These new recording methods benefit to biodiversity monitoring schemes by enabling a strong increase in the quantity and quality of the data collected. Such digital technologies are noninvasive, objective, and have minimal observation bias. In particular, some of these technologies can be coupled with automated signal recognition to allow species identification and individual count (Blumstein et al., 2011; Jeliazkov et al., 2016) that further reduce both observer bias and the time spent on data analysis. Interestingly, such digital recordings also allow subsequent data validation by taxonomic experts and can be reanalysed later using novel techniques and novel knowledge.

The type of data set varies greatly and so does the type of questions that can be addressed. Large-scale standardized monitoring schemes of species assemblages have been shown highly relevant and powerful to quantify changes in community composition across space and time (for example, breeding bird surveys), and linking those variations to environmental variables to investigate the ecological mechanisms involved. Such highly standardized schemes may be less relevant to detect the first stages of an introduction or to survey the beginning of the expansion of an invasive species as the spatial resolution or coverage of such scheme may not be adequate. In that case, opportunistic data can be useful. Widespread Internet access has indeed favoured the development of several extensive inventory projects involving massive networks of volunteers who provide observations following relatively unstructured protocols (e.g. Gallo and Waitt, 2011; Roberts et al., 2007). If the increase in sample size presents several advantages, the use of such data requires great care and good knowledge of the limitations (bias) of the data (Kremen et al., 2011; Snäll et al., 2011).

Data from citizen science programmes are increasingly used to assess and predict the distribution of invasive species. In their review, Dickinson et al. (2010) present several examples of the use of citizen science data to track changes in the distribution of invasive birds in North America and Europe. Compared to traditional academic surveys, citizen science can appear as a cost-effective way to monitor the spatial distribution of invasive species (Goldstein et al., 2014), and they are now used for a variety of taxa, from plants (e.g. Blois et al., 2011) to insects (e.g. Kadoya et al., 2009), mammals (e.g. Goldstein et al., 2014), or fishes (e.g. Zenetos et al., 2013).

The use of species distribution models (SDMs) (Franklin, 2013) allows linking these distributional observations with environmental data, such as land use and climate data, to better assess mechanisms of invasion and predict invasion risk and future distribution (Barbet-Massin et al., 2013; Kadoya et al., 2009). Beyond the use of data collected in the invaded zone, there is a real interest to use the species distribution data in species native area to predict future distribution as including data from the entire (native and nonnative) distribution of invasive species may help to better characterise its suitable habitat. This allows a better forecast of the potential for invasion in space and time, for example, under climate change (Beaumont et al., 2009).

Václavík and Meentemeyer (2012) using time series data on the invasion of a plant pathogen also showed that SDMs calibrated with data from later stages of the invasion performed better to estimate the potential range of invasive pathogen compared with those of the early stages of invasion.

Data from various citizen science programmes have been used to track the cascading effects invasive species have on the species of invaded communities. An important benefit of long-term monitoring of communities over large spatial scale is that they predate invasion, thereby allowing comparisons between invaded and uninvaded communities in space and time regardless of whether invasion events are predictable. Such comparison can provide almost "real-time" assessment of the impact of invasive species have on local communities.

Bird communities are among the first to be monitored by citizen scientists and clearly exemplify the benefits of such programmes (Dickinson et al., 2010). They can provide natural experimental set-up with contrasting situations to dissect the actual mechanisms, or interactions, linking invasive species to others. One good example is distinguishing between direct or apparent competition between house finches and house sparrows in North America by analyzing pattern of covariation in abundances among species (Cooper et al., 2007). Citizen science programmes also allow investigating how change in density of an invasive species can percolate to higher trophic levels within communities. Several cases highlighting the impacts invasive prey have on the density and distribution of local predators indeed come from the analysis of citizen science monitoring programmes (Barber et al., 2008; Koenig et al., 2013).

Such long-term monitoring programmes do also exist for other communities such as insects and have proven to be able to provide good quality data in a cost-effective way (Gardiner et al., 2012). For example, one of the rare well-documented cases of species decline following invasion comes from ladybird monitoring programmes in Belgium, Britain, and Switzerland, where several abundant local species have been shown declining after the arrival of *Harmonia axyridis* (Roy et al., 2012), a rapidly expanding invasive ladybird first introduced a biological control agent (Pell et al., 2008).

More recent programmes have taken advantages of new technologies available. For example, the programme eMammal in the United States uses camera traps to sample mammals over six contiguous states (McShea et al., 2016). Investigating the impact of domestic cats on wild fauna, they found that coyotes exclude domestic cats, and thereby protect natural areas by concentrating cat activity in urban areas where coyotes are rare (Kays et al., 2015).

Citizen science offers several advantages to monitor invasive species and invaded communities. If opportunistic data are useful to survey changes in distribution, the absence of information on effort developed and on its distribution in space and time limits the possibilities of using such data to survey changes in invaded communities. In contrast, in standardized monitoring programmes (using relevant sampling design in space and time), the potential bias can easier be detected and taken into account. The relevance of monitoring schemes to assess the state and trends of biodiversity across scales depends on both the quantity and the quality of data because the precision of estimates depends on the sample size (i.e. the number of sampling units available for estimation) and the natural variation of the measured parameter in time and space (i.e. variation within or among years, variation among populations or among habitats). A key issue is thus to increase the quantity and quality of locally collected data and to sustain the sampling effort through time (e.g. Dumbrell et al., 2016).

One way to increase data quantity is to try to increase the number of observers (e.g. the number of volunteers in citizen programmes), for example, by improving the communication about the existence and the general value of such programmes or by facilitating the participation with dedicated and user-friendly apps and websites. Regarding data quality, it is possible to reduce the observer error such as misidentification or poor species detection (Lotz and Allen, 2007; Strand, 1996; Thompson and Mapstone, 1997) through a better training of field staff and volunteers (Gallo and Waitt, 2011) and/or by using digital technologies (new recording methods and automatic recording devices). Finally, whatever the types of data used, citizen science data present analysis-related challenges (e.g. sampling bias, observer variability, and detection probability) that are not easily addressed with classical statistical approaches and implied the development of new approaches (Weir et al., 2005). Citizen science research has hence recently resulted in the development of new computational approaches for analysis of large, complex data sets (Bird et al., 2014; Isaac and Pocock, 2015; Kelling et al., 2009). New statistical models are developed to take into account the expertise of volunteers. For example, eBird project, which is one of the largest citizen science programmes in existence providing opportunistic data, allows birders to upload observations of bird species to an online database even if they have various levels of expertise. Modelling the expertise of birders improves the accuracy of predicting observations of a bird species at a site. Such models can also be used for predicting birder expertise given their history of eBird checklists and identifying bird species that are difficult for novices to detect (Yu et al., 2010).

As in other fields, these approaches are revolutionizing the ways in which ecologists analyse large-scale patterns and visualize change at large geographic scales.

2.1.4 Molecular Techniques

Molecular techniques based on the detection of DNA directly from organisms or from their traces (i.e. environmental DNA, also known as eDNA) are rapidly popularizing among ecologists and managers as cost- and time-effective tools for reliable monitoring of nonnative species and/or their impact on native communities. Indeed, the development of molecular techniques gained an important momentum during the last decade through efforts to build standardized DNA sequence reference databases, protocols, and analytical pipelines (e.g. Armstrong and Ball, 2005; Boyer et al., 2016; Coissac et al., 2012; Eichmiller et al., 2016; King et al., 2008; Ratnasingham and Hebert, 2007; Valentini et al., 2009b), thus rendering the approach more accurate and more user-friendly. As a consequence, molecular techniques are now very broadly applicable and have been successfully employed for the detection of a growing number of invasive organisms from eDNA samples comprising vertebrates (Adrian-Kalchhauser and Burkhardt-Holm, 2016; Jerde et al., 2011; Piaggio et al., 2014; Secondi et al., 2016), invertebrates (Ardura et al., 2015; Goldberg et al., 2013; Tréguier et al., 2014), plants (Scriver et al., 2015), pathogens (Hall et al., 2016; Lamarche et al., 2015; Schmidt et al., 2013a,b), and even invasive genotypes (Uchii et al., 2016).

According to the degree of taxonomic precision and the level of a priori knowledge about the studied system, eDNA approach could encompass a range of distinct techniques, varying in their technical requirements (Table 1). For instance, a simple polymerase chain reaction (PCR) could be one very cost-effective tool for diagnosing a single-target species, either directly from collected specimens or indirectly from its traces (e.g. excreta, fur, skin, eggs, etc.; Naaum et al., 2014). This approach requires only the development of species-specific and sensitive enough primers, allowing the unambiguous detection of the target DNA, even in very low concentrations. PCR could also be tailored to very specific, allowing the differential detection of distinct social forms (e.g. the invasive fire ant Solenopsis invicta; Yang et al., 2009) or sexes (e.g. the feral fox Vulpes vulpes; Berry et al., 2007), within the same species. For an optimal protocol, targeting short fragments

DNA Template	DNA Revealing Technique	Singleplex PCR — Single or Multiple Group-Specific or "Universal" primers	Multiplex PCR — Multiple Species- or Group-Specific Primers	Singleplex and Multiplex qPCR — Single or Multiple Species- or Group-Specific Primers Fluorescent probes	No PCR
Single whole specimen or eDNA	Gel electrophoresis	Presence/absence of a species — Prinsloo et al. (2002); Ramsey et al. (2015)	Presence/absence of a species — Thalinger et al. (2016)		
One known species		Prey attack rates — Boreau de Roincé et al. (2013); Gomez-Polo et al. (2015)	Prey attack rates — Hatteland et al. (2011); Pianezzola et al. (2013)		
		Parasitism rates — Agusti et al. (2005); Prinsloo et al. (2002)	Parasitism rates — Gariepy et al. (2005, 2008); Traugott et al. (2008)		
		Species taxonomical identity — Deng et al. (2015); Dupas et al. (2006)	Interaction network — Balmer et al. (2013); Sint and Traugott (2015)		
			Symbiont/parasite community — Kurata et al. (2016)		
				Presence/absence of one or several species — Domaizon et al. (2013); Tréguier et al. (2014)	
Single or multiple whole specimens or eDNA	Melt curves				
One or multiple known species				Prey attack rates — Gomez-Polo et al. (2015); Lundgren and Fergen (2011)	
				Parasitism rates — Liang et al (2015)	
				Interaction network — Campos-Herrera et al. (2011); Lundgren et al. (2009)	

Continued

Table 1 Summary of Major PCR Techniques Currently Used eDNA and Molecular Diet Analysis Studies—cont'd

DNA Template	DNA Revealing Technique	Singleplex PCR — Single or Multiple Group-Specific or "Universal" primers	Multiplex PCR — Multiple Species- or Group-Specific Primers	Singleplex and Multiplex qPCR — Single or Multiple Species- or Group-Specific Primers Fluorescent probes	No PCR
				Diet — Bowles et al. (2011); Deagle and Tollit (2007)	
				Relative biomass[a] — Deagle and Tollit (2007); Takahara et al. (2012)	
Single whole specimen or eDNA	Sanger sequencing	Presence/absence of a species — Adrian-Kalchhauser and Burkhardt-Holm (2016); Rougerie et al. (2011)			
One known or unknown species		Prey attack rates — Gorokhova (2006); Zarzoso-Lacoste et al. (2016)			
		Parasitism rates — Derocles et al. (2012); Traugott et al. (2006)			
		Species taxonomical identity — Kaartinen et al. (2010); Smith et al. (2011)			
		Phylogenetic signal — Kasper et al. (2004); Peralta et al. (2014)			
		Interaction network — Derocles et al. (2014); Peralta et al. (2014)			
		Diet — Wilson et al. (2009); Zarzoso-Lacoste et al. (2016)			

Multiple whole specimens or eDNA	High-throughput sequencing			
Multiple known or unknown species	Presence/absence of a species	Cannon et al. (2016); Zaiko et al. (2015b)	Presence/absence of one or several species	Andújar et al. (2015); Tang et al. (2015); Zhou et al. (2013)
	Prey attack rates	Mollot et al. (2014)	Prey attack rates	
	Parasitism rates		Parasitism rates	
	Species taxonomical identity[a]	Shokralla et al. (2015)	Species taxonomical identity	Gillett et al. (2014); Tang et al. (2015); Zhou et al. (2013)
	Phylogenetic signal[a]	See Lemmon and Lemmon (2013)	Phylogenetic signal	Andújar et al. (2015); Gillett et al. (2014)
	Interaction network	Ibanez et al. (2013); Mollot et al. (2014)	Interaction network	
	Diet	Boyer et al. (2013); Kartzinel et al. (2015)	Diet	Paula et al. (2015); Srivathsan et al. (2015, 2016)
	Relative biomass[a]	Murray et al. (2011)	Relative biomass[a]	Srivathsan et al. (2015); Tang et al. (2015); Zhou et al. (2013)
	Symbiont/parasite community	Gibson et al. (2014)	Symbiont/parasite community	Paula et al. (2015); Srivathsan et al. (2015, 2016)
	Intraspecific genetic variation		Intraspecific genetic variation	Srivathsan et al. (2016)

[a]Indicates cases where the validity of the data is debateable in the literature (e.g. quantitative estimates derived from sequence data or confident phylogenetic estimates from very short DNA fragments, targeted in metabarcoding studies).

Techniques are ordered according to the degree of a priori knowledge their require. The scope and the precision of generated data are provided. Empty cases indicate that no study using the corresponding technique is available to date.

of multicopy DNA is usually a leading choice as this optimizes the detection of rare, highly degraded eDNA molecules. Because of its simplicity, diagnostic PCR is implemented by a growing number of public agencies as an easily standardized tool for the routine surveillance of invasive species or for justifying the establishment of management actions. For instance, the Tasmanian government has adopted a sequential PCR approach (i.e. singleplex diagnostic PCR, followed by a Sanger sequencing) in the context of a programme aiming the eradication of the red fox from Tasmania (*V. vulpes*, Berry et al., 2007; Ramsey et al., 2015). Likewise, the State of Utah Division of Wildlife Resources in the United States is relying on PCR diagnostic as a monitoring tool of at-risk water bodies. Positive PCR detections are consequently used for the trigger off of appropriate management actions (for more details, see Darling and Mahon, 2011).

Multiple species- or group-specific primer pairs could be used together within a single PCR (multiplex PCR). This is a much more cost- and time-saving alternative to multiple separate singleplex PCRs, in cases where the simultaneous diagnosis of multiple species is required (e.g. Cooke et al., 2012; Láruson et al., 2012; Mackie et al., 2012; Nakamura et al., 2013). Multiplexed primer pairs usually need to target DNA fragments of contrasting lengths in order to allow the recovery of each taxon identity. Corresponding amplicons may or may not be Sanger sequenced for confirming species identity. Recent improvements in multiplexing protocols (Sint et al., 2012; Staudacher et al., 2016) as well as the availability of adapted reagents at very attractive rates (e.g. large choice of ready-to-use multiplex PCR commercial kits) could help achieving greater standardization, thus allowing comparison across multiplex PCR assays, regardless of the studied species or the target genes. For example, one interesting application of multiplex PCR is the concomitant detection and identification of invasive species and their symbionts or pathogens. Recently, Kurata et al. (2016) proposed a multiplex PCR approach for the diagnosis of *Bemisia tabaci* and its symbionts. Apparently, *B. tabaci* forms a species complex comprising several genetic groups that vary in their pest potential (pesticide resistance) and invasion impact (the ability to transmit pathogens to local crop plants), partly as a consequence of differences in the key symbionts each genetic group harbours.

PCR-based approaches could be complemented by a high-throughput sequencing of amplified DNA molecules (i.e. metabarcoding; Taberlet et al., 2012b). This approach allows the simultaneous taxonomic assignment of multiple species present in complex environmental samples (water, soil, faeces, gut contents, etc.), requiring no or little a priori knowledge about the biodiversity under examination (Table 1). Two recent studies used this eDNA metabarcoding approach for assessing the whole biodiversity from marine and freshwater bodies—ecosystems for which traditional methods provide very limited insight. In both studies, the metabarcoding approach showed to be very successful in detecting rare, cryptic species that have been missed with visual surveys (Cannon et al., 2016; Port et al., 2016). More interestingly, Cannon et al. (2016) were able to successfully detect not only aquatic and semiaquatic taxa but also terrestrial species occurring nearby the riverbanks of the Cuyahoga River they surveyed. The DNA from the invasive Asian carp was also detected, while this invader was not known to be present in the Cuyahoga River. These encouraging results suggest that DNA metabarcoding could be a sensitive tool for monitoring present biodiversity, even in previously uncharacterized environments. When coupled with relevant estimates about temporal scales, the metabarcoding approach also allows reconstructing past biodiversity dynamics, based on sedimentary ancient DNA (e.g. Domaizon et al., 2013; Pansu et al., 2015a; Thomsen and Willerslev, 2015) or DNA trapped in deep ice cores (Willerslev et al., 2007). Metabarcoding approach based on ancient DNA helped thus reveal major and often unexpected shifts in plant and animal assemblages through large geological periods (e.g. Domaizon et al., 2013; Pansu et al., 2015a; Willerslev et al., 2014) as well as long-lasting impact caused by invasive species on native communities (Pansu et al., 2015b).

Furthermore, biodiversity estimates based on metabarcoding data have been shown to provide more accurate and, most importantly, auditable biodiversity estimates compared to traditional taxonomic surveys, strongly encouraging their use as source for policymaking and ecosystem-level management (Baird and Hajibabaei, 2012; Ji et al., 2013; Taberlet et al., 2012b; Valentini et al., 2016), so why not also as a management tool within the context of biological invasions? For instance, the two first ever studies using metabarcoding for detecting widespread invasive marine plankton species recommended a DNA metabarcoding approach, in combination with visual observations, for the routine surveillance of marine invasions (Zaiko et al., 2015a,b).

Probably the most promising application of PCR-based approaches is the possibility to derive quantitative estimation about the biomass or the density of the target species directly from the quantity of DNA retrieved from environmental samples (Rees et al., 2014). Currently, our understanding of such estimates comes mainly from aquatic ecosystems, where several studies found positive correlations between DNA concentration and biomass, density, and/or abundance (Erickson et al., 2016;

Pilliod et al., 2013; Takahara et al., 2012; Thomsen et al., 2012). The quantification of eDNA is usually achieved using a quantitative PCR (qPCR), allowing to infer the number of molecules present in the DNA template or the number of sequences yielded by high-throughput sequencing machines (Evans et al., 2016; Kelly et al., 2014).

Many of the methodological constraints associated with data production are now well documented (e.g. Pedersen et al., 2014). For example, the application of strict clean-lab procedures, including numerous negative and positive controls at each stage of the data production, is the very first step for achieving robust results. The use of sterile disposable labware and separate stations for pre- and post-PCR procedures is usually recommended (Cooper and Poinar, 2000; Pedersen et al., 2014). For palaeoenvironmental DNA analysis, particular care should be taken for avoiding contamination by modern DNA (e.g. Boere et al., 2011; Rizzi et al., 2012; Thomsen and Willerslev, 2015; Torti et al., 2015). Dealing with imperfect detection (Box 1) and PCR/sequencing errors in metabarcoding studies requires the incorporation of multiple technical and field sampling replicates (e.g. Roussel et al., 2015), as well as the use appropriate bioinformatic tools (Boyer et al., 2016; Schloss et al., 2011) and sequence analysis framework (De Barba et al., 2014).

The sole use of eDNA precludes obtaining information about the state of an organism, including its size, developmental stage, or state (dead vs alive). This implies a more close integration with other complementary techniques (traditional trappings or visual observations for example; Tréguier et al., 2014; Valentini et al., 2016). On the other hand, fast-growing advances in high-throughput sequencing, combined with functional analysis of gene expression, could appear as an excellent opportunity for building the next generation of eRNA molecular tools based on the analysis of the transcriptome from the very same environmental samples. For example, patterns of DNA methylation are now routinely used as marks of developmental history in animals (cf. Meehan, 2003) and could be adapted to meet the requirements of eDNA methods.

Some systems are faced with resolving more specific issues related to the spatiotemporal information delivered through eDNA. In lotic systems, for instance, effects of the downstream transportation and dilution of DNA need to be better accounted for in future studies (see Roussel et al., 2015). Understanding DNA degradability and turnover within water column and the soil is another important challenge for the monitoring of contemporaneous eDNA (cf. Turner et al., 2015; Yoccoz et al., 2012).

BOX 1 Dealing With Imperfect Detection in eDNA Studies

Imperfect detection is a common problem in ecological studies and refers to the uncertainty associated with the presence/absence of a species within a given environment or environmental sample (i.e. eDNA). It could be subdivided into two major types of errors: type I error is the detection of a taxon in the area where it is apparently absent (false positive); type II error occurs when a taxon fails to be detected, while it is actually present (false negative). This problem came into the spotlight recently in the context of eDNA and metabarcoding techniques as rapidly popularizing tools for biodiversity monitoring. Major concerns arose as the extreme (false positives) or the insufficient (false negatives) sensitivity of molecular tools could lead to an overinflation of degree of detection uncertainty. Moreover, because DNA techniques usually imply multiple steps—from field sampling/filtration to DNA extraction, PCR, sequencing, and bioinformatic analyses—the probability of false estimations could increase proportionally. As a potential way to minimize uncertainty in detection, the adoption of strict field and lab procedures preventing common sources of bias (e.g. contaminations, assay sensitivity, PCR and sequencing errors, etc.) has been proposed (e.g. Bohmann et al., 2014; Darling and Mahon, 2011). Among these procedures, multiple field, extraction, and PCR replicates as well as numerous positive and negative controls at each step of the processes are the rules. More recently, the importance of bioinformatic analyses, especially in the case of HTS approach, has been stressed for obtaining reliable species distribution data (de Barba et al., 2014). Such procedures showed to be satisfyingly efficient for preventing or managing false negative occurrences but do not really allow to account for false-positive detections. Furthermore, even minimized, a certain level of uncertainty persists and none of the above procedures could provide reliable estimations of detection probability or proportions of false detections. Fortunately, new statistical methods inspired by site occupancy models (SOMs) have been recently proposed (Ficetola et al., 2015; Schmidt et al., 2013a,b). SOMs and SODMs (for site occupancy detection models) have been adapted to meet assumptions of eDNA studies and could be calibrated using presence/absence data validated by multiple detection methods or by observations for closely related taxa (Lahoz-Monfort et al., 2016). SODMs can be used to estimate the proportion of samples where a species is present or the minimal number of samples/replicates necessary to obtain a confident detection estimates (within a defined confidence interval). SODMs appear as a promising tool to estimate detection and species distribution probabilities despite imperfect detection and might appear as a powerful unifying framework for eDNA analysis that needs to be further developed.

Finally, the metabarcoding approach implies the taxonomic identification of multiple unknown taxa by comparison with reference library of taxonomically annotated sequences. The fine taxonomic assignment of highly diverse and taxonomically difficult organisms (e.g. Tara Oceans, palaeoenvironments) could be challenging, simply by lack of comprehensive reference databases. On the other hand, the short DNA fragments targeted in metabarcoding studies with the objective to maximize the capture of highly degraded eDNA often preclude species-level taxonomic identification even in more common study systems. Additionally, a single short DNA metabarcoding sequence could be assigned simultaneously with different levels of taxonomic precision (due to sequencing/PCR errors) (Yoccoz, 2012), therefore contributing to inflate the uncertainty in biodiversity estimates derived from metabarcoding data. One possible way to deal with imperfect taxonomic assignments is combining metabarcoding and traditional field surveys, which could significantly improve the taxonomic resolution of detected biodiversity (cf. Yoccoz et al., 2012). However as visual surveys are not always feasible, another way to deal with the problem, besides the continuous completion of public databases, is to improve taxonomical assignments even in the absence of complete reference database. Several approaches have been proposed recently (e.g. Munch et al., 2008a,b; Zhang et al., 2012).

Molecular methods offer a large array of applications: from the detection and the accurate taxonomic assignment of single specimens at various life stages to the large-scale spatial and temporal monitoring of invading and invaded communities. One particular interest of using DNA techniques as monitoring tool for biological invasions is their capacity to detect cryptic and rare, low-abundance species with greater precision and minimal investment in taxonomic expertise and sampling effort (e.g. Ji et al., 2013; Yoccoz, 2012). The high taxonomic resolution, accessed through molecular data, could help identifying divergent taxa, even at the very early stages of genetic divergence. This information, combined for example with historical data about the origin or the time since introduction, could inform about ongoing events of speciation/diversification (Folino-Rorem et al., 2009) as well as about their prospective impact on local species (e.g. hybridization, competition, etc.). Molecular data can also help distinguishing between morphologically cryptic species and species exhibiting phenotypic plasticity for taxonomically relevant morphological characters (Folino-Rorem et al., 2009; Stoof-Leichsenring et al., 2012), which is one important prerequisite for identifying possible mechanisms of invasion success. DNA approaches could also reveal coevolution patterns between taxa (e.g. host–parasite; Kyle et al., 2015). Moreover, as no diagnostic features are available for the large majority of the past and present biodiversity, DNA offers a unique opportunity to increase the number of species that can potentially be studied, including those retained in the sediment and permafrost records (Coolen et al., 2013; Domaizon et al., 2013) or in highly diverse microscopic soil or plankton communities (e.g. Tara Oceans; Karsenti et al., 2011). Specimens with imperfectly preserved morphology, old museum specimens, and different egg or larval stages are also accessible for diagnosis via DNA. This offers an unprecedented potential to analyse large-scale changes in the composition and structure of complex biological assemblages (Capo et al., 2015) and identify some general patterns. Beyond species distribution and genetic variation, molecular techniques could provide more functional insight of past and present communities, by targeting the expression of functional genes, such as the gene encoding for a cyanotoxin, which is produced by aquatic cyanobacterial communities (Savichtcheva et al., 2011). In turn, this allows a more direct estimation of a positive or negative impact within the invaded ecosystem. Other possible applications of DNA techniques for measuring the functional impact of biological invasions are discussed in Section 4.

Finally, the high sensibility of molecular techniques potentially allows the detection of nonnative species at the very early stages of the invasion process (i.e. at low species density), maximizing the chances for early intervention and successful management (e.g. Anderson, 2005). Therefore, DNA-based methods provide a robust and cost- and time-efficient methodological framework that has the potential to become the new institutional norm in terms of invasive species surveillance and management (e.g. Handley, 2015).

2.2 Indirect Methods for Reconstructing Past and Current Invasion History

2.2.1 Population Genetics

Studying invasions requires temporal data. These can be acquired by physically monitoring species abundances over time building upon long-term data sets or palaeorecords. However, good approximation of temporal data could also be derived from historical signatures in DNA, i.e., using population genetics. By studying intraspecific variability at presumably neutral DNA markers, one may infer the patterns of relatedness between current individuals. And these patterns of relatedness may inform on the recent past history of populations. Based on these principles, phylogeography, the use of phylogenetic information in relation to geography (Avise, 2000), has been much used in the context of bioinvasions. For instance, phylogeography has proved powerful to detecting multiple introductions (e.g. Facon et al., 2003). The study of patterns of within-population diversity (i.e. allelic richness, heterozygosity, private variability) and the patterns of population differentiation are classically used to detect expanding populations, populations that have gone through a bottleneck event and identify populations that have been the source of bioinvasions. Once the empirical expertise acquired, this approach has been statistically formalized using various statistical methods; Approximate Bayesian Computation (ABC) is one such method (Beaumont, 1999). ABC methods allow confronting concurrent complex demographic and evolutionary scenarios to genetic data within a statistically grounded framework. In ABC the limiting step of very complex likelihood computations is replaced by a simulation procedure. The principle is to simulate numerous data sets under various considered scenarios and, for each scenario, under many different values of historical and genetic parameters (e.g. bottleneck intensity and duration, mutation rate, etc.). Scenario choice and parameter inference are then made by studying the frequencies of scenarios and the distribution of parameter values among the simulated data sets that are closest to the observed data (Beaumont et al., 2002). The similarity between simulated and real data is generally evaluated based on diverse summary statistics picked from the classical repertoire of diversity indices (allelic richnesses, heterozygosities, differentiation indices, etc.). ABC methods have already been successfully used to reconstruct the introduction pathways of several invasive species (e.g. Barrès et al., 2012; Dutech et al., 2012; Guillemaud et al., 2011; Miller et al., 2005).

In addition to reconstructing scenarios of invasion, population genetics may allow inferring demographic changes. Population declines and expansion, recent or more ancient, may in principle be detected and quantified using adequate molecular markers (Gilbert and Whitlock, 2015).

All these possibilities are however limited by the informativeness of molecular markers. The rate of evolution of markers makes them adequate to study processes that have occurred at a specific temporal scale. Slowly evolving markers (such as some mitochondrial sequences for instance) will be useful to reconstruct ancient phenomena. Recent demographic changes will only be detected with the help of very resolutive, thus very rapidly evolving markers (such as microsatellites). A precise inference of population demography therefore requires a dedicated development of markers with adequate resolution, large samples from several populations, and the use of appropriate dedicated statistical inference methods. As a consequence, this type of approach has only been applied to species on a one-by-one basis. One can hardly imagine how it could be applied to networks of interacting species. But before discarding population genetics from the set of useful methods for network monitoring, one should consider the help of such methodologies in the absence of long-term abundance data (which is the case for a large proportion of nonnative species). With only contemporary samples and data, population genetics provides a window on past demographic changes. The ideal combined use of network analysis and population genetics could for instance consist in detecting keystone or indicator species of the networks and develop a dedicated population genetic approach for some of them.

A second case where population genetics has a natural place among the methods for monitoring invasions in networks is the one of rapidly evolving organisms, such as microbes. For organisms like viruses, bacteria, and many fungi, generation times are so short and populations so large that the timescales at which demographic and genetics processes occur could be confounded. Population genetics tools provide a direct access to demography. In the case of pathogens, this observation has led to the approach called molecular epidemiology. Classically molecular epidemiology uncovers invading pathogen strains in a network of resident, endemic. For instance, these invasive, epidemic strains can be associated with "star-like" networks of haplotypes, where a common haplotype is connected to numerous rare single locus variants (Achtman, 2008; Vernière et al., 2014). They may also be detected using phylogeny-based approach as a monophyletic clade emerging from the phylogenetic tree (Avise, 2000).

With the advent of high-throughput sequencing, population genetics can now be conducted on whole genomes (population genomics). The number of markers is no longer a limit and ample information on different temporal scales can be retrieved. Moreover, environmental samples such as faeces or gut contents could be used for inferring the genetic structure of interacting communities and monitor how it changes following invasion. This may consist, for example, in tracking the propagating effects of invading locus variants across trophic levels or quantifying the relative importance of intraspecific variability in key traits determining the degree of invasibility in local communities.

2.2.2 Chemical Fingerprinting

Preserved metabolically inert tissues such as skeletons or shells can be another valuable source for tracing back the establishment of a nonnative species and possibly its geographical origin. By incorporating various chemical compounds from the environment, usually with a very low turnover, metabolically inert tissues "imprint" the more or less subtle variations in local chemistry over long time periods. Therefore, such tissues could be used as "environmental recorders" for tracking more or less accurately, any changes in habitat use by a species, throughout a lifecycle or across multiple generations. For example, the geochemical characterization of single or multiple trace elements (i.e. chemical fingerprinting) of preserved contemporaneous or historical tissues could be used for revealing the ancient geographic origin of an invasive species or for reconstructing transgenerational migration routes (e.g. Elsdon and Gillanders, 2003; Rubenstein and Hobson, 2004). This approach has shown to be particularly useful for retrospectively tracking migration routes and habitat use in teleost fishes because their otoliths (fish ear stones) grow continuously throughout the fish life cycle, with little or no reabsorption of material incorporated into their structure. Fish otoliths thus reflect the local chemical composition of water at the moment of the incorporation (Campana and Tzeng, 2000). For instance, lifetime variations in strontium (Sr) or strontium/calcium (Sr/Ca) ratios were frequently used for investigating fish migration patterns (e.g. Carpenter et al., 2003; Clarke et al., 2009; Tanner et al., 2011; Thorrold et al., 1997; Townsend et al., 1995) as they showed to be good proxies for environmental salinity (Campana, 1999). If different habitats with very similar microchemistry are to be discriminated, multielemental otolith fingerprints (i.e. the simultaneous analysis of multiple trace elements from the same tissue; Clarke et al., 2009; Forrester and Swearer, 2002; Mercier et al., 2011) could be used for increasing the resolution of analysis. Microchemistry fingerprinting could therefore be a powerful tool for tracking the arrival, and the life-history characteristics of nonnative species that are otherwise difficult to monitor (extinct, elusive, or small-sized invertebrates, aquatic organisms, etc., Carpenter et al., 2003; Zazzo et al., 2006). Moreover, because otoliths are usually well represented in fossil records (Nolf, 1994), the analysis of otolith trace elements could also be used for tracing past environmental changes, concomitant to the establishment of a nonnative species, thus providing functional understanding about prospective invasives' impact and environmental conditions that could have enhanced it (e.g. Jones and Campana, 2009). More recently, multielemental fingerprinting also showed to be a valuable tool for elucidating complex connectivity patterns among species as investigated by López-Duarte et al. (2012) in an impressive 12-year study.

Yet, for some trace elements, there could be great inter- and intraspecific variation for species originating from the same geographic locations that needs to be accounted for, especially in studies involving comparisons among multiple species (Sturrock et al., 2012). For example, factors such as temperature, age, or speed of growth have all been shown to influence elemental incorporation in fish otoliths (reviewed in Elsdon and Gillanders, 2003; Sturrock et al., 2012). Despite the existence of relatively well-characterized large-scale gradients in water chemistry, multiple confounding factors as local fine-scale heterogeneity or temporal variations in elemental composition for the same site could exist (Elsdon and Gillanders, 2004, 2006). This implies the collection of multiple samples per location and per time point for each site in order to obtain stable signatures. As for other approaches (e.g. molecular, stable isotopes) an experimental validation might be required in order to disentangle among intra- and interspecific variability, environmental heterogeneity, and methodological issues in detection (e.g. Elsdon and Gillanders, 2004). Finally, as indicated by López-Duarte et al. (2012), trace-elemental fingerprinting for multiple species could be costly and logistically demanding. The combination of different complementary methods (e.g. real-time species tracking or molecular markers; Cook et al., 2007) could provide more predictive power.

Alongside physical or molecular tracking methods, stable isotope ratios of oxygen ($\delta^{18}O$) and hydrogen (δD) are increasingly used for tracking species movement of terrestrial and aquatic organisms using their metabolically inert tissues (Gannes et al., 1998; Hobson, 1999; Rubenstein and Hobson, 2004). $\delta^{18}O$ and δD isotope ratios could be informative, for example, about the origin of an invasive or its migration dynamics across large, continental scales (Farmer et al., 2008). As such, $\delta^{18}O$ and δD isotopes could provide useful, complementary information to DNA-based population genetics tools in the case of recent introductions of species for which we lack samples from the geographical area of origin.

3. IDENTIFYING AND MONITORING ECOLOGICAL INTERACTIONS OF AND WITH INVASIVE SPECIES

3.1 Trophic Interactions

3.1.1 Stable Isotopes

Chemical elements with more than one stable isotopic form have been used as natural biomarkers in ecosystems. The two most commonly used to address questions about feeding interactions are stable isotopes of carbon (^{13}C) and nitrogen (^{15}N). These two elements and their stable isotope compositions are necessarily acquired from an animal's diet and incorporated into that animal's tissues (DeNiro and Epstein, 1978, 1981). Primary producers can often be discriminated from their δ^{13}C (measured as the ratio of ^{13}C to ^{12}C relative to a reference standard) because of the strong imprint of photosynthetic modes on plants' δ^{13}C. Thus, an animal's δ^{13}C composition can potentially identify the origin of the carbon in its tissues, which can be used to infer foraging location(s) and behaviours (Bearhop et al., 2004). For example, δ^{13}C naturally varies between the low trophic level organisms in the nearshore and offshore habitats of lakes (France, 1995), and the δ^{13}C composition of freshwater fish is used to estimate the relative amount of carbon derived from these habitats (e.g. Vander Zanden and Vadeboncoeur, 2002; Vander Zanden et al., 1999). On the other hand, an animal's δ^{15}N composition has been used as a surrogate for a continuous assessment of its trophic position within a food chain (as reviewed by Fry, 2006). Animals become progressively enriched in ^{15}N at higher trophic levels due to preferential retention of the heavier isotope during protein metabolism, which results in a stepwise enrichment of ^{15}N between consumer and resource.

The time frame to which stable isotope information is relevant in animal tissue directly depends on its turnover. For instance, once synthesized, the amino acids in feathers, scales, and hair are hardly remobilized. Their isotopic composition could be thought of as a documented record of their dietary past. In comparison, blood, muscle, or liver tissues have higher (albeit variable and possibly irregular) turnover rates that provide evidence of recent dietary history at varying time windows (Hesslein et al., 1993) or seasons (Perga and Gerdeaux, 2005). Not only can stable isotope measurements be taken from multiple tissues that provide dietary information at different timescales, but because some of these tissues can resist degradation (e.g. archived in collections of fish scales (Gerdeaux and Perga, 2006) or zooplankton exoskeleton extracted from sediment cores (Perga et al., 2010)), stable isotope analysis (SIA) has the potential to retrospectively reconstruct temporal changes in dietary characteristics of a given species/community, including before and after invasion. Sampling of feathers, blood, or fin clips can also be a good nonlethal alternative with the obvious advantage of sparing vulnerable populations of rare species (Kelly et al., 2006).

If the potentialities of SIA for questions related to invasions have been highlighted several times, these have been clearly underexploited so far.

They have yet proven invaluable to document invasive dietary position and niche width, how invaders might alter the overall food web structure, with subsequent consequences on key processes, such as nutrient cycling.

Being a complement rather than a surrogate to classical or molecular gut content analyses, SIA can evaluate on which local, native organisms an invasive one might potentially feed on. Because SIA is not time consuming, it allows repeatable comparisons between situations, documenting how the predation activities from the invasive species might change depending on the local availability of prey (Caut et al., 2008). Because SIA provides a time-integrated assessment of individual feeding habits, they also offer the option of quantifying the trophic niche of a species (Bearhop et al., 2004). This isotope-based life-history trait allows a good comparison of how nonnative and native species may compete, as shown for invasives as crayfish (Olsson et al., 2009) or fish species in UK waters (Britton et al., 2010).

The impact of an invasive species can expand beyond its own trophic level and consequently affect the overall food web structure. The seminal study by Vander Zanden et al. (1999) documented a reduced trophic position and a dietary shift from littoral to pelagic habitat in native lake trout (*Salvelinus namaycush*) in North American boreal shield lakes following the invasion by two nonnative predators, namely smallmouth bass (*Micropterus dolomieu*) and rock bass (*Ambloplites rupestris*). SIA also provides the opportunity to follow through long-term impact of invasive species at the food web scale. For example, the invading carnivorous Argentine ant (*Linepithema humile*) occupied a similar trophic level as ants in their native habitats. However, once established, the ants shifted to a lower trophic position as they consumed more plant material following severe reductions in native ant prey populations (Tillberg et al., 2007). Based on SIA, Inger et al. (2010) highlighted that the nonnative bream (*Abramis brama*) got caught in Lough Neagh food web, in which it sustains the river lamprey diets. Because SIA data have been piling up over the last decades, large meta-analyses allow testing for more generalistic patterns on the impacts of invasive within food web, as did Cucherousset et al. (2012), suggesting that invasions might promote a greater trophic variability within food webs.

Because primary producers' stable isotope composition is tightly connected to how carbon and nitrogen nutrients circulate within the given ecosystem, SIA may be able to address how these altered trophic interactions finally affect larger biogeochemical processes. A terrestrial example of this application is the study of Carey et al. (2004). They used SIA and physiological measurements to document how carbon parasitism via arbuscular mycorrhizae may be an important mechanism explaining the success of spotted knapweed (*Centaurea maculosa*), out-competing its native prairie neighbours.

Nevertheless, other sources of variation, not specifically related to diet and trophic behaviour, could influence isotopic fractionation or data interpretation (reviewed by Boecklen et al., 2011; Vanderklift and Ponsard, 2003) and therefore need to be considered (Spence and Rosenheim, 2005). Most of these factors have already been well documented and could now be accounted for in many statistical methods (Kadoya et al., 2012; Parnell et al., 2013; Phillips and Koch, 2002; Phillips et al., 2014; Post, 2002; Post et al., 2007; Ward et al., 2011; Yeakel et al., 2011), while the integration of compound-specific stable isotope ratios (e.g. Chikaraishi et al., 2011, 2014; see also Traugott et al., 2013) could help improving the accuracy of stable isotope data. Finally, growing body of the literature advocates the incorporation of additional isotopic tracers (Jaouen et al., 2016; Vander Zanden et al., 2016), therefore opening perspectives to extend SIA to species with large diet spectra.

3.1.2 Molecular Techniques

DNA-based molecular methods for diet analysis emerge as a valuable complementary approach to SIAs. Not only do they provide the opportunity to get a detailed insight into the menu of a species, but also the taxonomic resolution of identified prey is increasingly improving with the swell of public reference databases. Historically, the use of molecular techniques for diet reconstruction derived from the enzyme-linked immunosorbent assays (ELISA) using monoclonal antibodies that react with an antigen of the target prey in a very specific manner (reviewed by Symondson, 2002). Although this technique has been progressively replaced by the PCR-based methods, monoclonal antibodies offer a range of advantages for tracking prey and parasitoid detection with high precision and sensibility (Greenstone, 1996; Stuart and Greenstone, 1990). The method can be specific enough to discriminate between the different developmental stages of the prey consumed (e.g. insect egg, larvae, or adults; Crook et al., 1996). Such high specificity greatly outperforms all currently available molecular techniques. Moreover, monoclonal antibodies could be designed in a way that increases prey detection success over time (e.g. Harwood et al., 2001; Symondson, 2002), thus maximizing prey detection especially in fast-metabolizing small-sized organisms. An additional advantage is that the probability of detecting secondary predation with monoclonal antibodies (i.e. detection of a prey within the gut of the target prey organism under study) is very low (e.g. Harwood et al., 2001), compared to a PCR approach (e.g. Sheppard et al., 2005). Monoclonal antibodies thus offer interesting opportunities for characterizing trophic interactions of nonnative species. For example, this method showed very useful for detecting trophic linkages between two invasive insect species: the coccinellid *H. axyridis* feeding upon eggs from the leafhopper pest *Homalodisca coagulata* (Fournier et al., 2006). In a neat experimental study, Lundgren et al. (2013) used ELISA and showed that dandelion seeds (*Taraxacum officinale*, Asteraceae) marked with rabbit monoclonal antibody are consumed by a much broader community of arthropods in their nonnative range of distribution.

In general, monoclonal antibodies could be the easiest method for detecting very specific or unknown trophic linkages (e.g. different developmental stages) while allowing for without a priori screening of large number of predators. Yet, the monoclonal antibody approach is very limited for studying broad-spectrum diets, mainly because developing such a large array of distinct antibodies is prohibitively expensive. For more details see Symondson (2002).

Currently, PCR-based techniques are probably the most versatile and cost-effective molecular methods for the characterization of trophic interactions (Table 1). Indeed, PCR diet analysis could be seen as an alternative for assigning morphologically unidentifiable prey remains to specific prey organisms (e.g. Bartley et al., 2015; Kasper et al., 2004; Pérez-Sayas et al., 2015), bypassing most of the limitations of traditional visual surveys of feeding behaviour or gut content analyses (e.g. cryptic feeding, liquid prey, etc.). Singleplex diagnostic PCR, for example, could be one very cheap and robust approach for tracking the consumption of a target species by a wide range of predators. Egeter et al. (2015) used diagnostic PCR and demonstrated that the DNA of an endangered frog species could be detected in the stomachs and faecal pellets of a range of invasive rodents with PCR, showing overwhelmingly superior results compared to the traditional morphological identification of prey remains. Similar results were achieved in another recent study where the diagnostic PCR allowed higher detectability and more accurate taxonomic identification of ingested prey by invasive mammal species in French Polynesia (Zarzoso-Lacoste et al., 2016).

Besides predation, PCR methods are also a reliable tool for disentangling host–parasitoid interactions. Insect parasitoids (Hymenoptera, Diptera) are frequently introduced into new habitats as a part of biological control programmes or could be efficient control agents of introduced species within their nonnative ranges. But parasitoids are also small-sized and a taxonomically challenging group. The monitoring of their trophic interactions by traditional rearing methods could therefore be laborious (Gariepy et al., 2008). As an alternative, Gariepy et al. (2014) combined a singleplex PCR and Sanger sequencing for detecting Scelionidae parasitoids within eggs of the invasive brown marmorated stink bug (*Halyomorpha halys*). The developed assay was sensitive enough to detect with 100% efficiency parasitoid DNA from parasitized eggs at different time periods (and as soon as 1 h after oviposition), and with lesser success from empty eggs after parasitoid emergence. When applied to field collected egg masses, the assay was also successful in identifying cases of hyperparasitism. This is one noticeable illustration of how eDNA (DNA shed on insect egg masses) could be used for retrieving trophic interactions without the physical disruption of local food webs opening promising avenues for the direct reconstruction of invaded food webs in a cost- and time-efficient way. One particularly revolutionizing application of PCR methods concerns planktonic assemblages, where the small size of the organisms involved usually precludes the use of other methods for quantifying predator–prey interactions (Maloy et al., 2013; Riemann et al., 2010; Sotton et al., 2014; Troedsson et al., 2009). For example, the presence of the toxic cyanobacteria *Planktothrix rubescens* in the gut contents of various zooplanktonic taxa (*Daphnia*, *Bosmina*, and *Chaoborus*) was estimated by qPCR, showing that these cyanobacterial cells constitute a part of food resource for herbivorous zooplanktonic taxa during metalimnetic cyanobacteria bloom periods (Sotton et al., 2014). As a consequence, zooplanktonic herbivores by diel vertical migration act as vectors of cyanotoxins by encapsulating grazed cyanobacteria and contribute to the contamination of zooplanktonic predators (Sotton et al., 2014).

More recently, the scope of PCR techniques for diet analysis has been significantly broadened by the development of the DNA metabarcoding as this approach introduces the possibility to examine the full diet spectrum of a species, while requiring very little a priori knowledge. DNA metabarcoding for diet analysis implies the use of general primers, for amplifying prey DNA from food remains, present in a dietary sample (e.g. gut contents or faeces). For dietary samples from unknown species, an additional set of species-specific primers could be used in order to confirm/reveal the identity of the target predator species (e.g. Shehzad et al., 2012). The use of unique identifiers ("tags") allows recovering data after sequencing from each individual consumer using bioinformatic approach (Binladen et al., 2007;

Boyer et al., 2016). DNA metabarcoding could be used for unravelling trophic interactions in herbivorous (Ait Baamrane et al., 2012; Ibanez et al., 2013; McClenaghan et al., 2015; Quéméré et al., 2013; Valentini et al., 2009a) and carnivore organisms (Boyer et al., 2013; Leray et al., 2015; Mata et al., 2016; Vesterinen et al., 2016). DNA metabarcoding for diet analysis has not been applied yet in the context of biological invasions but two recent studies indicate that this approach seems to be advanced enough to be applied within a more explicitly hypothesis-testing context. By combining SIA and DNA metabarcoding, Kartzinel et al. (2015) investigated the fine-scale trophic partitioning in a community of large mammalian herbivores, while Craine et al. (2015) showed how dietary changes, induced by climate warming, could potentially cause nutritional stress in native North American bisons (*Bison bison*).

Molecular techniques for diet analysis are one very promising tool and offer numerous research opportunities in invasion ecology and management. However, as previously mentioned, they are far from being perfect. For instance, Lundgren et al. (2013) noticed that the antibodies used for marking the sentinel prey for their ELISA analysis seemingly altered prey palatability and consequently food preferences of some predators, leading to biased estimations of prey attacks. They also found that marker (antibody) stability in the environment could be relatively short (90% of the marker was lost within 4 days), which could be an important source of misinterpretation, if not quantified prior the study. In PCR-based studies, the existence of multiple non-dietary sources of variation often preclude the comparison of dietary data obtained with multiple distinct primers, using different dietary samples or from different species. The existence of such bias ideally requires setting up complex multispecies, multifactorial experimental studies where the different sources of variation could be quantified at once, and hierarchized according to the relative importance of the bias they introduce. Experimental data could in turn be used for building general, corrective models similar to the species occupancy detection models (SODMs), which are currently used for optimizing the number of replicates necessary to minimize the probability of false positive or negative detections in eDNA studies (cf. Box 1).

When using DNA metabarcoding for a diet assessment, additional constraints apply. For example, as for eDNA biodiversity monitoring, highly conserved general primers are required to guarantee the successful amplification of multiple, phylogenetically diverse taxa (Taberlet et al., 2012b). In the case of carnivorous species that are closely related to their prey (e.g. a mammal predator consuming mammal prey), using general primers might lead to the preferential amplification of the highly concentrated, non-digested predator DNA. In some cases, this could be avoided by specifically preventing the amplification of the predator/host DNA. Several methods for target DNA enrichment do exist (reviewed by O'Rorke et al., 2012), with the most popular including the use of predator-specific endonuclease restriction enzymes (Blankenship and Yayanos, 2005; Dunshea, 2009) or blocking primers (Deagle et al., 2009; Shehzad et al., 2012; Vestheim and Jarman, 2008). PCR techniques are also particularly prone to detecting secondary predation because of their high sensitivity (Sheppard et al., 2005). However, in some cases, secondary predation could be interpreted as an interesting opportunity for quantifying tri-trophic links or intraguild predation (Sheppard et al., 2005) from a single dietary sample. As for eDNA biodiversity monitoring, deriving quantitative information about ingested prey numbers or biomass remains challenging as biological processes like differential digestion rates or the different gene copy numbers between food species appear to distort relative species proportions (Deagle and Tollit, 2007; Deagle et al., 2005, 2010, 2013). However, recent findings encouragingly suggest that some sources of variation could be controlled for by using appropriate correction factors (Thomas et al., 2014, 2016a). Finally, another important yet unresolved methodological issue in diet analysis in general but particularly in the context of molecular techniques, as they are rapidly generalizing among ecologists, is the capacity to distinguish between active and passive predation (i.e. scavenging). Indeed, carrion could be a valuable and easily available resource and rates of scavenging are expectedly high (e.g. Foltan et al., 2005; von Berg et al., 2012), with significant but yet underestimated impact on food web dynamics (Wilson and Wolkovich, 2011). In terms of biological invasions, quantifying the rates of passive feeding is important in order to estimate realistic impacts on local food webs (e.g. Brown et al., 2015) or invasion success (e.g. Wilson-Rankin, 2015). However, experimental attempts to distinguish fresh from carrion prey with PCR techniques show that ingested decaying carrion prey is detected as efficiently as any of the fresh prey items offered (Foltan et al., 2005; Heidemann et al., 2011; Juen and Traugott, 2005; von Berg et al., 2012). As a possible solution, Juen and Traugott (2005) proposed the use of isoenzyme electrophoresis technique, which offers the opportunity to target specific enzymes, known to persist in dead corpses long after death, without being destroyed during the digestion process (i.e. have high retention times). To our knowledge, this approach has not yet been empirically tested. Another promising but still unexplored approach has been proposed by Wilson et al. (2010b).

Authors took inspiration from techniques used by forensic pathologists to determine the putative causes and time since death by relying on predictable changes in various physiological properties such as muscle pH and water loss rates. In an experimental setting focusing on the invasive predatory western yellowjacket (*Vespula pensylvanica*), Wilson et al. (2010b) showed that they were able to identify with 88% success which of the yellowjacket prey was carrion and which was killed by active predation, based on the physiological imprint of stress levels induced by predation. And even if this method has not been tested yet on prey subjected to digestion, it opens exciting new opportunities to explore in the near future.

Overall, molecular methods provide very straightforward presence/ absence diet data. A semiquantitative approach is possible if the proportion of individuals positive for a given trophic link is considered. In such case, weighted trophic networks could be built and their properties examined. From the management perspective, molecular techniques could be useful for tracking dynamic changes in trophic behaviour following invasion as well as the successful integration of invasive species into local food webs (e.g. Gorokhova, 2006). Moreover, the growing numbers of empirical studies provide encouraging examples of how molecular diet analysis could possibly support decision making and management (e.g. Hatteland et al., 2011; Pianezzola et al., 2013). The best illustration for this comes from intensively managed agroecosystems, where in the context of biological control programmes, molecular methods could be a valuable tool for rapid large-scale-estimations of attack rates (e.g. Derocles et al., 2012; Traugott et al., 2008) as well as rates of incidental intraguild predation among predators (e.g. Davey et al., 2013; Traugott and Symondson, 2008). For example, multiplex PCR showed to be a valuable approach for identifying key predators and their attack rates on invasive slug species, which are important crop pests in Europe (Hatteland et al., 2011; Pianezzola et al., 2013). Bohan et al. (2000) further demonstrated that ELISA tests could be used for tracking dynamic predation of earthworms and slugs by a generalist carabid predator (*Pterostichus melanarius*). They showed that changing spatial associations between the predator and its prey were mostly driven by their respective density-dependent distributions rather than by agronomical factors such as crop or soil characteristics. By using diagnostic PCR for gut content analyses, Bell et al. (2010) extended this approach to a multispecies community of invertebrate predators and concluded that the relationship between predator–prey co-occurrences and feeding behaviour is fairly consistent across species, and therefore spatiotemporal community dynamics could be manipulated in order to optimize pest regulation (in this case, slugs). Finally, DNA metabarcoding (Mollot et al., 2014; Vacher et al., 2016) has been successfully applied for characterizing the trophic behaviour of key arthropod pest predators, allowing for recommendations about relevant management practices aiming to enhance the biological control potential within arable fields.

3.1.3 Other Methods

For particular study questions, some nonconventional or less popular techniques could be a valuable source of trophic information. For example, Smith and Gardiner (2013) used video cameras for comparing field rates of egg predation between native and exotic coccinellid species. Sloggett et al. (2009) employed gas chromatography–mass spectrometry to track egg predation of native ladybird species *Hippodamia convergens* within the guts of exotic ladybird *H. axyridis*.

3.2 Mutualistic Interactions

There is now a vast body of literature investigating the impact of biological invasions on mutualistic networks (plant pollinators, seed dispersers) (Traveset and Richardson, 2006), but the majority of empirical data have been produced using direct observations (Giovanetti et al., 2015; Tiedeken and Stout, 2015) or experimental approaches (Chung et al., 2014; Russo et al., 2014), while tools such as molecular and SIAs have been still surprisingly underused. Studies suggest for instance the successful integration of invasive plant or pollinators within native interaction networks (e.g. Vilà et al., 2009) that could sometimes lead to important changes in the network structure and dynamics (Albrecht et al., 2014; Spotswood et al., 2012). However, it is still not clear if these changes are translated into functional impact, for example, in terms of gene or energy flow across and among trophic levels. For example, Bartomeus et al. (2008) showed that despite apparently high rates of pollen transfer between native and invasive plant species, the actual pollen deposition on native plants was really low. These findings highlight the importance of incorporating DNA and stable isotope analysis into the study of mutualistic networks, as they provide a direct and time-integrated measure of fluxes (energy, genes) and therefore of potential functional impacts following invasion. Such unique opportunity for gleaning both taxonomic and functional data at once offers an efficient and cost-effective methodological alternative for assessing the functional impact of invasive species that has the full potential to become a novel paradigm for invasives' biomonitoring (e.g. Jackson et al., 2016). For example, a DNA metabarcoding could be a very straightforward approach for characterizing mutualistic networks. First, this type of interactions involves partners that are phylogenetically distantly related, thus facilitating the design of well-conserved, group-specific primers. Second, their DNA is more readily accessible as usually it is not degraded (compared to say prey DNA in antagonist interactions). Pollen DNA is generally of higher quality than digested prey items from faeces or gut contents, and there is no need for prior visual sorting or separation of pollen grains as for traditional palynology surveys (which could be time and effort demanding as well as subject to observer's bias; Richardson et al., 2015a,b). Moreover, molecular tools are much more sensitive in detecting rare, low-abundant plant species compared to visual observations (Richardson et al., 2015a,b). They are also suitable for detecting ancient pollen DNA in pollinators' crops from historical museum collections (e.g. Wilson et al., 2010a). The additional advantage of using highly sensitive detection techniques is the potential they open in terms of biodiversity monitoring as pollinators could be reliable "environmental recorders", thanks to their capacity to sample low-abundant nonnative plant species (Galimberti et al., 2014) that could be undetectable by other means. Moreover, several molecular markers for plants have already been developed, and their efficiency compared (e.g. Galimberti et al., 2014; Richardson et al., 2015a,b; Wilson et al., 2010a), which is a valuable resource allowing the design of further studies. Molecular data could also be complemented by SIA in order to reveal patterns of seasonal switching between an insectivorous and frugivorous diet in facultative pollinators (Frick et al., 2014), tracking community-level nutritious carbon pathways across heterogeneous habitats (Herrera Montalvo et al., 2013) or evidencing multiple hidden facilitative interactions between nonnative species, likely promoting invasion success (Lach et al., 2010).

Mutualistic interactions involving symbiotic bacteria or mycorrhizal fungi are another important type of facilitative interactions, and increasingly shown to influence life-history traits and fitness of nonnative species in invaded environments (e.g. Bogar et al., 2015; Himler et al., 2011). However, understanding direct and indirect effects of these symbiotic interactions on invasiveness requires the combination of multiples methods. First, bacteria or fungal symbionts are taxonomically challenging groups and most of them are not culturable. Second, because of the potentially intricate interactions between symbionts and their host, and/or environment, quantifying symbiont-related impacts might be delicate. Fortunately, in terms of taxonomic diagnosis of multispecies symbiont communities, molecular techniques show encouraging results (e.g. Bansal et al., 2014; Cotton et al., 2015; Kurata et al., 2016; Thierry et al., 2015; Vasanthakumar et al., 2008) and should therefore be explored further in a more network-explicit context. For example, Hansen et al. (2007) found that the prevalence of secondary symbionts in the invasive psyllid *Glycaspis brimblecombei* was strongly correlated to parasitism rates by its main parasitoid, indicating that symbiont community could have multitrophic cascading effects.

3.3 Parasitic Interactions

The introduction of nonnative species in new habitats carries the risk of concomitant introduction of other "passenger" organisms such as parasites and various microbial pathogens. Such collateral introductions could significantly contribute to invasion success (Roy et al., 2012) or exacerbate impact on local communities, which usually lack suitable defences against exotic pathogens (Vilcinskas, 2015; Vilcinskas et al., 2013). Because of the disease risk collateral introductions involve, the characterization of parasitic interactions in invasive species is comparatively well documented. Particularly, molecular techniques have been a valuable diagnostic tool for detecting pathogens and their spread in a variety of aquatic and terrestrial invaders (e.g. Collins et al., 2014; Grabner et al., 2015; Lester et al., 2015; Spikmans et al., 2013; Vilcinskas et al., 2013). However, as for symbiont communities, the explicit consideration of nonnative parasite interactions within a multispecies network context is rare (Emde et al., 2014, 2016; Sato et al., 2011, 2012; Thieltges et al., 2013). A growing body of literature highlights the importance of integrating parasite interactions into ecological network models (Dunne et al., 2013; Lafferty et al., 2006). This is particularly relevant in the context of biological invasions, as nonnative parasites have been shown to mediate complex, unexpected changes in local interaction networks (Sato et al., 2011, 2012). One very neat example is the abovementioned SIA study, showing the importance of carbon parasitism via arbuscular mycorrhizae as mechanism explaining the success of the invasive spotted knapweed (Carey et al., 2004). On the other hand, the everincreasing power of HTS techniques allows the simultaneous detection of prey and parasite communities within the same individual sample (Berry et al., 2015; Srivathsan et al., 2015, 2016; Tiede et al., 2016). Additionally, HTS has the power to detect cases of coinfections by multiple pathogens within a single invasive host and has thus tremendous potential to reveal complex indirect effects, alongside direct trophic impact, eventually promoting invasiveness (e.g. Lu et al., 2010; Zhao et al., 2013).

4. MEASURING THE IMPACT OF BIOLOGICAL INVASIONS ON ECOSYSTEM FUNCTIONS

Throughout the chapter, we already provided several examples how different techniques and methods can contribute to estimate the impact of invasive species—through changes in the local community composition and biodiversity, to ecosystem functions. In this section, and without providing too much methodological details, we try to put a specific focus on measuring impact on ecosystem-level processes. The need for broad-scale, ecosystem-level approaches for understanding and predicting biological invasions impact has been recognized more than a decade ago (e.g. Vander Zanden et al., 1999). However, assessing ecosystem-level processes is methodologically challenging, especially considering our constantly evolving perception of biological complexity. Still, the rising number of methodological opportunities brings the assessment and the integration of the ecosystem-level impact into tangible reach. For example, progress in molecular techniques has been paramount for the discovery of all the unseen biodiversity of microbial communities as well as their functional implications in ecological processes (Bik et al., 2012; Lansdown et al., 2016; Thompson et al., 2015). Molecular methods have been also essential in the characterization of past biodiversity dynamics, helping to understand general mechanisms behind long-term ecosystem functioning (Pedersen et al., 2014). This offers multiple opportunities for assessing and monitoring the impact of biological invasions. Accordingly, molecular techniques currently start playing a prominent role for estimating short-term or contemporaneous impact of invasion. For example, HTS allows the simultaneous diet assessment of multiple species, thus providing information about the degree of niche partitioning (e.g. Kartzinel et al., 2015) or habitat coupling (e.g. Soininen et al., 2014), both mechanisms playing an important role in ecosystem functioning and stability. Additionally, by yielding trophic data, molecular techniques allow the estimation of metrics indicative about the functional role and the degree of integration of an invader within the local interaction network. This is important as nonnative species have been shown to rapidly integrate into local ecosystems, sometimes as key network "hubs", establishing a high number of strong and weak linkages to local species (Aizen et al., 2008; Memmott and Waser, 2002; Vacher et al., 2010).

Consequently, if not properly quantified, their removal following (partly informed) management actions could cause unpredictable disrupting changes in local ecosystem functioning (e.g. surprise effect; Caut et al., 2009; Courchamp et al., 2003).

Finally, the fast-growing field of functional metagenomics—which seeks to connect an organism to its respective function in an environment—opens new so far unexplored opportunities. Methods such as metatranscriptomics (the analysis of the transcripts isolated from a community of organisms) and meta-proteomics (the analysis of the protein profiles expressed by a community) are very complementary approaches to DNA sequence-based methods.

First, these methods allow the detection of a functionally active species that does not need to be characterized morphologically nor taxonomically. For example, a metatranscriptomics approach has recently been used for discovering functionally active viral natural enemies without the need to physically culture them and these viruses could be used for controlling non-native species (Valles et al., 2012, 2013). Another possible application is the profiling of protein expression in natural communities and across environ-mental gradients in order to monitor how the community functional structure changes following invasion. This could be a powerful tool for monitoring invaded communities or/and invader's symbiont and parasite communities, helping to identify key genes determining invader's success and impact (e.g. Dlugosch et al., 2013; Scully et al., 2013). Second, partic-ularly the metaproteomics approach allows the discovery of novel proteins/enzymes (and therefore ecological functions) that could not be predicted based on DNA sequences alone (Chistoserdova, 2009). As such, it has a promising avenue in discovering new functions that have not existed previously or have been disrupted by nonnative species.

In perspective, functional metagenomics might open an entirely new framework, where natural communities could be manipulated accordingly to the function that needs to be preserved/optimized. Consequently, man-agement strategies will not aim in removing a certain species anymore but rather in optimizing its integration in local ecosystems, while minimizing its impact on ecosystem functioning. Nevertheless, the broad-scale ecological application of functional metagenomics is still in its infancy and further methodological challenges related to clonal library preparation, enzyme activity expression, or genome annotation need to be addressed (Chistoserdova, 2009; Lam et al., 2015). Moreover, as the function of great majority of genes is still unknown, predicting the function of newly discov-ered proteins is not always possible (Lam et al., 2015).

SIA is another practical method for measuring the impact of biological invasions on ecosystem functioning thanks to the synchronous and diachronous comparisons of ecosystems they allow. Its main advantage in this context resides in the existence of numerous data sets, already available worldwide. For example, Sagouis et al. (2015) used SIA data from the published literature for investigating the impact of nonnative fish species on the food web structure across 496 freshwater fish communities worldwide. For this study, historical collections of fish tissues, a part of long-term-monitoring programmes of the Laurentian Great Lakes, have been analysed with SIA over a period of more than 2 years in order to see how the introduction of nonnative species has influenced the local food web structure (e.g. Paterson et al., 2014; Rennie et al., 2009; Rush et al., 2012). The results show a profound impact of nonnative species on food web structure, energy flow, and stability within all these aquatic ecosystems, emphasizing importance of SIA as a sensitive approach in this context.

Additionally, recent development of new quantitative metrics for SIA, inspired by those used in functional ecology, might allow more efficient and comprehensive assessment of existing stable isotope data, thus providing new perspectives in terms of functional impact of invasive species (Cucherousset and Villéger, 2015; Layman et al., 2007).

5. USING EMPIRICAL DATA FOR IMPROVING OUR PREDICTIVE CAPABILITY THROUGH MODELLING AND MACHINE-LEARNING APPROACHES

We provided numerous arguments throughout this chapter about increasing capacity to produce large ecological data sets and large-scale environmental monitoring thanks to the emergence of a variety of complementary techniques. However, what has been missing, to date, are methods that could repurpose these data sets to build or reconstruct ecological networks that have not been empirically observed. Here, we present two major approaches for network reconstruction: predictive models and machine learning. The use of models predicting interactions between pairs of species or helping to construct interaction networks from massive amounts of correlative data (machine learning) can be useful tools for the assessment of the impact and potential success of exotic species. Combining these modelling approaches with new data acquisition protocols, such as the ones highlighted in the previous sections of this chapter, will pave the way for more reliable and rapid assessments of invasion trajectories and impact on the ecosystem.

5.1 Predictive Models

5.1.1 Principle

Predicting novel interactions, i.e., interactions between species that have never been observed co-occurring in the same locations, is a challenge that has been tackled a few times, using different methods, in the past 10 years. For instance, Pearse and Altermatt (2013) have been among the first to propose a model to predict how native caterpillars might interact, through herbivory, with nonnative plants based on the phylogenetic proximity of native and nonnative plant species. One intrinsic difficulty associated with this type of model consists in defining a strategy to make efficient use of the available information, i.e., a statistical regression problem. In species interaction networks, information can come in different guises:

(i) one can make use of information on species traits, i.e., use information on the nodes of the networks to infer potential links based on existing links and node-related information of already known parts of the network. This is typically the approach followed when predicting food webs from the size of organisms (Gravel et al., 2013; Petchey et al., 2008);

(ii) one can make use of information on "distances" or "similarities" among species (i.e. dyadic or relational information among nodes) to infer their role in the network, e.g., by assuming that closely related species tend to interact in a very similar fashion due to phylogenetic conservatism or, on the contrary, tend not to share certain interaction partners given the competitive exclusion principle. Phylogeny is one obvious way of defining similarity in an eco-evolutionary context, but other means are available (e.g. distances in the space of carbon and nitrogen isotope ratios, or distances based on co-occurrence in different patches);

(iii) finally, one can make use of latent trait variables associated to nodes and/or to node relations. Latent traits or relations are, by definition, not measured per se, but can be estimated indirectly through data on the emerging network. When a two-way relation exists between latent traits and the probability of an interaction actually existing, i.e., when traits predict the interaction and the observation of interactions serves to infer the latent traits, then inference on latent traits in a "known part" of the network might help in predicting interactions in an "unknown" part of the same network. For instance, the methods developed by Allesina and Pascual (2008), Eklöf et al. (2013a,b),

Rohr et al. (2016), Williams et al. (2010), or Dalla Riva and Stouffer (2016) make use of latent traits that determine predator–prey interactions based on match–mismatch (e.g. Dalla Riva and Stouffer, 2016) or hierarchical relations (e.g. Allesina and Pascual, 2008; Williams et al., 2010), or even more complicated combinations such as match–mismatch and generalist–specialist information at the same time (Rohr et al., 2016).

5.1.2 Types of Models

Grossly caricaturing the current state of the art of models predicting interactions between pairs of species, we can classify them along two orthogonal axes: on the one hand, models can be based either on measured variables (in general, species traits such as body size or phenology) or on latent variables that are inferred from a "learning network" (i.e. part of the network to predict or a different one with some overlap of ecological communities); on the other hand, models can be divided based on their use of traits (be they measured or latent) through an "intervality" principle (as in the niche model), a "matching" principle (to interact, the vulnerability trait of the prey must closely match the foraging trait of the predator, e.g., Rohr et al., 2016), or a "generality" principle (species have different degrees, and so interact with different number of species, but somehow randomly; e.g. Pearse and Altermatt, 2013). Table 2 is an attempt at describing the various methods encountered in the literature. Except in the case of recent studies (Dalla Riva and Stouffer, 2016; Rohr et al., 2016), there are very few comparisons of the various models in terms of predictability or goodness of fit.

5.2 Machine Learning

5.2.1 Principle

At the core of network reconstruction is the idea that in the sample data there are the impressions of past interactions. Taxa that have interacted will have correlated values in the sample data, and there will be identifiable patterns or motifs between groups of interacting taxa. Such "ghosts of interactions past" can be searched for in the data and machine-learning methods used to reconstruct the ecological network in which they occurred (Vacher et al., 2016).

Network reconstruction has only a relatively short history in Ecology and has typically used either models based on finding using formal logic links in the data or mixed learning that combines statistical inference methods, such as Bayesian approaches, with logic to learn network structure. In one of the first applications of learning to network reconstruction in

Table 2 Summary of Existing Models Used to Predict Species Interactions

Paper	Type of Network	Information Used	Summary
Allesina and Pascual (2008)	All types	(Partial) network topology to obtain latent variables	The model predicts interactions using a minimal number of latent variables (the dimensionality of the model) using a variation of the food web niche model of Williams and Martinez (2000). Intervality in the initial niche model is broken when there is more than one dimension. Learning the model on part of the interactions may predict the other part using the latent variable estimated for each species.
Bartomeus (2013)	Plant–pollinator networks	Habitat (flower density, etc.) and plant species (flower morphology, etc.) covariates	Simultaneously estimate regression coefficients with measured covariates and detection probability for observed interactions. The associated hierarchical model can be used to predict novel interactions (or unobserved ones) based on covariates. Model selection is used to cut out unnecessary covariates from the regression.
Beckerman et al. (2006); Petchey et al. (2008)	Food webs	Body sizes, allometries	Using optimal foraging theory (i.e. the ratio of energy gained to time spent handling prey), the model generates probabilities for the various interactions to exist based on their profitability. The allometric version of the model makes use of the functional dependence between species sizes and handling time.
Dalla Riva and Stouffer (2016)	Food webs	(Partial) network topology to obtain latent variables	Network topology is used to obtain "latent traits" through a PCA-like approach, to summarize the position of species as preys and predators. Inferring traits can be performed on a partial network, thus allowing to predict interactions between certain species based on their interactions with others.

Reference	Network type	Data used	Description
Dehling et al. (2016)	Plant–bird network (bipartite)	Various traits measured in both species groups	Each species of plant and bird is projected onto their respective functional space through a PCA. Each species is also projected onto the partner functional space through the centroid of its partner species. Procrustes rotation is used to find the morphism linking both functional spaces (i.e. to predict which functional position in birds matches with which functional position in plants).
Eklöf et al. (2013a,b)	All types	Different types of traits (phenology, size, habitat, etc.)	Interactions are predicted based on intervals of "matching traits" required by the focal species (e.g. a predator could only eat preys within a certain size interval, occurring over a certain habitat interval, during a certain phenology interval, etc.).
Gravel et al. (2013)	Food webs	Body sizes	The niche food web model of Williams and Martinez (2000) is the basis for a model of interaction prediction based on body sizes. Linear regression outputs, notably the coefficient of regression and the 95% quantiles, yield the variables used to fit the niche model.
Guimerà and Sales-Pardo (2009)	All types	Network topology	The model implements a series of stochastic block models (SBMs) to infer missing and spurious links within the observed network. At the end of the procedure, the model predicts interactions based on the SBM inference.
Ovaskainen et al. (2016)	Co-occurrence network	Measures of habitat characteristics (available resources), co-occurrence with other species	The model implements a co-species distribution modelling framework that makes use of both measured covariates and latent variables measuring species interactions.

Continued

Table 2 Summary of Existing Models Used to Predict Species Interactions—cont'd

Paper	Type of Network	Information Used	Summary
Pearse and Altermatt (2013)	Plant–herbivore network (bipartite)	Partial network topology, phylogenetic similarity among plants	The model to infer interactions "learns" on a part of the network is validated on another part and is used to predict novel interactions with exotic plants. The model is a generalized linear model that makes use of herbivore degree in the network and of phylogenetic similarity among plant species.
Rohr et al. (2010)	Food webs	Partial network topology, species body sizes	To explain interactions between species, the model makes use of body size and latent variables referring to vulnerability and foraging breadth, i.e., generality traits. Latent variables can in turn be correlated to other variables like phylogeny.
Rohr et al. (2016)	All types	Partial network topology	Latent trait variables are used to model interaction patterns. Two types of latent variables are used: matching traits and generality traits. The nested or modular nature of the network can be more efficiently captured by generality (respective matching) traits so that the estimation of the regression coefficients associated with these latent traits also informs on the position of the network along the nested–modular continuum.
Williams et al. (2010)	Food webs	(Partial) network topology to obtain latent variables, possibly using species size as a direct proxy for these variables	The model is the probabilistic version of the food web niche model of Williams and Martinez (2000). It can feed on either latent trait variables, which must be estimated from a part of the network, or species traits (such as size) that are suspected to drive the food web niche model.

ecology, Bohan et al. (2011) demonstrated that machine learning has the potential to construct realistic agricultural food webs, using a logic-based approach called abductive/inductive logic programming (A/ILP). A/ILP was used to generate plausible and testable networks from field sample data of the abundance of taxa (network nodes) alone. Importantly, this process did not just recover the obvious links that we already known from observation, but also suggested surprising and apparently illogical link. Spiders were consistently inferred by the machine learning as prey, despite being obligate predators. High probability links were also hypothesized for intraguild predation that might destabilize the network. Importantly, the learning reconstruction pinpointed a much lower number of test links necessary to validate the network than would have been required to build the network, using a classical approach, from scratch. A review of the literature revealed that many of the high probability links in the model had already been independently observed or suggested for this system. Moreover, the apparently illogical links to and the position of predatory spiders in the network were subsequently demonstrated to be correct using molecular testing for prey spiders in the guts of carabid beetles (Davey et al., 2013). This would suggest not only that learning and network reconstruction methods can produce plausible ecological networks (food webs) from sample data, but that by hypothesizing verifiable new links the learning is actually doing genuinely novel science.

5.2.2 Applications

Learning methods are being developed to reconstruct/hypothesize network interactions from the abundance patterns and additional background information, such as functional traits or meta-data associated with the samples (e.g. Bohan et al., 2011; Deng et al., 2012; Faust and Raes, 2012; Kurtz et al., 2015). Of enormous interest is the potential of HTS techniques as source for raw data for network reconstruction (Evans et al., 2016; Vacher et al., 2016). HTS platforms can generate several millions of DNA sequences for a few hundred dollars (Liu et al., 2012; Quail et al., 2012), and the price of this is reducing all the time. The approach is also quite general, allowing the characterization of DNA diversity using ostensibly the same methods across a broad range of complex environments (e.g. in air, soil, water, faeces, and gut contents, and on/in plant tissues, etc.), producing sample data containing many hundreds of taxa. Increasing numbers of these sequences can now be identified at the species level thanks to expanding taxonomic databases (see Abarenkov et al., 2010; DeSantis et al., 2006; Kõljalg et al., 2005; Quast et al., 2013; Ratnasingham and Hebert, 2007).

Developments in the use of HTS data for reconstructing networks of ecological interaction have recently begun to be made. Vacher et al. (2016) described the reconstruction of a microbial network on the surface (phyllosphere) of oak leaves, using a mixed, statistical, and logical approach from pure HTS data on microorganism OTU co-occurrences. This network, which was studied to understand the behaviour of a pathogenic fungi, *Erysiphe alphitoides*, revealed striking patterns of connectivity once this pathogen has invaded the phyllosphere community. The pathogen was connected to the rest of the network through strong and negative links that suggested that *E. alphitoides* might be associated with the absence, and possible removal, of other microorganisms.

6. PERSPECTIVES AND CHALLENGES AHEAD

Conceptual and technical progress in terms of analytical methods over the last decade has been remarkable. To date, ecologists have a comprehensive toolkit for investigating, describing, and understanding biodiversity. We review a vast spectrum of methods covering most aspects of invasion ecology, and they all have their respective strengths and weaknesses. However, their full or partial combination constitutes a powerful synergy and offers the potential to revolutionise our understanding of biological invasions either by providing deeper insight into general ecological processes associated with invasion or by allowing the effective monitoring of their spread and impact. In this section, we try to provide some visionary insight into the future avenues that this potential offers in terms of methodological improvements but also some societal and conceptual challenges associated, for example, with the implementation of efficient management actions.

6.1 Methodological Challenges and Perspectives
6.1.1 Molecular Techniques
Increasing production of molecular and DNA sequence data will raise important challenges in terms of comparison and integration across studies. Currently, the majority of studies use a vast array of primers and protocols that are optimized for a specific question or model taxon but lack standardization in terms of detection sensitivity thresholds and/or taxonomic coverage. This means that merging disparate data from various studies could be questionable. A general methodological framework is imperative if we want to take full

advantage from molecular data advent. Nevertheless, the raising awareness about major drawbacks of molecular techniques has led to an increasing number of experimental studies aiming to document and quantify their effects on final estimates. Still, this increasing amount of information needs to be systematically incorporated into molecular data analyses. In this respect, molecular ecologists could find inspiring examples from stable isotope ecology for how complex confounding factors could be incorporated into data analyses. Encouraging efforts have been made recently in terms of the integration of false-positive/negative signals through the development of appropriate models (Box 1). The use of molecular tools is still taxon- and ecosystem-biased. For instance, the majority of eDNA studies cover aquatic environments and/or vertebrate species, while terrestrial and/or invertebrate and plant communities have been overlooked so far. Increasing the range of applications of molecular tools will not only provide more insights about methodological bias and limitations but will also create opportunities for methodological innovations and new research questions. Accordingly, more and more creative initiatives are emerging such as monitoring biodiversity through eDNA retrieved from air samples (Kraaijeveld et al., 2015; Taberlet et al., 2012a), or nationwide eDNA citizen science campaigns (Biggs et al., 2015), which will definitely broaden the scope of molecular tools. From this standpoint, the integration of molecular tools and citizen science offers tremendous potential in terms of biodiversity surveillance. Encouragingly, an increasing number of such of initiatives are being successfully launched (e.g. http://malaiseprogram. ca/; http://studentdnabarcoding.org/). The rapid development of portable, cheap HTS devices (Box 3) promises even more unexplored opportunities that could be harnessed in the context of biological invasions. For instance, significant improvements in cost might help bringing molecular tools closer to research and policymaking institutions in low-income countries, where invasives' pressure is probably the least documented.

In terms of network reconstruction, building large-scale ecological networks from molecular data is still anecdotal, given the huge potential offered by these techniques, especially with regard to plant–herbivore and plant–pollinator networks, where methodological constraints are relatively few. But we believe that the ever-increasing accessibility of high-throughput molecular techniques will promote their use by a wider community of ecologists, helping to take full advantage of these methods rapidly.

Finally, molecular methods have a promising future as a powerful tool integrating multiple organization levels. Very soon, it will be possible, with a single sequencing event, to collect information about species' distributional ranges, abundance, phylogeny, population and functional genetics,

BOX 2 Shotgun Sequencing for Biodiversity Assessment

Shotgun sequencing refers to the direct sequencing of genomic DNA. According to the matrix used to extract the DNA, shotgun sequencing has multiple applications. It could be used for whole genome sequencing when DNA is extracted from a specimen or for biodiversity assessment when DNA extraction is made from environmental samples. Considering the last case, this approach can bridge the main current limitations of DNA metabarcoding, namely PCR amplification bias (Coissac et al., 2016; Taberlet et al., 2012b; Zhou et al., 2013). Taking advantage of the unprecedented expansion of sequencing capacity provided by high-throughput sequencers, shotgun sequencing has the full potential to draw up new dimensions in biodiversity research (Papadopoulou et al., 2015). The direct sequencing of genomic DNA from bulk or environmental samples allows the parallel acquisition of a large array of genetic information including multiple mitochondrial, plastid, and nuclear markers, useful for robust and auditable taxonomic assignation and phylogenetic inference within a single sampling event (e.g. Gillett et al., 2014; Tang et al., 2015; Zhou et al., 2013), while available information could be used for inferring specimens' evolutionary history (Besnard et al., 2014), community assembly (Andújar et al., 2015), and diet or symbiont community (Paula et al., 2015; Srivathsan et al., 2015, 2016). Moreover, with the routine sequencing of genomes, scientists will no more be limited in their initial choice of loci. The same effort to build reference databases will provide, from the same specimens, markers useful for taxonomists, ecologists, and phylogeneticists, thus unifying their efforts for describing biodiversity (Coissac et al., 2016).

Shotgun sequencing could be realized directly from DNA extracts fragmented prior sequencing (i.e. genome skimming) or from DNA extracts that have been previously enriched using oligonucleotide probes or "baits" (for more details about the different DNA capture techniques, see Ávila-Arcos et al., 2011; Horn, 2012). While the genome skimming results in the low-coverage random subsampling of a small proportion of the total genomic DNA (Dodsworth, 2015), target enrichment consists in selectively targeting user-defined sequences across plastid or nuclear genomes (Ávila-Arcos et al., 2011). Because of their usually high-copy numbers, the nuclear ribosomal cistron, mitochondrial, and plastid genomes are fully sequenced with high sequencing depth. Most of the time, their complete sequence can be reconstructed/assembled from a genome skimming data set, while providing multiple useful markers for robust phylogenetic inference (Dodsworth, 2015). However, the large majority of reads belongs to the nuclear genome. Even if the data they provide cannot allow the assembly of full nuclear genomes, they can still provide some additional phylogenetic information. On the other hand, selective enrichment prior sequencing enables an increased sequencing depth and lower cost over regions of interest, very appropriate for large-scale biodiversity assessment via mass sequencing of bulk samples. This is because the capture of desired genomic regions takes more advantage of the sequencing depth enabling the recovery even of low-abundant target taxa (e.g. Ávila-Arcos et al., 2011). The current cost per gigabase ranges between $40 (target enrichment; Zhou et al., 2013) and $80 (genome skimming; Coissac et al., 2016). But while the shotgun sequencing discovery of a single species within a bulk sample is estimated to about $20 (i.e. approximately the price of outsourced Sanger sequencing barcoding; Zhou et al., 2013), the preparation cost of high-throughput sequencing library is still relatively high, especially when it comprises only one specimen ($200; Coissac et al., 2016). This price excludes the cost related to computational power and data storage infrastructures. Therefore, further challenges include cost reduction as well as the development of appropriate bioinformatic tools for high-throughput genomic data analysis. Nevertheless, several recent projects demonstrated as a proof of a principle that upscalling the shotgun sequencing approach is feasible (Coissac et al., 2016; Ribeiro et al., 2012; Stull et al., 2013; Tang et al., 2015). The perspective of bringing high-throughput shotgun sequencing directly to the field via the portable MinION™ personal sequencing device (Box 3) is the very next step that will revolutionize the taxonomical and functional analysis of biodiversity.

symbiont and parasite communities, as well as its trophic interactions with other species (Table 1; Box 2; see also Barnes and Turner, 2016; Hajibabaei et al., 2007). This opens important perspectives for conceptual and analytical advancements in order to be able to integrate all these data in a common framework (cf. Kéfi et al., 2016).

6.1.2 Stable Isotopes

SIA is now a well-established and increasingly applied method in ecological studies, calling for a further expansion of the isotopic toolbox towards research questions incorporating more biological complexity and over larger set of spatiotemporal scales. The current development of new analytical frameworks and models (e.g. Healy et al., 2016; Phillips et al., 2014; Stock and Semmens, 2016; Yeakel et al., 2016) offers the possibility to derive an increasing spectrum of biological information from a limited number of isotopic elements including physiological, ecological, and environmental factors that shape the dynamics of the isotopic space. Therefore, it would

BOX 3 Oxford Nanopore MinION™

The MinION™ device is the first miniaturized portable real-time DNA sequencing device and also the first to offer no limits for the length of DNA fragments to be sequenced. The core of the MinION™ sequencer is a single-use flow cell composed by an array of protein nanopores, embedded in a synthetic polymer membrane and high-salt buffer. During the sequencing reaction, a voltage is applied across the membrane, which makes a current of ions to flow through the pores. The trans-location of a single-strand DNA molecule through each pore disrupts the ionic flow in a sequence-specific manner. The signal (squiggle plot defined by the duration and the value of the disrupted current) is directly linked to the pattern of the DNA sequence present inside the pore (i.e. base calling). Each strand of a double-stranded DNA molecule is sequenced separately, but the two strands are needed for the base calling in order to generate one final consensus sequence (2D read). The running time for a single flow cell could be adjusted according to the need or the length of the DNA molecules but usually does not exceed 48 h, which corre-sponds to the maximal duration of the protein nanopore activity. Once all the pores have been deactivated, the flow cell needs to be replaced. For example, the R7.3 model of Nanopore Flow Cell available until mid-2016 has allowed the simulta-neous read of up to 512 DNA molecules for an average of 115 million double-strand bases or approximately 20,000 reads (Ip et al., 2015). Considering this sequence yield and a minimal cost of $500 for a single flow cell, the current cost per gigabase could be estimated at about $5500. The required sequence coverage with the Min-ION™ will depend on two parameters: (i) the complexity of the target genome; (ii) the analytical method used. For example, a coverage of 29 × was required to obtain the complete 4.6 megabases de novo assembled bacterial genome (Loman et al., 2015). However, at this stage of technological advancement, the amount of MinION™ generated data was still too low to allow the sequencing of large genomes.

Concerning the library preparation, various kits have already been released by Oxford Nanopore Technologies according to the nature of the DNA to be sequenced (e.g. amplicon, shotgun, or RNA sequencing). The price per kit varies between $400 and $500 and usually allows the preparation of up to six libraries. The minimum amount of DNA material required ranges between 20 ng and >1 μg, depending on the size of the DNA fragments to be sequenced, with usu-ally, a higher DNA input improving the overall sequence yield. In terms of amplicon sequencing, between 12 and 96 distinct libraries can be pooled and sequenced at the same time, while the RNA sequencing requires the preliminary preparation of a complementary DNA.

The MinION™ device is connected to a laptop or a desktop computer through a USB connection. A control software, MinKNOW, allows the manage-ment of sequencing core tasks and related parameters during the sequencing run. At the end of the run, the base calling is activated through a cloud-based software Metrichor, provided by Oxford Nanopore. Raw sequencing data are first uploaded on a cloud cluster and then downloaded and analysed as fast5 format data files on the local host computer. There are now several different bioinformatic tools and pipelines allowing the exploitation fast5 format data. Poretools (Loman and Quinlan, 2014) and poRe (Watson et al., 2015) are useful for converting and visualizing the raw data. MinoTour is a software for pre- and postalignment analysis of MinION™ sequencing data, and for determining sequence quality and error profiles. Some software packages were specifically developed to allow genome assembly that account for the high error rate of the raw data (currently >10%). These softwares include NanoCORR (Goodwin et al., 2015), NaS (Madoui et al., 2015), Nanopolish (Loman et al., 2015), and poreSeq (Szalay and Golovchenko, 2015), while the newest mapping algorithm for Nanopore sequencing reads, GraphMap, has been released this year (Sović et al., 2016). The first open-source base callers for Nanopore sequencing have also been released this year, DeepNano (Boža et al., 2016) and Nanocall (David et al., 2016).

In terms of applications, MinION™ was used for the real-time genomic surveillance in the resource-limited context of the Ebola virus outbreak (Quick et al., 2016). Several small bacteria and yeast genomes have been already sequenced, some combining MinION™ and Illumina data (Goodwin et al., 2015; Madoui et al., 2015; Risse et al., 2015), while other only relying on the MinION™ (Loman et al., 2015). MinION™ could find useful applications in metagenomics as well. It has been used for instance for identifying viral pathogens in complex clinical samples (Greninger et al., 2015), or unknown bacteria and viruses by amplicon sequencing (Kilianski et al., 2015). The characterization of highly diverse microbial community using the 16S rRNA gene is also possible (Benitez-Paez et al., 2016). Other applications include the analysis of methylated DNA bases (Simpson et al., 2016; Rand et al., 2016) or complementary DNA (Oikonomopoulos et al., 2016), while Li et al. (2016) proposed a pragmatic approach for circumvent the high single read error rate. Finally, the research group of Massimo Delledonne successfully used the MinION™ for sequencing the DNA of a wild frog species during a field expedition in Tanzania (https://publications.nanoporetech.com/2015/05/15/minions-and-nanofrogs/).

The newest flow cell models, R9 and then R9.4, were launched by Oxford Nanopore Technologies in May and October 2016, respectively. Their DNA base calling accuracy mostly depends on the speed of DNA translocation through the pores and the performance of the base caller itself. The R9 and R9.4 versions allow a faster translocation rate (up to 450 DNA bases per second instead of the 70 with the R7.3 version), and higher sequence yield with lower error rates (from >10% to ~5%), thus reducing the cost per Gb. An upcoming device called VolTRAX™—a palm-sized cartridge—promises for an automated DNA library preparation without the need for human intervention.

This opens promising avenues for the adoption of the MinION™ device as a portable, versatile, diagnostic tool by practitioners and even citizen scientists in order to ensure the large-scale, real-time mapping of dynamical changes in biodiversity and species distributional ranges.

be possible to hierarchize the importance of individual vs community-level processes that determine the impact of invasives on food web structure and ecosystem functioning. If applied to data from fossil or museum records, this could be done in temporally explicit context in order to see how fine-scale variations in the functional or foraging diversity—based on individual isotopic signatures—correlate with the degree of invasibility in a given community or location. Furthermore, growing advancements in our understanding about the relative incorporation of dietary vs environmental sources of isotopic elements in animal tissues could allow in the near future to characterise both the diet and habitat of an individual by using compound-specific SIA from the same sample (e.g. amino acids; Fogel et al., 2016).

SIA could also benefit from a greater integration with molecular data. For example, the combination of SIA and DNA analysis could be particularly valuable for revealing the taxonomic identity, the ecological process, and the underlying functional impact of a nonnative species as elegantly demonstrated by Matsuzaki et al. (2010). On the other hand, techniques such as DNA/RNA-stable-isotope probing used in microbiology (Manefield et al., 2002; Neufeld et al., 2007; Radajewski et al., 2000), targeting the analysis of incorporated isotope-labelled specific compounds into nucleic acids using functional metagenomics, showed particularly valuable for characterizing the functional roles of a large array of unculturable microorganisms by revealing their unique biochemical pathways (e.g. Krause et al., 2010; Lueders et al., 2004). This type of methods could find multiple interesting applications in the context of biological invasions where the existence and the spread of cryptic invasive functions or genes within a given ecosystem could be characterized without the need to accessing the taxonomical identity of the invader. Finally, the development of artificial diet tracers, integrating the advantages of both DNA and stable isotope methods within a single analysis (e.g. silica particles with encapsulated DNA, Mora et al., 2015), could be an original and probably cost-effective alternative to the combination of multiple techniques. Further investigations will show how applicable is this approach in more realistic ecological context.

6.2 Perspectives and Challenges for Network Reconstruction
6.2.1 Interaction Network Models
The models presented under Section 5.1 aim either at reconstructing interaction networks from incomplete data or indirect evidence, or to predict

novel interactions between partners that have never been in contact before. In both cases, such models must deal with statistical issues linked to the amount of data and the number of parameters estimated used to make inferences—with big data, parsimony, and predictive power being serious issues (Giraud, 2014). In most cases, these models are also phenomenological in nature, i.e., they do not rely on a well-understood theoretical model, but rather make inferences based on correlations (Mouquet et al., 2015). As it has been the case for SDMs, phenomenological approaches have to evolve from comparisons of goodness-of-fit indicators to predictive power comparison and considerations about divergence of predictions among models (Araújo and New, 2007; Gritti et al., 2013; Peterson et al., 2011; Thuiller et al., 2008). While such comparisons have already been undertaken for some machine-learning approaches aimed at uncovering network structure from abundance data (Faisal et al., 2010), this has yet to be done on a systematic basis for models predicting novel interactions from the food web structure (but see Dalla Riva and Stouffer, 2016).

From a more biological point of view and again taking the example of SDMs, interaction network models will have to account for factors such as species evolution and dispersal between biogeographical units (Saltré et al., 2015; Thuiller et al., 2013). In the case of antagonistic interactions such as those between predator and prey or host and parasite, rapid coevolution through either frequency-dependent selection or an evolutionary arms race is expected, following Red Queen dynamics (Decaestecker et al., 2007; Gandon et al., 2008; Kerfoot and Weider, 2004; Salathé et al., 2008). Theoretical models investigating the evolution of plant defences have shown that, depending on plant dispersal rate and overall ecosystem productivity, one expects the evolution of little defence, specialized cheap defences, or high, costly defence (Loeuille and Leibold, 2008). Such considerations will have to be taken into account when modelling interaction networks on large spatial and temporal scales because they entail potential shifts in species role among interaction networks sampled at different times and locations (see also Poisot et al., 2012).

Theoretically linking interaction network characteristics to their invasibility and the potential impact of invaders is still in its infancy. While the amount of work linking network structure to network stability and robustness to species extinction is important (Allesina and Tang, 2012; Allesina et al., 2015; Astegiano et al., 2015; Dunne et al., 2002; Eklöf et al., 2013a,b; Goldstein and Zych, 2016; Kokkoris et al., 2002; Lehman and Tilman, 2000; May, 1973; Santamaría et al., 2016; Tang and

Allesina, 2014; Tang et al., 2014), similar explorations need to be conducted to link the network structure with invasion success and impacts, following the seminal work of Romanuk and colleagues (Romanuk et al., 2009; see also Romanuk et al., 2017).

6.2.2 Machine Learning

The future development of HTS-based reconstruction of networks has the potential to allow us to develop networks faster and more cheaply than previously, allowing comparison of networks across a great range of situations. Consequently, it becomes possible to imagine a situation where HTS methods may be used to sample ecosystems continuously, for the detection of invasions into reconstructed networks. Importantly, this would in addition probably be more sensitive than current methods of detecting invasion. To achieve this potential, however, it will be necessary to demonstrate clearly the methodological validity of network reconstruction from HTS data by: (i) identifying ecosystems with known and well-characterized networks; (ii) sampling these ecosystems using HTS; and (iii) reconstructing networks from the HTS sample data for comparison and testing them against the already existing networks.

6.3 Societal Challenges and Perspectives for Management

The development of new tools and the data acquisition they allow open multiple new challenges far beyond the sole area of scientific research. For example, throughout this chapter, we argued that a complete toolbox is an important prerequisite in biological invasions' management. Nevertheless, translating multiple type of data into decision-making programmes could be challenging as it requires a good understanding of the distinct advantages and disadvantages, as well as the level of uncertainty, associated with each method (Table 3). This is particularly relevant with regard to rising eDNA techniques for species surveillance, where a general, robust framework for data interpretation is still lacking. As advocated by Darling and Mahon (2011) an open and transparent discussion between multiple stakeholders will be necessary in order to negotiate the trade-offs between various sources of potential errors for each method and associated cost–benefits in terms of invasion management.

Another important challenge is related to the implementation of network perspective in management programmes (Kaiser-Bunbury and Blüthgen, 2015). Despite substantial advances in fundamental knowledge, it is still not clear how to use network theory for decision making as the

Table 3 Overview of the Advantages and Disadvantages Associated With the Different Tools and Methods Presented in This Paper As Well As Their Potential Applications in Invasion Ecology

Type of Method	Advantages	Disadvantages	Type of Information
1. Visual analysis of fossil records	• Easy and straightforward analyses • Cost effective	• Unsuitable for detecting the absence of a species • Not sensitive to short-term changes • Limited to organisms with fossilizing parts • Coarse taxonomic assignation	• Species occurrence • Passenger vs invasive organisms • Spatiotemporal changes in invasives distributions • Source of invasion • Invasion routes • Number of introduction events
2. Long-term data series	• Easy and straightforward analyses • Cost effective • Sensitive to short-term changes	• Few existing data sets • Limited to specific taxonomic groups • Drawbacks of long-term monitoring programmes (e.g. inertia, trade-offs on observed data, etc.)	• Passenger vs invasive organisms • Comparison of pre- and postinvasion • Spatiotemporal extent of invasions
3. Population genetics	• Historical inference from present-day data only	• Need for developing of markers with adequate resolution • Need for large samples from multiple populations	• Phylogenetic origin of the invader • Population size of founders • Population differentiation • Demographic changes • Genetic diversity • Hybridization events • Invasion routes and dispersal • Single vs multiple introductions
4. Citizen science	• Covers large spatiotemporal scales • Cost effective • Increasing number of taxa covered • Real-time monitoring of invasion	• Varying levels of protocol standardization • Need for simplified sampling designs and strong coordination • Need for advanced statistical tools	• Spatiotemporal changes in invasives distributions • Species occurrence • Community structure • Measures of individual traits • Predict invasion risk and future distributions • Comparison of pre- and postinvasion • Impact on higher trophic levels

Continued

Table 3 Overview of the Advantages and Disadvantages Associated With the Different Tools and Methods Presented in This Paper As Well As Their Potential Applications in Invasion Ecology—cont'd

Type of Method	Advantages	Disadvantages	Type of Information
5. Stable isotope analysis	• Easy and straightforward analyses • Rapid screening of large number of samples • Quantitative estimation of the proportional contribution of multiple prey • Time-integrative measure of diet	• Existence of multiple nondietary sources of variation • Limited to a narrow number of food sources	• Nutrients and energy flow • Trophic links • Food web structure • Trophic level • Habitat use • Invasion routes and dispersal
6. Chemical fingerprinting	• Fine-scale resolution of organism life history • Time-integrative measure of organism life history	• Need for very high precision • Need for (local) calibration	• Invasion routes and dispersal • Geographic origin of the invader • Habitat switch
7. PCR– and sequence-based DNA methods	• Versatile tool adaptable for different research questions and needs • Clues of species occurrence or interactions even when the species itself cannot be observed • Potentially large taxonomic coverage • The taxonomic specificity could be adjusted according to the research context • Rapid and cost-effective screening of large number of samples	• Sensitive to DNA–cross contamination • Sensitive to false positives and false negatives • Existence of multiple nondietary sources of variation • No quantitative estimation of the proportional contribution of prey	• Species occurrence • Predict invasion risk and future distributions • Trophic links • Food web structure • Predation and parasitism rates • Diet breadth • Impact on higher trophic levels
8. Predictive models	• Cost effective • Using different types of data	• Need for tailoring models based on available data • Possible indeterminacy (different possible mode is for the same observed data) • Possible diverging, nonrobust results (changing the model drastically change the predictions)	• Potential interactions between species • Inference of interaction networks from various data (traits, co-occurrence time series, etc.) • Changes in interactions due to invasion/ environmental changes

choice of relevant metrics is potentially large. There is clearly need for more a posteriori validation (when practically feasible) of management impact in order to evaluate how invaded networks respond to different management scenarios (Courchamp et al., 2003). Beyond the financial arguments, getting prior feedback on management actions is important in order to prevent cascading "surprise effects" following management that could further jeopardize ecosystem functioning and services. So far such experimental validations are rare. Make a thorough use of above-mentioned set of surveillance techniques, as well as a closer collaboration between scientists and practitioners could help advancing in this direction (Kelly et al., 2014).

ACKNOWLEDGEMENTS

We thank the CESAB working group COREIDS for opportunities to develop this work, as well as TOTAL and Fondation pour la Recherche sur la Biodiversité for funding COREIDS. F.M. is supported by two French ANR projects: AFFAIRS project in the BIOADAPT programme (P.I. P. David—Grant No. 12-ADAP-005) and ARSENIC project (P.I. F. Massol—Grant No. 14-CE02-0012). S.K. is funded by the Atlantic Canada Opportunities Agency (Project 2.2.3 "Barcoding: Innovative DNA-based diagnostic for spruce budworm, its natural enemies, and other conifer-feeding species", P.I. Alex Smith and Eldon Eveleigh). A.L. is supported by the Genoscope, the Commissariat à l'Energie Atomique et aux Energies Alternatives (CEA) and France Génomique (ANR-10-INBS-09-08). V.R. acknowledges financial support from The European Union (ERDF), Conseil Régional de La Réunion, and the French Agropolis Foundation (Labex Agro—Montpellier, E-SPACE project number 1504-004). I.D. is supported by French EC2CO projects (REPLAY: paleoecological reconstruction of lacustrine biodiversity from sedimentary DNA).

GLOSSARY

Palaeoecology

Allozyme (alloenzyme) diversity Diversity of variant forms of an enzyme that are coded by different alleles at the same locus, as characterized by gel electrophoresis. A method (pre-next-generation sequencing era) to quantify genetic diversity.

Diatom frustules Hard and porous cell wall of diatoms, made of silica. They are preserved in the sediment and their typical morphological features can be used for taxonomic identification.

Metalimnetic cyanobacterial bloom Outbreak of cyanobacterial biomass occurring at the metalimnion depth, at the interface between the surface, warm and deeper, colder water layers of a lake.

Phytoplankton cysts Resting spores produced by some phytoplankton species. They are preserved in the sediment and their typical morphological features can be used for taxonomic identification.

Stratigraphic analysis Stratigraphy is a branch of geology, which studies rock layers (strata) and layering (stratification). It is primarily used in the study of sedimentary and layered volcanic rocks. A stratigraphic analysis focusses on archived components along a rock layer or sediment sequence as a way to go back in time.

Citizen science

Automated signal recognition Software based on classification algorithms used to identify digital records of species such as audio or pictures.

Detection probability Probability to detect a species when it is present. Such probabilities can vary among species, among environmental conditions of recordings, and among observers.

Fundamental niche The full range of conditions (biotic and abiotic) and resources in which a species could survive and reproduce.

Species distribution models (SDMs) Algorithms used to predict the distribution of species in space based on their known distribution in environmental space. The environmental space is most often characterized by climatic variables (e.g. temperature, precipitation), but can also include other variables such as soil type, water depth, and land cover.

Molecular ecology

Base-caller Algorithm that analyses the raw data produced by automated sequencers to predict the individual DNA bases.

Blocking primer This is a unique primer specifically designed to block the amplification of particular DNA sequences when universal primers are used in metabarcoding studies. Blocking primers are of particular interest for studies implying the HTS of ancient DNA or diet analysis to prevent the preferential amplification of modern or consumer DNA over target ancient or food DNA.

Diagnostic PCR A PCR assay that is used to test samples for the presence/absence of DNA from a single-target species or from a particular taxonomic group.

DNA template This is a matrix of DNA molecules that contains the target sequence the primers bind to.

Enzyme-linked immunosorbent assay (ELISA) This is an assay using an antibody specific to a particular antigen. The binding reaction between antigen and antibody is detected thanks to an enzymatic reaction provoking colour change in the assay substrate.

Environmental DNA (eDNA) This is related to the DNA molecules trapped in environmental samples like water, soil, or faeces.

Gel electrophoresis Method allowing the visualization of DNA fragments based on their size and charge. It implies the application of an electric field inducing negatively charged DNA molecules to move through a porous matrix (usually agarose gel). The method could be automatized and allow the simultaneous separation of multiple DNA fragments for multiple samples with great precision (e.g. capillary electrophoresis). The single-stranded conformation polymorphism (SSCP) is a particular case of electrophoresis, which allows separating DNA fragments that differ in their nucleotide sequence without sequencing (Sunnucks et al., 2000). The SSCP technique could offer a relatively simple and inexpensive alternative to sequencing in some cases (e.g. Varennes et al., 2014).

High-throughput sequencing (HTS) Technologies that parallelise the sequencing process by generate millions of sequences at the same time. There are several different high-throughput sequencing platforms, with currently the most popular being Illumina, Ion Torrent, and Oxford Nanopore. Regular updates about different platform specifications and cost could be found at http://www.molecularecologist.com/next-gen-fieldguide-2014/.

k-mers This is a term that refers to all the possible subsequences (of length k) from a read obtained through DNA sequencing.

Melt curves Analysis is used to determine the specificity and the sensitivity of a qPCR reaction. It refers to the temperature-dependent denaturation of the double-strand DNA measured by intercalating fluorescent probe.

Metabarcoding A method for the characterization of biodiversity recovered from complex environmental (soil, water, faeces, etc.) or bulk (Malaise traps) samples. It relies on simultaneous amplification of multiple species via PCR universal primers, high-throughput sequencing, and bioinformatic analysis.

Nested PCR It refers to a PCR reaction that involves two successive steps using the same or two different sets of primers, where the second step aims to amplify the PCR product of the first step. Nested PCR is used for amplifying secondary target gene regions or for enhancing the amplification of recalcitrant target regions.

Pair-end sequencing This is the high-throughput sequencing of both ends of the same target DNA fragment, allowing the high-quality alignment of sequence data.

Phylogenetic inference This refers to methods that provide estimates about phylogenetic (evolutionary) relationships among organisms based on observed heritable traits like morphology or DNA sequences.

Quantitative PCR (qPCR) Where the regular PCR primer set is combined with a specific fluorescent probe allowing a quantitative estimation of the number of molecules present in the template by comparison with a reference threshold.

Sanger sequencing This is a method of DNA sequencing developed by Frederick Sanger and colleagues in 1977. It refers to the selective incorporation of chain-terminating dideoxynucleotides by DNA polymerase during in vitro DNA replication, generating sequences between 100 and 1000 bp. In contrast to HTS, Sanger sequencing does not allow the simultaneous sequencing of multiple DNA molecules.

Sedimentary ancient DNA (sedaDNA) This is a metabarcoding method for the reconstruction of past communities and biodiversity dynamics based on ancient DNA trapped in sediments or ice cores.

Sequence reads These are millions of usually short DNA sequences, produced by high-throughput sequencing machines.

Singleplex/multiplex PCR A polymerase chain reaction (PCR) where one pair of primers is used to amplify one specific PCR fragment is called a singleplex PCR, whereas in multiplex PCR more than one primer pair is employed to simultaneously amplify several PCR fragments within one reaction.

Squiggle plot This is a graphical interpretation of the fluctuating electrical signals generated by the DNA translocation through the nanopore used in the MinION™ sequencing device.

Tag This is a unique short sequence added to the $5'$-end of a primer allowing the downstream sequence sorting and sample assignation. Tags usually range between 8 and 12 bp. Samples with unique tags can be pooled and sequenced in the same sequencing run (=multiplexing). Sequences are later assigned to samples via bioinformatic pipelines.

Population genetics

Allelic richness This is the average number of alleles per locus. Allelic richness is used as a measure of population genetic diversity.

Approximate Bayesian Computation A class of model-based likelihood-free methods for statistical inference. Based on principles of Bayesian statistics, ABC algorithms provide a way to identify the models and model parameters that are most congruent with data. To do so, they quantify the probability of the observed data under a particular model, that is, likelihood, and seek the model and parameters for which likelihood is maximal. But the likelihood function can easily be mathematically derived and numerically evaluated only for reasonably simple models. In ABC, the limiting step of very complex likelihood computations is replaced by an approximation of the likelihood, which is obtained by simulating (millions of) data sets under considered models and studying the distributions of models and model parameter values among the simulated data sets that are closest to the observed data using a regressive approach on summary statistics (Beaumont et al., 2002).

Genetic bottleneck A reduction in population effective size.

Haplotype This is a group of genes or alleles that progeny inherited from one parent.

Heterozygosity This is a measure of genetic variation within a population. Observed heterozygosity is defined as the percentage of heterozygous individuals per locus. Expected heterozygosity (also called gene diversity) is the expected number of heterozygotes given allele frequencies (Nei, 1987). Observed heterozygosity can only be computed for diploid genomes, while expected heterozygosity can always be computed.

Microsatellite A DNA sequence containing repeats of a certain motif (generally ranging from 2 to 5 base pairs). Microsatellites belong to the Variable Number of Tandem Repeats (VNTR) markers and also named short tandem repeats (STRs) or simple sequence repeats (SSRs).

Haplotype network A diagram representing genetic relationships between haplotypes. In a haplotype network, each haplotype is represented by a circle, with size proportional to the number of individuals belonging to that haplotype, connected to the most similar haplotype by a line. Each circle may be divided in slices of pie, in which the colour indicates sampling localities.

Neutral DNA marker A DNA sequence that is not subject to selection. Noncoding regions of DNA are traditionally supposed neutral, although for some of them, deleterious effects have been detected.

Private variability This is the part of within-population diversity that is unique to a given population. Private alleles are only found in one population.

REFERENCES

Abarenkov, K., Nilsson, R.H., Larsson, K.H., Alexander, I.J., Eberhardt, U., Erland, S., Hoiland, K., Kjoller, R., Larsson, E., Pennanen, T., Sen, R., Taylor, A.F.S., Tedersoo, L., Ursing, B.M., Vralstad, T., Liimatainen, K., Peintner, U., Koljalg, U., 2010. The UNITE database for molecular identification of fungi—recent updates and future perspectives. New Phytol. 186, 281–285.

Achtman, M., 2008. Evolution, population structure, and phylogeography of genetically monomorphic bacterial pathogens. Annu. Rev. Microbiol. 62, 53–70. http://dx.doi.org/10.1146/annurev.micro.62.081307.162832.

Adrian-Kalchhauser, I., Burkhardt-Holm, P., 2016. An eDNA assay to monitor a globally invasive fish species from flowing freshwater. PLoS One 11, e0147558. http://dx.doi.org/10.1371/journal.pone.0147558.

Aguirre-Gutiérrez, J., Kissling, W.D., Carvalheiro, L.G., WallisDeVries, M.F., Franzén, M., Biesmeijer, J.C., 2016. Functional traits help to explain half-century long shifts in pollinator distributions. Sci. Rep. 6, 24451. http://dx.doi.org/10.1038/srep24451.

Agusti, N., Bourguet, D., Spataro, T., Delos, M., Eychenne, N., Folcher, L., Arditi, R., 2005. Detection, identification and geographical distribution of European corn borer larval parasitoids using molecular markers. Mol. Ecol. 14, 3267–3274. http://dx.doi.org/10.1111/j.1365-294X.2005.02650.x.

Ait Baamrane, M.A., Shehzad, W., Ouhammou, A., Abbad, A., Naimi, M., Coissac, E., Taberlet, P., Znari, M., 2012. Assessment of the food habits of the Moroccan dorcas gazelle in M'Sabih Talaa, West Central Morocco, using the trnL approach. PLoS One 7, e35643. http://dx.doi.org/10.1371/journal.pone.0035643.

Aizen, M.A., Morales, C.L., Morales, J.M., 2008. Invasive mutualists erode native pollination webs. PLoS Biol. 6, e31. http://dx.doi.org/10.1371/journal.pbio.0060031.

Albrecht, M., Padron, B., Bartomeus, I., Traveset, A., 2014. Consequences of plant invasions on compartmentalization and species' roles in plant-pollinator networks. Proc. R. Soc. B Biol. Sci. 281, 20140773. http://dx.doi.org/10.1098/rspb.2014.0773.

Allesina, S., Pascual, M., 2008. Network structure, predator–prey modules, and stability in large food webs. Theor. Ecol. 1, 55–64. http://dx.doi.org/10.1007/s12080-007-0007-8.

Allesina, S., Tang, S., 2012. Stability criteria for complex ecosystems. Nature 483, 205–208. http://dx.doi.org/10.1038/nature10832.

Allesina, S., Grilli, J., Barabás, G., Tang, S., Aljadeff, J., Maritan, A., 2015. Predicting the stability of large structured food webs. Nat. Commun. 6, 7842. http://dx.doi.org/10.1038/ncomms8842.

Alric, B., Möst, M., Domaizon, I., Pignol, C., Spaak, P., Perga, M.-E., 2016. Local human pressures influence gene flow in a hybridizing *Daphnia* species complex. J. Evol. Biol. 29, 720–735. http://dx.doi.org/10.1111/jeb.12820.

Amsellem, L., Brouat, C., Duron, O., Porter, S.S., Vilcinskas, A., Facon, B., 2017. Importance of microorganisms to macroorganisms invasions: is the essential invisible to the eye? (The little prince, A. de Saint-Exupéry, 1943). Adv. Ecol. Res. 57, 99–146.

Anderson, L.W.J., 2005. California's reaction to Caulerpa taxifolia: a model for invasive species rapid response. Biol. Invasions 7, 1003–1016. http://dx.doi.org/10.1007/s10530-004-3123-z.

Anderson, P.K., Cunningham, A.A., Patel, N.G., Morales, F.J., Epstein, P.R., Daszak, P., 2004. Emerging infectious diseases of plants: pathogen pollution, climate change and agrotechnology drivers. Trends Ecol. Evol. 19, 535–544. http://dx.doi.org/10.1016/j.tree.2004.07.021.

Andújar, C., Arribas, P., Ruzicka, F., Crampton-Platt, A., Timmermans, M.J.T.N., Vogler, A.P., 2015. Phylogenetic community ecology of soil biodiversity using mitochondrial metagenomics. Mol. Ecol. 24, 3603–3617. http://dx.doi.org/10.1111/mec.13195.

Araújo, M.B., New, M., 2007. Ensemble forecasting of species distributions. Trends Ecol. Evol. 22, 42–47. http://dx.doi.org/10.1016/j.tree.2006.09.010.

Ardura, A., Zaiko, A., Martinez, J.L., Samulioviene, A., Semenova, A., Garcia-Vazquez, E., 2015. eDNA and specific primers for early detection of invasive species—a case study on the bivalve Rangia cuneata, currently spreading in Europe. Mar. Environ. Res. 112, 48–55. http://dx.doi.org/10.1016/j.marenvres.2015.09.013.

Armstrong, K.F., Ball, S.L., 2005. DNA barcodes for biosecurity: invasive species identification. Philos. Trans. R. Soc. B Biol. Sci. 360, 1813–1823. http://dx.doi.org/10.1098/rstb.2005.1713.

Astegiano, J., Massol, F., Vidal, M.M., Cheptou, P.-O., Guilmarães Jr., P.R., 2015. The robustness of plant-pollinator assemblages: linking plant interaction patterns and sensitivity to pollinator loss. PLoS One 10, e0117243. http://dx.doi.org/10.1371/journal.pone.0117243.

Ávila-Arcos, M.C., Cappellini, E., Romero-Navarro, J.A., Wales, N., Moreno-Mayar, J.V., Rasmussen, M., Fordyce, S.L., Montiel, R., Vielle-Calzada, J.-P., Willerslev, E.,

Gilbert, M.T.P., 2011. Application and comparison of large-scale solution-based DNA capture-enrichment methods on ancient DNA. Sci. Rep. 1, 74. http://dx.doi.org/10.1038/srep00074.

Avise, J.C., 2000. Phylogeography: The History and Formation of Species. Harvard University Press, Cambridge.

Baird, D.J., Hajibabaei, M., 2012. Biomonitoring 2.0: a new paradigm in ecosystem assessment made possible by next-generation DNA sequencing. Mol. Ecol. 21, 2039–2044.

Balmer, O., Pfiffner, L., Schied, J., Willareth, M., Leimgruber, A., Luka, H., Traugott, M., 2013. Noncrop flowering plants restore top-down herbivore control in agricultural fields. Ecol. Evol. 3, 2634–2646. http://dx.doi.org/10.1002/ece3.658.

Bansal, R., Mian, M.A.R., Michel, A.P., 2014. Microbiome diversity of *Aphis glycines* with extensive superinfection in native and invasive populations: microbiome diversity in soybean aphid. Environ. Microbiol. Rep. 6, 57–69. http://dx.doi.org/10.1111/1758-2229.12108.

Barber, N.A., Marquis, R.J., Tori, W.P., 2008. Invasive prey impacts the abundance and distribution of native predators. Ecology 89, 2678–2683.

Barbet-Massin, M., Rome, Q., Muller, F., Perrard, A., Villemant, C., Jiguet, F., 2013. Climate change increases the risk of invasion by the Yellow-legged hornet. Biol. Conserv. 157, 4–10. http://dx.doi.org/10.1016/j.biocon.2012.09.015.

Barnes, M.A., Turner, C.R., 2016. The ecology of environmental DNA and implications for conservation genetics. Conserv. Genet. 17, 1–17. http://dx.doi.org/10.1007/s10592-015-0775-4.

Barnosky, A.D., Hadly, E.A., Bascompte, J., Berlow, E.L., Brown, J.H., Fortelius, M., Getz, W.M., Harte, J., Hastings, A., Marquet, P.A., Martinez, N.D., Mooers, A., Roopnarine, P., Vermeij, G., Williams, J.W., Gillespie, R., Kitzes, J., Marshall, C., Matzke, N., Mindell, D.P., Revilla, E., Smith, A.B., 2012. Approaching a state shift in Earth's biosphere. Nature 486, 52–58. http://dx.doi.org/10.1038/nature11018.

Barrès, B., Carlier, J., Seguin, M., Fenouillet, C., Cilas, C., Ravigné, V., 2012. Understanding the recent colonization history of a plant pathogenic fungus using population genetic tools and approximate Bayesian computation. Heredity 109, 269–279.

Bartley, T.J., Braid, H.E., McCann, K.S., Lester, N.P., Shuter, B.J., Hanner, R.H., 2015. DNA barcoding increases resolution and changes structure in Canadian boreal shield lake food webs. DNA Barcodes 3, 30–43.

Bartomeus, I., 2013. Understanding linkage rules in plant-pollinator networks by using hierarchical models that incorporate pollinator detectability and plant traits. PLoS One 8, e69200. http://dx.doi.org/10.1371/journal.pone.0069200.

Bartomeus, I., Bosch, J., Vila, M., 2008. High invasive pollen transfer, yet low deposition on native stigmas in a Carpobrotus-invaded community. Ann. Bot. 102, 417–424. http://dx.doi.org/10.1093/aob/mcn109.

Bearhop, S., Adams, C.E., Waldron, S., Fuller, R.A., Macleod, H., 2004. Determining trophic niche width: a novel approach using stable isotope analysis. J. Anim. Ecol. 73, 1007–1012.

Beaumont, M.A., 1999. Detecting population expansion and decline using microsatellites. Genetics 153, 2013–2029.

Beaumont, M.A., Zhang, W., Balding, D.J., 2002. Approximate Bayesian computation in population genetics. Genetics 162, 2025–2035.

Beaumont, L.J., Gallagher, R.V., Thuiller, W., Downey, P.O., Leishman, M.R., Hughes, L., 2009. Different climatic envelopes among invasive populations may lead to underestimations of current and future biological invasions. Divers. Distrib. 15, 409–420. http://dx.doi.org/10.1111/j.1472-4642.2008.00547.x.

Beckerman, A.P., Petchey, O.L., Warren, P.H., 2006. Foraging biology predicts food web complexity. Proc. Natl. Acad. Sci. U.S.A. 103, 13745–13749. http://dx.doi.org/10.1073/pnas.0603039103.

Bell, J.R., Andrew King, R., Bohan, D.A., Symondson, W.O.C., 2010. Spatial co-occurrence networks predict the feeding histories of polyphagous arthropod predators at field scales. Ecography 33, 64–72. http://dx.doi.org/10.1111/j.1600-0587.2009.06046.x.

Benitez-Paez, A., Portune, K., Sanz, Y., 2016. Species-level resolution of 16S rRNA gene amplicons sequenced through MinION™ portable nanopore sequencer. GigaScience 5, 4. http://dx.doi.org/10.1186/s13742-016-0111-z. eCollection 2016.

Ben Rais Lasram, F., Tomasini, J.A., Romdhane, M.S., Do Chi, T., Mouillot, D., 2008. Historical colonization of the Mediterranean Sea by Atlantic fishes: do biological traits matter? Hydrobiologia 607, 51–62. http://dx.doi.org/10.1007/s10750-008-9366-4.

Berry, O., Sarre, S.D., Farrington, L., Aitken, N., 2007. Faecal DNA detection of invasive species: the case of feral foxes in Tasmania. Wildl. Res. 34, 1. http://dx.doi.org/10.1071/WR06082.

Berry, O., Bulman, C., Bunce, M., Coghlan, M., Murray, D., Ward, R., 2015. Comparison of morphological and DNA metabarcoding analyses of diets in exploited marine fishes. Mar. Ecol. Prog. Ser. 540, 167–181. http://dx.doi.org/10.3354/meps11524.

Besnard, G., Christin, P.-A., Male, P.-J.G., Lhuillier, E., Lauzeral, C., Coissac, E., Vorontsova, M.S., 2014. From museums to genomics: old herbarium specimens shed light on a C3 to C4 transition. J. Exp. Bot. 65, 6711–6721. http://dx.doi.org/10.1093/jxb/eru395.

Biggs, J., Ewald, N., Valentini, A., Gaboriaud, C., Dejean, T., Griffiths, R.A., Foster, J., Wilkinson, J.W., Arnell, A., Brotherton, P., Williams, P., Dunn, F., 2015. Using eDNA to develop a national citizen science-based monitoring programme for the great crested newt (Triturus cristatus). Biol. Conserv. 183, 19–28. http://dx.doi.org/10.1016/j.biocon.2014.11.029.

Bik, H.M., Porazinska, D.L., Creer, S., Caporaso, J.G., Knight, R., Thomas, W.K., 2012. Sequencing our way towards understanding global eukaryotic biodiversity. Trends Ecol. Evol. 27, 233–243. http://dx.doi.org/10.1016/j.tree.2011.11.010.

Binladen, J., Gilbert, M.T., Bollback, J.P., Panitz, F., Bendixen, C., Nielsen, R., Willerslev, E., 2007. The use of coded PCR primers enables high-throughput sequencing of multiple homolog amplification products by 454 parallel sequencing. PLoS One 2 (2), e197.

Bird, T.J., Bates, A.E., Lefcheck, J.S., Hill, N.A., Thomson, R.J., Edgar, G.J., Stuart-Smith, R.D., Wotherspoon, S., Krkosek, M., Stuart-Smith, J.F., Pecl, G.T., Barrett, N., Frusher, S., 2014. Statistical solutions for error and bias in global citizen science datasets. Biol. Conserv. 173, 144–154. http://dx.doi.org/10.1016/j.biocon.2013.07.037.

Blankenship, L.E., Yayanos, A.A., 2005. Universal primers and PCR of gut contents to study marine invertebrate diets: marine invertebrate diet analysis. Mol. Ecol. 14, 891–899. http://dx.doi.org/10.1111/j.1365-294X.2005.02448.x.

Blois, J.L., Williams, J.W., Grimm, E.C., Jackson, S.T., Graham, R.W., 2011. A methodological framework for assessing and reducing temporal uncertainty in paleovegetation mapping from late-Quaternary pollen records. Q. Sci. Rev. 30, 1926–1939. http://dx.doi.org/10.1016/j.quascirev.2011.04.017.

Blumstein, D.T., Mennill, D.J., Clemins, P., Girod, L., Yao, K., Patricelli, G., Deppe, J.L., Krakauer, A.H., Clark, C., Cortopassi, K.A., Hanser, S.F., McCowan, B., Ali, A.M., Kirschel, A.N.G., 2011. Acoustic monitoring in terrestrial environments using microphone arrays: applications, technological considerations and prospectus: acoustic monitoring. J. Appl. Ecol. 48, 758–767. http://dx.doi.org/10.1111/j.1365-2664.2011.01993.x.

Boecklen, W.J., Yarnes, C.T., Cook, B.A., James, A.C., 2011. On the use of stable isotopes in trophic ecology. Annu. Rev. Ecol. Evol. Syst. 42, 411–440. http://dx.doi.org/10.1146/annurev-ecolsys-102209-144726.

Boere, A.C., Sinninghe Damsté, J.S., Rijpstra, W.I.C., Volkman, J.K., Coolen, M.J.L., 2011. Source-specific variability in post-depositional DNA preservation with potential implications for DNA based paleoecological records. Org. Geochem. 42, 1216–1225. http://dx.doi.org/10.1016/j.orggeochem.2011.08.005.

Bogar, L.M., Dickie, I.A., Kennedy, P.G., 2015. Testing the co-invasion hypothesis: ectomycorrhizal fungal communities on *Alnus glutinosa* and *Salix fragilis* in New Zealand. Divers. Distrib. 21, 268–278. http://dx.doi.org/10.1111/ddi.12304.

Bohan, D.A., Bohan, A.C., Glen, D.M., Symondson, W.O.C., Wiltshire, C.W., Hughes, L., 2000. Spatial dynamics of predation by carabid beetles on slugs. J. Appl. Ecol. 69, 367–379.

Bohan, D.A., Boursault, A., Brooks, D.R., Petit, S., 2011. National-scale regulation of the weed seedbank by carabid predators: carabid seed predation. J. Appl. Ecol. 48, 888–898. http://dx.doi.org/10.1111/j.1365-2664.2011.02008.x.

Bohan, D.A., et al., 2016. Networking our way to better ecosystem service provision. Trends Ecol. Evol. 31, 105–115. http://dx.doi.org/10.1016/j.tree.2015.12.003.

Bohmann, K., Evans, A., Gilbert, M.T.P., Carvalho, G.R., Creer, S., Knapp, M., Yu, D.W., de Bruyn, M., 2014. Environmental DNA for wildlife biology and biodiversity monitoring. Trends Ecol. Evol. 29, 358–367. http://dx.doi.org/10.1016/j.tree.2014.04.003.

Boreau de Roincé, C., Lavigne, C., Mandrin, J.-F., Rollard, C., Symondson, W.O.C., 2013. Early-season predation on aphids by winter-active spiders in apple orchards revealed by diagnostic PCR. Bull. Entomol. Res. 103, 148–154. http://dx.doi.org/10.1017/S0007485312000636.

Bowles, E., Schulte, P.M., Tollit, D.J., Deagle, B.E., Trites, A.W., 2011. Proportion of prey consumed can be determined from faecal DNA using real-time PCR. Mol. Ecol. Resour. 11, 530–540. http://dx.doi.org/10.1111/j.1755-0998.2010.02974.x.

Boyer, S., Wratten, S.D., Holyoake, A., Abdelkrim, J., Cruickshank, R.H., 2013. Using next-generation sequencing to analyse the diet of a highly endangered land snail (Powelliphanta augusta) feeding on endemic earthworms. PLoS One 8, e75962. http://dx.doi.org/10.1371/journal.pone.0075962.

Boyer, F., Mercier, C., Bonin, A., Le Bras, Y., Taberlet, P., Coissac, E., 2016. OBITools: a Unix-inspired software package for DNA metabarcoding. Mol. Ecol. Resour. 16, 176–182. http://dx.doi.org/10.1111/1755-0998.12428.

Boža, V., Brejová, B., Vinař, T., 2016. DeepNano: deep recurrent neural networks for base calling in MinION nanopore reads. arXiv:1603.09195v1.

Brede, N., Sandrock, C., Straile, D., Spaak, P., Jankowski, T., Streit, B., 2009. The impact of human-made ecological changes on the genetic architecture of Daphnia species. Proc. Natl. Acad. Sci. U.S.A. 106, 4758–4763. http://dx.doi.org/10.1073/pnas.0807187106.

Britton, J.R., Davies, G.D., Harrod, C., 2010. Trophic interactions and consequent impacts of the invasive fish Pseudorasbora parva in a native aquatic foodweb: a field investigation in the UK. Biol. Invasions 12, 1533–1542. http://dx.doi.org/10.1007/s10530-009-9566-5.

Brose, U., Hillebrand, H., 2016. Biodiversity and ecosystem functioning in dynamic landscapes. Philos. Trans. R. Soc. B Biol. Sci. 371, 20150267. http://dx.doi.org/10.1098/rstb.2015.0267.

Brown, M.B., Schlacher, T.A., Schoeman, D.S., Weston, M.A., Huijbers, C.M., Olds, A.D., Connolly, R.M., 2015. Invasive carnivores alter ecological function and enhance complementarity in scavenger assemblages on ocean beaches. Ecology 96, 2715–2725.

Butchart, S.H.M., Walpole, M., Collen, B., van Strien, A., Scharlemann, J.P.W., Almond, R.E.A., Baillie, J.E.M., Bomhard, B., Brown, C., Bruno, J., Carpenter, K.E., Carr, G.M., Chanson, J., Chenery, A.M., Csirke, J., Davidson, N.C., Dentener, F., Foster, M., Galli, A., Galloway, J.N., Genovesi, P., Gregory, R.D., Hockings, M., Kapos, V., Lamarque, J.-F., Leverington, F., Loh, J., McGeoch, M.A., McRae, L., Minasyan, A., Morcillo, M.H., Oldfield, T.E.E., Pauly, D., Quader, S., Revenga, C., Sauer, J.R., Skolnik, B., Spear, D., Stanwell-Smith, D., Stuart, S.N., Symes, A., Tierney, M., Tyrrell, T.D., Vie, J.-C., Watson, R., 2010. Global biodiversity: indicators of recent declines. Science 328, 1164–1168. http://dx.doi.org/10.1126/science.1187512.

Campana, S.E., 1999. Chemistry and composition of fish otoliths: pathways, mechanisms and applications. Mar. Ecol. Prog. Ser. 188, 263–297.

Campana, S.E., Tzeng, W.-N., 2000. Otolith composition. Fish. Res. 46, 287–288.

Campos-Herrera, R., El-Borai, F.E., Stuart, R.J., Graham, J.H., Duncan, L.W., 2011. Entomopathogenic nematodes, phoretic Paenibacillus spp., and the use of real time quantitative PCR to explore soil food webs in Florida citrus groves. J. Invertebr. Pathol. 108, 30–39. http://dx.doi.org/10.1016/j.jip.2011.06.005.

Cannon, M.V., Hester, J., Shalkhauser, A., Chan, E.R., Logue, K., Small, S.T., Serre, D., 2016. In silico assessment of primers for eDNA studies using PrimerTree and application to characterize the biodiversity surrounding the Cuyahoga River. Sci. Rep. 6, 22908. http://dx.doi.org/10.1038/srep22908.

Capo, E., Debroas, D., Arnaud, F., Domaizon, I., 2015. Is planktonic diversity well recorded in sedimentary DNA? Toward the reconstruction of past protistan diversity. Microb. Ecol. 70, 865–875. http://dx.doi.org/10.1007/s00248-015-0627-2.

Carey, E.V., Marler, M.J., Callaway, R.M., 2004. Mycorrhizae transfer carbon from a native grass to an invasive weed: evidence from stable isotopes and physiology. Plant Ecol. 172, 133–141.

Carpenter, S.J., Erickson, J.M., Holland, F.D., 2003. Migration of late Cretaceous fish. Nature 423, 70–74. http://dx.doi.org/10.1038/nature01575.

Carvalheiro, L.G., Barbosa, E.R.M., Memmott, J., 2008. Pollinator networks, alien species and the conservation of rare plants: Trinia glauca as a case study. J. Appl. Ecol. 45, 1419–1427. http://dx.doi.org/10.1111/j.1365-2664.2008.01518.x.

Carvalheiro, L.G., Buckley, Y.M., Memmott, J., 2010. Diet breadth influences how impacts propagate through food webs. Ecology 91, 1063–1074.

Caut, S., Angulo, E., Courchamp, F., 2008. Dietary shift of an invasive predator: rats, seabirds and sea turtles. J. Appl. Ecol. 45, 428–437. http://dx.doi.org/10.1111/j.1365-2664.2007.01438.x.

Caut, S., Angulo, E., Courchamp, F., 2009. Avoiding surprise effects on Surprise Island: alien species control in a multitrophic level perspective. Biol. Invasions 11, 1689–1703. http://dx.doi.org/10.1007/s10530-008-9397-9.

Chauvet, E., Ferreira, V., Giller, P.S., McKie, B.G., Tiegs, S.D., Woodward, G., Elosegi, A., Dobson, M., Fleituch, T., Graça, M.A.S., Gulis, V., Hladyz, S., Lacoursière, J.O., Lecerf, A., Pozo, J., Preda, E., Riipinen, M., Rîşnoveanu, G., Vadineanu, A., Vought, L.B.-M., Gessner, M.O., et al., 2016. Litter decomposition as an indicator of stream ecosystem functioning at local-to-continental scales: insights from the European RivFunction project. Adv. Ecol. Res. 55, 99–182.

Chikaraishi, Y., Ogawa, N.O., Doi, H., Ohkouchi, N., 2011. 15N/14N ratios of amino acids as a tool for studying terrestrial food webs: a case study of terrestrial insects (bees, wasps, and hornets). Ecol. Res. 26, 835–844. http://dx.doi.org/10.1007/s11284-011-0844-1.

Chikaraishi, Y., Steffan, S.A., Ogawa, N.O., Ishikawa, N.F., Sasaki, Y., Tsuchiya, M., Ohkouchi, N., 2014. High-resolution food webs based on nitrogen isotopic composition of amino acids. Ecol. Evol. 4, 2423–2449. http://dx.doi.org/10.1002/ece3.1103.

Chistoserdova, L., 2009. Functional metagenomics: recent advances and future challenges. Biotechnol. Genet. Eng. Rev. 26, 335–352. http://dx.doi.org/10.5661/bger-26-335.

Chung, Y.A., Burkle, L.A., Knight, T.M., 2014. Minimal effects of an invasive flowering shrub on the pollinator community of native forbs. PLoS One 9, e109088. http://dx. doi.org/10.1371/journal.pone.0109088.

Clarke, L., Walther, B., Munch, S., Thorrold, S., Conover, D., 2009. Chemical signatures in the otoliths of a coastal marine fish, Menidia menidia, from the northeastern United States: spatial and temporal differences. Mar. Ecol. Prog. Ser. 384, 261–271. http:// dx.doi.org/10.3354/meps07927.

Coissac, E., Riaz, T., Puillandre, N., 2012. Bioinformatic challenges for DNA metabarcoding of plants and animals. Mol. Ecol. 21, 1834–1847. http://dx.doi.org/ 10.1111/j.1365-294X.2012.05550.x.

Coissac, E., Hollingsworth, P.M., Lavergne, S., Taberlet, S., 2016. From barcodes to genomes: extending the concept of DNA barcoding. Mol. Ecol. 25, 1423–1428.

Collins, L.M., Warnock, N.D., Tosh, D.G., McInnes, C., Everest, D., Montgomery, W.I., Scantlebury, M., Marks, N., Dick, J.T.A., Reid, N., 2014. Squirrelpox virus: assessing prevalence, transmission and environmental degradation. PLoS One 9, e89521. http:// dx.doi.org/10.1371/journal.pone.0089521.

Conrad, C.C., Hilchey, K.G., 2011. A review of citizen science and community-based environmental monitoring: issues and opportunities. Environ. Monit. Assess. 176, 273–291. http://dx.doi.org/10.1007/s10661-010-1582-5.

Cook, B.D., Bunn, S.E., Hughes, J.M., 2007. Molecular genetic and stable isotope signatures reveal complementary patterns of population connectivity in the regionally vulnerable southern pygmy perch (Nannoperca australis). Biol. Conserv. 138, 60–72. http://dx. doi.org/10.1016/j.biocon.2007.04.002.

Cooke, G.M., King, A.G., Miller, L., Johnson, R.N., 2012. A rapid molecular method to detect the invasive golden apple snail Pomacea canaliculata (Lamarck, 1822). Conserv. Genet. Resour. 4, 591–593. http://dx.doi.org/10.1007/s12686-011-9599-9.

Coolen, M.J.L., van de Giessen, J., Zhu, E.Y., Wuchter, C., 2011. Bioavailability of soil organic matter and microbial community dynamics upon permafrost thaw. Environ. Microbiol. 13, 2299–2314.

Coolen, M.J.L., Orsi, W.O., Balkema, C., Quince, C., Harris, K., Sylva, S.P., Filipova-Parinova, M., Giosan, L., 2013. Evolution of the plankton paleome in the Black Sea from the Deglacial to Anthropocene. Proc. Natl. Acad. Sci. U.S.A. 110, 8609–8614. http:// dx.doi.org/10.1073/pnas.1219283110.

Cooper, A., Poinar, H.N., 2000. Ancient DNA: do it right or not at all. Science 289, 1139. http://dx.doi.org/10.1126/science.289.5482.1139b.

Cooper, C.B., Hochachka, W.M., Dhondt, A.A., 2007. Contrasting natural experiments confirm competition between house finches and house sparrows. Ecology 88, 864–870.

Cotton, T.E.A., Fitter, A.H., Miller, R.M., Dumbrell, A.J., Helgason, T., 2015. Fungi in the future: interannual variation and effects of atmospheric change on arbuscular mycorrhizal fungal communities. New Phytol. 205, 1598–1607.

Courchamp, F., Chapuis, J.-L., Pascal, M., 2003. Mammal invaders on islands: impact, control and control impact. Biol. Rev. 78, 347–383. http://dx.doi.org/10.1017/ S1464793102006061.

Couvet, D., Jiguet, F., Julliard, R., Levrel, H., Teyssedre, A., 2008. Enhancing citizen contributions to biodiversity science and public policy. Interdisc. Sci. Rev. 33, 95–103. http://dx.doi.org/10.1179/030801808X260031.

Craine, J.M., Towne, E.G., Miller, M., Fierer, N., 2015. Climatic warming and the future of bison as grazers. Sci. Rep. 5, 16738. http://dx.doi.org/10.1038/srep16738.

Crall, A.W., Newman, G.J., Jarnevich, C.S., Stohlgren, T.J., Waller, D.M., Graham, J., 2010. Improving and integrating data on invasive species collected by citizen scientists. Biol. Invasions 12, 3419–3428. http://dx.doi.org/10.1007/s10530-010-9740-9.

Cristescu, M.E., 2015. Genetic reconstructions of invasion history. Mol. Ecol. 24, 2212–2225. http://dx.doi.org/10.1111/mec.13117.

Crook, A.M.E., Keane, G., Solomon, M.G., 1996. Production and selection of monoclonal antibodies for use in detecting vine weevil *Otiorhynchus sulcatus* (Coleoptera: Curculionidae). In: BCPC Symposium Proceedings 65: Diagnostics in Crop Production.

Cucherousset, J., Villéger, S., 2015. Quantifying the multiple facets of isotopic diversity: new metrics for stable isotope ecology. Ecol. Indic. 56, 152–160. http://dx.doi.org/10.1016/j.ecolind.2015.03.032.

Cucherousset, J., Bouletreau, S., Martino, A., Roussel, J.-M., Santoul, F., 2012. Using stable isotope analyses to determine the ecological effects of non-native fishes. Fish. Manag. Ecol. 19, 111–119. http://dx.doi.org/10.1111/j.1365-2400.2011.00824.x.

Dalla Riva, G.V., Stouffer, D.B., 2016. Exploring the evolutionary signature of food webs' backbones using functional traits. Oikos 125, 446–456. http://dx.doi.org/10.1111/oik.02305.

Darling, J.A., Mahon, A.R., 2011. From molecules to management: adopting DNA-based methods for monitoring biological invasions in aquatic environments. Environ. Res. 111, 978–988. http://dx.doi.org/10.1016/j.envres.2011.02.001.

Davey, J.S., Vaughan, I.P., Andrew King, R., Bell, J.R., Bohan, D.A., Bruford, M.W., Holland, J.M., Symondson, W.O.C., 2013. Intraguild predation in winter wheat: prey choice by a common epigeal carabid consuming spiders. J. Appl. Ecol. 50, 271–279. http://dx.doi.org/10.1111/1365-2664.12008.

David, M., Dursi, L.J., Yao, D., Boutros, P.C., Simpson, J.T., 2016. Nanocall: an open source basecaller for Oxford Nanopore sequencing data. Bioinformatics. http://dx.doi.org/10.1093/bioinformatics/btw569.

David, P., Thébault, E., Anneville, O., Duyck, P.-F., Chapuis, E., Loeuille, N., 2017. Impacts of invasive species on food webs: a review of empirical data. Adv. Ecol. Res. 56, 1–60.

De Barba, M., Miquel, C., Boyer, F., Mercier, C., Rioux, D., Coissac, E., Taberlet, P., 2014. DNA metabarcoding multiplexing and validation of data accuracy for diet assessment: application to omnivorous diet. Mol. Ecol. Resour. 14, 306–323. http://dx.doi.org/10.1111/1755-0998.12188.

Deagle, B.E., Tollit, D.J., 2007. Quantitative analysis of prey DNA in pinniped faeces: potential to estimate diet composition? Conserv. Genet. 8, 743–747. http://dx.doi.org/10.1007/s10592-006-9197-7.

Deagle, B.E., Tollit, D.J., Jarman, S.N., Hindell, M.A., Trites, A.W., Gales, N.J., 2005. Molecular scatology as a tool to study diet: analysis of prey DNA in scats from captive Steller sea lions. Mol. Ecol. 14, 1831–1842. http://dx.doi.org/10.1111/j.1365-294X.2005.02531.x.

Deagle, B.E., Kirkwood, R., Jarman, S.N., 2009. Analysis of Australian fur seal diet by pyrosequencing prey DNA in faeces. Mol. Ecol. 18, 2022–2038. http://dx.doi.org/10.1111/j.1365-294X.2009.04158.x.

Deagle, B.E., Chiaradia, A., McInnes, J., Jarman, S.N., 2010. Pyrosequencing faecal DNA to determine diet of little penguins: is what goes in what comes out? Conserv. Genet. 11, 2039–2048. http://dx.doi.org/10.1007/s10592-010-0096-6.

Deagle, B.E., Thomas, A.C., Shaffer, A.K., Trites, A.W., Jarman, S.N., 2013. Quantifying sequence proportions in a DNA-based diet study using Ion Torrent amplicon sequencing: which counts count? Mol. Ecol. Resour. 13, 620–633. http://dx.doi.org/10.1111/1755-0998.12103.

Decaestecker, E., Gaba, S., Raeymaekers, J.A.M., Stoks, R., Van Kerckhoven, L., Ebert, D., De Meester, L., 2007. Host–parasite "Red Queen" dynamics archived in pond sediment. Nature 450, 870–873. http://dx.doi.org/10.1038/nature06291.

Dehling, D.M., Jordano, P., Schaefer, H.M., Böhning-Gaese, K., Schleuning, M., 2016. Morphology predicts species' functional roles and their degree of specialization in plant–frugivore interactions. Proc. R. Soc. B Biol. Sci. 283, 20152444. http://dx.doi.org/10.1098/rspb.2015.2444.

Deng, Y., Jiang, Y.-H., Yang, Y., He, Z., Luo, F., Zhou, J., 2012. Molecular ecological network analyses. BMC Bioinf. 13, 113–133.

Deng, J., Wang, X.-B., Yu, F., Zhou, Q.-S., Bernardo, U., Zhang, Y.-Z., Wu, S.-A., 2015. Rapid diagnosis of the invasive wax scale, Ceroplastes rusci Linnaeus (Hemiptera: Coccoidea: Coccidae) using nested PCR. J. Appl. Entomol. 139, 314–319. http://dx. doi.org/10.1111/jen.12155.

DeNiro, M.J., Epstein, S., 1978. Influence of diet on the distribution of carbon isotopes in animals. Geochim. Cosmochim. Acta 42, 495–506.

DeNiro, M.J., Epstein, S., 1981. Influence of diet on the distribution of carbon isotopes in animals. Geochim. Cosmochim. Acta 45, 341–351.

Derocles, S.A.P., Plantegenest, M., Simon, J.-C., Taberlet, P., Le Ralec, A., 2012. A universal method for the detection and identification of Aphidiinae parasitoids within their aphid hosts. Mol. Ecol. Resour. 12, 634–645. http://dx.doi.org/10.1111/j.1755-0998.2012.03131.x.

Derocles, S.A.P., Le Ralec, A., Besson, M.M., Maret, M., Walton, A., Evans, D.M., Plantegenest, M., 2014. Molecular analysis reveals high compartmentalization in aphid-primary parasitoid networks and low parasitoid sharing between crop and noncrop habitats. Mol. Ecol. 23, 3900–3911. http://dx.doi.org/10.1111/mec.12701.

DeSantis, T.Z., Hugenholtz, P., Larsen, N., Rojas, M., Brodie, E.L., Keller, K., Huber, T., Dalevi, D., Hu, P., Andersen, G.L., 2006. Greengenes, a chimera-checked 16S rRNA gene database and workbench compatible with ARB. Appl. Environ. Microbiol. 72, 5069–5072. http://dx.doi.org/10.1128/AEM.03006-05.

Desprez-Loustau, M., Robin, C., Buee, M., Courtecuisse, R., Garbaye, J., Suffert, F., Sache, I., Rizzo, D., 2007. The fungal dimension of biological invasions. Trends Ecol. Evol. 22, 472–480. http://dx.doi.org/10.1016/j.tree.2007.04.005.

Dickinson, J.L., Zuckerberg, B., Bonter, D.N., 2010. Citizen science as an ecological research tool: challenges and benefits. Annu. Rev. Ecol. Evol. 41, 149–172.

Didham, R.K., Tylianakis, J.M., Hutchison, M.A., Ewers, R.M., Gemmell, N.J., 2005. Are invasive species the drivers of ecological change? Trends Ecol. Evol. 20, 470–474.

Dlugosch, K.M., Lai, Z., Bonin, A., Hierro, J., Rieseberg, L.H., 2013. Allele identification for transcriptome-based population genomics in the invasive plant Centaurea solstitialis. G3 (Bethesda) 3, 359–367. http://dx.doi.org/10.1534/g3.112.003871.

Dodsworth, S., 2015. Genome skimming for next-generation biodiversity analysis. Trends Plant Sci. 20, 525–527. http://dx.doi.org/10.1016/j.tplants.2015.06.012.

Domaizon, I., Savichtcheva, O., Debroas, D., Arnaud, F., Villar, C., Pignol, C., Alric, B., Perga, M.E., 2013. DNA from lake sediments reveals the long-term dynamics and diversity of Synechococcus assemblages. Biogeosciences 10, 3817–3838. http://dx.doi.org/10.5194/bg-10-3817-2013.

Duffy, M.A., Perry, L.J., Kearns, C.M., Weider, L.J., 2000. Paleogenetic evidence for a past invasion of Onondaga Lake, New York, by exotic Daphnia curvirostris using mtDNA from dormant eggs. Limnol. Oceanogr. 45, 1409–1414.

Dumbrell, A.J., Kordas, R.L., Woodward, G., 2016. Large-scale ecology: model systems to global perspectives—preface. Adv. Ecol. Res. 55, xix–xxiv.

Dunne, J.A., Williams, R.J., Martinez, N.D., 2002. Network structure and biodiversity loss in food webs: robustness increases with connectance. Ecol. Lett. 5, 558–567.

Dunne, J.A., Lafferty, K.D., Dobson, A.P., Hechinger, R.F., Kuris, A.M., Martinez, N.D., McLaughlin, J.P., Mouritsen, K.N., Poulin, R., Reise, K., Stouffer, D.B., Thieltges, D.W., Williams, R.J., Zander, C.D., 2013. Parasites affect food web structure primarily through increased diversity and complexity. PLoS Biol. 11. e1001579. http://dx.doi.org/10.1371/journal.pbio.1001579.

Dunshea, G., 2009. DNA-based diet analysis for any predator. PLoS One 4. e5252. http://dx. doi.org/10.1371/journal.pone.0005252.

Dupas, S., Gitau, C., Le Rü, B., Silvain, J.-F., 2006. Single-step PCR differentiation of *Cotesia sesamiae* (Cameron) and *Cotesia flavipes* Cameron (Hymenoptera: Braconidae) using polydnavirus markers. Ann. Soc. Entomol. Fr. 42, 319–323. http://dx.doi.org/10.1080/00379271.2006.10697463.

Dutech, C., Barrès, B., Bridier, J., Robin, C., Milgroom, M.G., Ravigné, V., 2012. The chestnut blight fungus world tour: successive introduction events from diverse origins in an invasive plant fungal pathogen. Mol. Ecol. 21, 3931–3946. http://dx.doi.org/10.1111/j.1365-294X.2012.05575.x.

Egeter, B., Bishop, P.J., Robertson, B.C., 2015. Detecting frogs as prey in the diets of introduced mammals: a comparison between morphological and DNA-based diet analyses. Mol. Ecol. Resour. 15, 306–316. http://dx.doi.org/10.1111/1755-0998.12309.

Eichmiller, J.J., Miller, L.M., Sorensen, P.W., 2016. Optimizing techniques to capture and extract environmental DNA for detection and quantification of fish. Mol. Ecol. Resour. 16, 56–68. http://dx.doi.org/10.1111/1755-0998.12421.

Eklöf, A., Helmus, M.R., Moore, M., Allesina, S., 2012. Relevance of evolutionary history for food web structure. Proc. R. Soc. B Biol. Sci. 279, 1588–1596. http://dx.doi.org/10.1098/rspb.2011.2149.

Eklöf, A., Jacob, U., Kopp, J., Bosch, J., Castro-Urgal, R., Chacoff, N.P., Dalsgaard, B., de Sassi, C., Galetti, M., Guimarães, P.R., Lomáscolo, S.B., Martín González, A.M., Pizo, M.A., Rader, R., Rodrigo, A., Tylianakis, J.M., Vázquez, D.P., Allesina, S., 2013a. The dimensionality of ecological networks. Ecol. Lett. 16, 577–583. http://dx.doi.org/10.1111/ele.12081.

Eklöf, A., Tang, S., Allesina, S., 2013b. Secondary extinctions in food webs: a Bayesian network approach. Methods Ecol. Evol. 4, 760–770. http://dx.doi.org/10.1111/2041-210X.12062.

Elsdon, T.S., Gillanders, B.M., 2003. Reconstructing migratory patterns of fish based on environmental influences on otolith chemistry. Rev. Fish Biol. Fish. 13, 217–235.

Elsdon, T.S., Gillanders, B.M., 2004. Fish otolith chemistry influenced by exposure to multiple environmental variables. J. Exp. Mar. Biol. Ecol. 313, 269–284. http://dx.doi.org/10.1016/j.jembe.2004.08.010.

Elsdon, T.S., Gillanders, B.M., 2006. Temporal variability in strontium, calcium, barium, and manganese in estuaries: implications for reconstructing environmental histories of fish from chemicals in calcified structures. Estuar. Coast. Shelf Sci. 66, 147–156. http://dx.doi.org/10.1016/j.ecss.2005.08.004.

Emde, S., Kochmann, J., Kuhn, T., Plath, M., Klimpel, S., 2014. Getting what is served? Feeding ecology influencing parasite-host interactions in invasive round goby Neogobius melanostomus. PLoS One 9. e109971. http://dx.doi.org/10.1371/journal.pone.0109971.

Emde, S., Kochmann, J., Kuhn, T., Dörge, D.D., Plath, M., Miesen, F.W., Klimpel, S., 2016. Cooling water of power plant creates "hot spots" for tropical fishes and parasites. Parasitol. Res. 115, 85–98. http://dx.doi.org/10.1007/s00436-015-4724-4.

Erickson, R.A., Rees, C.B., Coulter, A.A., Merkes, C.M., McCalla, S.G., Touzinsky, K.F., Walleser, L., Goforth, R.R., Amberg, J.J., 2016. Detecting the movement and spawning activity of bigheaded carps with environmental DNA. Mol. Ecol. Resour. 16, 957–965. http://dx.doi.org/10.1111/1755-0998.12533.

Evans, D.M., Kitson, J.J.N., Lunt, D.H., Straw, N.A., Pocock, M.J.O., 2016. Merging DNA metabarcoding and ecological network analysis to understand and build resilient terrestrial ecosystems. Funct. Ecol.. http://dx.doi.org/10.1111/1365-2435.12659.

Facon, B., Pointier, J.-P., Glaubrecht, M., Poux, C., Jarne, P., David, P., 2003. A molecular phylogeography approach to biological invasions of the New World by parthenogenetic Thiarid snails. Mol. Ecol. 12, 3027–3039. http://dx.doi.org/10.1046/j.1365-294X.2003.01972.x.

Faisal, A., Dondelinger, F., Husmeier, D., Beale, C.M., 2010. Inferring species interaction networks from species abundance data: a comparative evaluation of various statistical and machine learning methods. Eco. Inform. 5, 451–464. http://dx.doi.org/10.1016/j.ecoinf.2010.06.005.

Farmer, A., Cade, B.S., Torres-Dowdall, J., 2008. Fundamental limits to the accuracy of deuterium isotopes for identifying the spatial origin of migratory animals. Oecologia 158, 183–192.

Faust, K., Raes, J., 2012. Microbial interactions: from networks to models. Nat. Rev. Microbiol. 10, 538–550. http://dx.doi.org/10.1038/nrmicro2832.

Ficetola, G.F., Pansu, J., Bonin, A., Coissac, E., Giguet-Covex, C., De Barba, M., Gielly, L., Lopes, C.M., Boyer, F., Pompanon, F., Rayé, G., Taberlet, P., 2015. Replication levels, false presences and the estimation of the presence/absence from eDNA metabarcoding data. Mol. Ecol. Resour. 15, 543–556. http://dx.doi.org/10.1111/1755-0998.12338.

Fisher, M.C., Henk, D.A., Briggs, C.J., Brownstein, J.S., Madoff, L.C., McCraw, S.L., Gurr, S.J., 2012. Emerging fungal threats to animal, plant and ecosystem health. Nature 484, 186–194. http://dx.doi.org/10.1038/nature10947.

Fogel, M.L., Griffin, P., Newsome, S.D., 2016. Hydrogen isotopes in individual amino acids reflect differentiated pools of hydrogen from food and water in Escherichia coli. Proc. Natl. Acad. Sci. USA 113 (32), E4648–E4653. http://dx.doi.org/10.1073/pnas.1525703113.

Folino-Rorem, N.C., Darling, J.A., D'Ausilio, C.A., 2009. Genetic analysis reveals multiple cryptic invasive species of the hydrozoan genus Cordylophora. Biol. Invasions 11, 1869–1882. http://dx.doi.org/10.1007/s10530-008-9365-4.

Foltan, P., Sheppard, S., Konvicka, M., Symondson, W.O.C., 2005. The significance of facultative scavenging in generalist predator nutrition: detecting decayed prey in the guts of predators using PCR. Mol. Ecol. 14, 4147–4158. http://dx.doi.org/10.1111/j.1365-294X.2005.02732.x.

Forrester, G.E., Swearer, S.E., 2002. Trace elements in otoliths indicate the use of open-coast versus bay nursery habitats by juvenile California halibut. Mar. Ecol. Prog. Ser. 241, 201–213.

Fournier, V., Hagler, J.R., Daane, K.M., de León, J.H., Groves, R.L., Costa, H.S., Henneberry, T.J., 2006. Development and application of a glassy-winged and smoke-tree sharpshooter egg-specific predator gut content ELISA. Biol. Control 37, 108–118. http://dx.doi.org/10.1016/j.biocontrol.2005.12.015.

France, R.L., 1995. Carbon-13 enrichment in benthic compared to planktonic algae: foodweb implications. Mar. Ecol. Prog. Ser. 124, 307–312.

Franklin, J., 2013. Species distribution models in conservation biogeography: developments and challenges. Divers. Distrib. 19, 1217–1223. http://dx.doi.org/10.1111/ddi.12125.

Frick, W.F., Shipley, J.R., Kelly, J.F., Heady, P.A., Kay, K.M., 2014. Seasonal reliance on nectar by an insectivorous bat revealed by stable isotopes. Oecologia 174, 55–65. http://dx.doi.org/10.1007/s00442-013-2771-z.

Fry, B., 2006. Stable Isotope Ecology. Springer, New York.

Galimberti, A., De Mattia, F., Bruni, I., Scaccabarozzi, D., Sandionigi, A., Barbuto, M., Casiraghi, M., Labra, M., 2014. A DNA barcoding approach to characterize pollen collected by honeybees. PLoS One 9. e109363. http://dx.doi.org/10.1371/journal.pone.0109363.

Gallo, T., Waitt, D., 2011. Creating a successful citizen science model to detect and report invasive species. Bioscience 61, 459–465. http://dx.doi.org/10.1525/bio.2011.61.6.8.

Gandon, S., Buckling, A., Decaestecker, E., Day, T., 2008. Host-parasite coevolution and patterns of adaptation across time and space. J. Evol. Biol. 21, 1861–1866. http://dx.doi.org/10.1111/j.1420-9101.2008.01598.x.

Gannes, L.Z., del Rio, C.M., Koch, P., 1998. Natural abundance variations in stable isotopes and their potential uses in animal physiological ecology. Comp. Biochem. Physiol. 119, 725–737.

Gardiner, M.M., Allee, L.L., Brown, P.M., Losey, J.E., Roy, H.E., Smyth, R.R., 2012. Lessons from lady beetles: accuracy of monitoring data from US and UK citizen-science programs. Front. Ecol. Environ. 10, 471–476. http://dx.doi.org/10.1890/110185.

Gariepy, T.D., Kuhlmann, U., Haye, T., Gillott, C., Erlandson, M., 2005. A single-step multiplex PCR assay for the detection of European *Peristenus* spp., parasitoids of *Lygus* spp. Biocontrol Sci. Tech. 15, 481–495. http://dx.doi.org/10.1080/09583150500086771.

Gariepy, T., Kuhlmann, U., Gillott, C., Erlandson, M., 2008. A large-scale comparison of conventional and molecular methods for the evaluation of host-parasitoid associations in non-target risk-assessment studies. J. Appl. Ecol. 45, 708–715. http://dx.doi.org/10.1111/j.1365-2664.2007.01451.x.

Gariepy, T.D., Haye, T., Zhang, J., 2014. A molecular diagnostic tool for the preliminary assessment of host-parasitoid associations in biological control programmes for a new invasive pest. Mol. Ecol. 23, 3912–3924. http://dx.doi.org/10.1111/mec.12515.

Gerdeaux, D., Perga, M.-E., 2006. Changes in whitefish scales $\delta^{13}C$ during eutrophication and reoligotrophication of subalpine lakes. Limnol. Oceanogr. 51, 772–780.

Geslin, B., Gauzens, B., Baude, M., Dajoz, I., Fontaine, C., Henry, M., Ropars, L., Rollin, O., Thébault, E., Vereecken. N.J., 2017. Massively introduced managed species and their consequences for plant-pollinator interactions. Adv. Ecol. Res. 57, 147–199.

Gibson, J., Shokralla, S., Porter, T.M., King, I., van Konynenburg, S., Janzen, D.H., Hallwachs, W., Hajibabaei, M., 2014. Simultaneous assessment of the macrobiome and microbiome in a bulk sample of tropical arthropods through DNA metasystematics. Proc. Natl. Acad. Sci. U.S.A. 111, 8007–8012. http://dx.doi.org/10.1073/pnas.1406468111.

Giguet-Covex, C., Pansu, J., Arnaud, F., Rey, P.-J., Griggo, C., Gielly, L., Domaizon, I., Coissac, E., David, F., Choler, P., Poulenard, J., Taberlet, P., 2014. Long livestock farming history and human landscape shaping revealed by lake sediment DNA. Nat. Commun. 5, 3211. http://dx.doi.org/10.1038/ncomms4211.

Gilbert, K.J., Whitlock, M.C., 2015. Evaluating methods for estimating local effective population size with and without migration. Evolution 69, 2154–2166. http://dx.doi.org/10.1111/evo.12713.

Gillett, C.P.D.T., Crampton-Platt, A., Timmermans, M.J.T.N., Jordal, B.H., Emerson, B.C., Vogler, A.P., 2014. Bulk de novo mitogenome assembly from pooled total DNA elucidates the phylogeny of weevils (Coleoptera: Curculionoidea). Mol. Biol. Evol. 31, 2223–2237. http://dx.doi.org/10.1093/molbev/msu154.

Giovanetti, M., Mariotti Lippi, M., Foggi, B., Giuliani, C., 2015. Exploitation of the invasive Acacia pycnantha pollen and nectar resources by the native bee Apis mellifera. Ecol. Res. 30, 1065–1072. http://dx.doi.org/10.1007/s11284-015-1308-9.

Giraud, C., 2014. Introduction to High-Dimensional Statistics. CRC Press (Taylor & Francis Group), Boca Raton, FL.

Goldberg, C.S., Sepulveda, A., Ray, A., Baumgardt, J., Waits, L.P., 2013. Environmental DNA as a new method for early detection of New Zealand mudsnails (*Potamopyrgus antipodarum*). Freshw. Sci. 32, 792–800. http://dx.doi.org/10.1899/13-046.1.

Goldstein, J., Zych, M., 2016. What if we lose a hub? Experimental testing of pollination network resilience to removal of keystone floral resources. Arthropod Plant Interact. 10, 263–271. http://dx.doi.org/10.1007/s11829-016-9431-2.

Goldstein, E.A., Lawton, C., Sheehy, E., Butler, F., 2014. Locating species range frontiers: a cost and efficiency comparison of citizen science and hair-tube survey methods for use in tracking an invasive squirrel. Wildl. Res. 41, 64. http://dx.doi.org/10.1071/WR13197.

Gomez-Polo, P., Alomar, O., Castañé, C., Lundgren, J.G., Piñol, J., Agustí, N., 2015. Molecular assessment of predation by hoverflies (Diptera: Syrphidae) in Mediterranean lettuce crops: molecular assessment of predation by hoverflies in lettuce. Pest Manag. Sci. 71, 1219–1227. http://dx.doi.org/10.1002/ps.3910.

Goodwin, S., Gurtowski, J., Ethe-Sayers, S., Deshpande, P., Schatz, M.C., McCombie, W.R., 2015. Oxford Nanopore sequencing, hybrid error correction, and de novo assembly of a eukaryotic genome. Genome Res. 25, 1750–1756. http://dx.doi.org/10.1101/gr.191395.115.

Gorokhova, E., 2006. Molecular identification of the invasive cladoceran Cercopagis pengoi (Cladocera: Onychopoda) in stomachs of predators. Limnol. Oceanogr. Methods 4, 1–6.

Grabner, D.S., Weigand, A.M., Leese, F., Winking, C., Hering, D., Tollrian, R., Sures, B., 2015. Invaders, natives and their enemies: distribution patterns of amphipods and their microsporidian parasites in the Ruhr Metropolis, Germany. Parasit. Vectors 8, 419. http://dx.doi.org/10.1186/s13071-015-1036-6.

Gravel, D., Poisot, T., Albouy, C., Velez, L., Mouillot, D., 2013. Inferring food web structure from predator-prey body size relationships. Methods Ecol. Evol. 4, 1083–1090. http://dx.doi.org/10.1111/2041-210X.12103.

Greenstone, M.H., 1996. Serological analysis of arthropod predation: past, present and future. In: Symondson, W.O.C., Liddell, J.E. (Eds.), The Ecology of Agricultural Pests: Biochemical Approaches. Chapman & Hall, London, pp. 265–300.

Greninger, A.L., Naccache, S.N., Federman, S., Yu, G., Mbala, P., Bres, V., Stryke, D., Bouquet, J., Somasekar, S., Linnen, J.M., Dodd, R., Mulembakani, P., Schneider, B.S., Muyembe-Tamfum, J.-J., Stramer, S.L., Chiu, C.Y., 2015. Rapid metagenomic identification of viral pathogens in clinical samples by real-time nanopore sequencing analysis. Genome Med. 7, 99. http://dx.doi.org/10.1186/s13073-015-0220-9.

Grigorovich, I.A., Colautti, R.I., Mills, E.L., Holeck, K., Ballert, A.G., MacIsaac, H.J., 2003. Ballast-mediated animal introductions in the Laurentian Great Lakes: retrospective and prospective analyses. Can. J. Fish. Aquat. Sci. 60, 740–756. http://dx.doi.org/10.1139/f03-053.

Gritti, E.S., Duputié, A., Massol, F., Chuine, I., 2013. Estimating consensus and associated uncertainty between inherently different species distribution models. Methods Ecol. Evol. 4, 442–452. http://dx.doi.org/10.1111/2041-210X.12032.

Grosholz, E., 2002. Ecological and evolutionary consequences of coastal invasions. Trends Ecol. Evol. 17, 22–27.

Grosholz, E.D., 2005. Recent biological invasion may hasten invasional meltdown by accelerating historical introductions. Proc. Natl. Acad. Sci. U.S.A. 102, 1088–1091.

Guillemaud, T., Ciosi, M., Lombaert, É., Estoup, A., 2011. Biological invasions in agricultural settings: insights from evolutionary biology and population genetics. C. R. Biol. 334, 237–246. http://dx.doi.org/10.1016/j.crvi.2010.12.008.

Guimerà, R., Sales-Pardo, M., 2009. Missing and spurious interactions and the reconstruction of complex networks. Proc. Natl. Acad. Sci. U.S.A. 106, 22073–22078. http://dx.doi.org/10.1073/pnas.090836610.

Hairston, N.G., Perry, L.J., Bohonak, A.J., Fellows, M.Q., Kearns, C.M., 1999. Population biology of a failed invasion: paleolimnology of *Daphnia exilis* in upstate New York. Limnol. Oceanogr. 44, 477–486.

Hairston, N.G., Ellner, S.P., Geber, M.A., Yoshida, T., Fox, J.A., 2005. Rapid evolution and the convergence of ecological and evolutionary time: rapid evolution and the convergence of ecological and evolutionary time. Ecol. Lett. 8, 1114–1127. http://dx.doi.org/10.1111/j.1461-0248.2005.00812.x.

Hajibabaei, M., Singer, G.A.C., Hebert, P.D.N., Hickey, D.A., 2007. DNA barcoding: how it complements taxonomy, molecular phylogenetics and population genetics. Trends Genet. 23, 167–172. http://dx.doi.org/10.1016/j.tig.2007.02.001.

Hall, E.M., Crespi, E.J., Goldberg, C.S., Brunner, J.L., 2016. Evaluating environmental DNA-based quantification of ranavirus infection in wood frog populations. Mol. Ecol. Resour. 16, 423–433. http://dx.doi.org/10.1111/1755-0998.12461.

Handa, I.T., Aerts, R., Berendse, F., Berg, M.P., Bruder, A., Butenschoen, O., Chauvet, E., Gessner, M.O., Jabiol, J., Makkonen, M., McKie, B.G., Malmqvist, B., Peeters, E.T.H.M., Scheu, S., Schmid, B., van Ruijven, J., Vos, V.C.A., Hättenschwiler, S., 2014. Consequences of biodiversity loss for litter decomposition across biomes. Nature 509, 218–221. http://dx.doi.org/10.1038/nature13247.

Handley, L.L., 2015. How will the 'molecular revolution' contribute to biological recording? Biol. J. Linn. Soc. 115, 750–766.

Hansen, A.K., Jeong, G., Paine, T.D., Stouthamer, R., 2007. Frequency of secondary symbiont infection in an invasive psyllid relates to parasitism pressure on a geographic scale in California. Appl. Environ. Microbiol. 73, 7531–7535. http://dx.doi.org/10.1128/AEM.01672-07.

Harwood, J.D., Phillips, S.W., Sunderland, K.D., Symondson, W.O.C., 2001. Secondary predation: quantification of food chain errors in an aphid–spider–carabid system using monoclonal antibodies. Mol. Ecol. 10, 2049–2057.

Hatteland, B.A., Symondson, W.O.C., King, R.A., Skage, M., Schander, C., Solhøy, T., 2011. Molecular analysis of predation by carabid beetles (Carabidae) on the invasive Iberian slug Arion lusitanicus. Bull. Entomol. Res. 101, 675–686. http://dx.doi.org/10.1017/S0007485311000034.

Healy, K., Kelly, S.B.A., Guillerme, T., Inger, R., Bearhop, S., Jackson, A.L., 2016. Predicting trophic discrimination factor using Bayesian inference and phylogenetic, ecological and physiological data. DEsIR: Discrimination Estimation in R, PeerJ 4, e1950v1.

Heidemann, K., Scheu, S., Ruess, L., Maraun, M., 2011. Molecular detection of nematode predation and scavenging in oribatid mites: laboratory and field experiments. Soil Biol. Biochem. 43, 2229–2236. http://dx.doi.org/10.1016/j.soilbio.2011.07.015.

Henneman, M.L., Memmott, J., 2001. Infiltration of a Hawaiian community by introduced biological control agents. Science 293, 1314–1316. http://dx.doi.org/10.1126/science.1060788.

Herrera Montalvo, L.G., Rodríguez Galindo, M., Ibarra López, M.P., 2013. Asymmetric contribution of isotopically contrasting food sources to vertebrate consumers in a subtropical semi-arid ecosystem. Biotropica 45, 357–364. http://dx.doi.org/10.1111/btp.12018.

Hesslein, R.H., Hallard, K.A., Ramlal, P., 1993. Replacement of sulfur, carbon, and nitrogen in tissue of growing broad whitefish (Coregonus nasus) in response to a change in diet traced by δ^{34}S, δ^{13}C, and δ^{15}N. Can. J. Fish. Aquat. Sci. 50, 2071–2076.

Himler, A.G., Adachi-Hagimori, T., Bergen, J.E., Kozuch, A., Kelly, S.E., Tabashnik, B.E., Chiel, E., Duckworth, V.E., Dennehy, T.J., Zchori-Fein, E., et al., 2011. Rapid spread of a bacterial symbiont in an invasive whitefly is driven by fitness benefits and female bias. Science 332, 254–256.

Hines, J., van der Putten, W.H., De Deyn, G.B., Wagg, C., Voigt, W., Mulder, C., Weisser, W.W., Engel, J., Melian, C., Scheu, S., Birkhofer, K., Ebeling, A., Scherber, C., Eisenhauer, N., 2015. Towards an integration of biodiversity-ecosystem functioning and food web theory to evaluate relationships between multiple ecosystem services. In: Woodward, G., Bohan, D.A. (Eds.), In: Advances in Ecological Research vol. 53. Academic Press, Oxford, pp. 161–199. ISBN: 978-0-12-803885-7.

Hobson, K.A., 1999. Tracing origins and migration of wildlife using stable isotopes: a review. Oecologia 120, 314–326. http://dx.doi.org/10.1007/s004420050865.

Hoffmann, B.D., Courchamp, F., 2016. Biological invasions and natural colonisations: are they that different? NeoBiota 29, 1–14. http://dx.doi.org/10.3897/neobiota.29.6959.

Hofreiter, M., Paijmans, J.L.A., Goodchild, H., et al., 2015. The future of ancient DNA: technical advances and conceptual shifts. BioEssays 37, 284–293.

Hooper, D.U., Chapin III, F.S., Ewel, J.J., Hector, A., Inchausti, P., Lavorel, S., Lawton, J.H., Lodge, D.M., Loreau, M., Naeem, S., Schmid, B., et al., 2005. Effects of biodiversity on ecosystem functioning: a consensus of current knowledge. Ecol. Monogr. 75, 3–35.

Horn, S., 2012. Target enrichment via DNA hybridization capture. Methods Mol. Bio. 840, 177–188.

Hulme, P.E., Bacher, S., Kenis, M., Klotz, S., Kühn, I., Minchin, D., Nentwig, W., Olenin, S., Panov, V., Pergl, J., Pyšek, P., Roques, A., Sol, D., Solarz, W., Vilà, M., 2008. Grasping at the routes of biological invasions: a framework for integrating pathways into policy. J. Appl. Ecol. 45, 403–414. http://dx.doi.org/10.1111/j.1365-2664.2007.01442.x.

Hulme, P.E., Pyšek, P., Jarošík, V., Pergl, J., Schaffner, U., Vilà, M., 2013. Bias and error in understanding plant invasion impacts. Trends Ecol. Evol. 28, 212–218. http://dx.doi.org/10.1016/j.tree.2012.10.010.

Ibanez, S., Manneville, O., Miquel, C., Taberlet, P., Valentini, A., Aubert, S., Coissac, E., Colace, M.-P., Duparc, Q., Lavorel, S., Moretti, M., 2013. Plant functional traits reveal the relative contribution of habitat and food preferences to the diet of grasshoppers. Oecologia 173, 1459–1470. http://dx.doi.org/10.1007/s00442-013-2738-0.

Inger, R., McDonald, R.A., Rogowski, D., Jackson, A.L., Parnell, A., Jane Preston, S., Harrod, C., Goodwin, C., Griffiths, D., Dick, J.T.A., Elwood, R.W., Newton, J., Bearhop, S., 2010. Do non-native invasive fish support elevated lamprey populations? J. Appl. Ecol. 47, 121–129. http://dx.doi.org/10.1111/j.1365-2664.2009.01761.x.

Ip, C.L., et al., 2015. MinION analysis and reference consortium: phase 1 data release and analysis. F1000 Res. 4, 1075. http://dx.doi.org/10.12688/f1000research.7201.1.

Isaac, N.J.B., Pocock, M.J.O., 2015. Bias and information in biological records. Biol. J. Linn. Soc. 115, 522–531.

Jackson, S.T., Overpeck, J.T., 2000. Responses of plant populations and communities to environmental changes of the late Quaternary. Paleobiology 26, 194–220. http://dx.doi.org/10.1666/0094-8373(2000)26[194:ROPPAC]2.0.CO;2.

Jackson, S.T., Sax, D.F., 2010. Balancing biodiversity in a changing environment: extinction debt, immigration credit and species turnover. Trends Ecol. Evol. 25, 153–160. http://dx.doi.org/10.1016/j.tree.2009.10.001.

Jackson, M.C., Weyl, O.L.F., Altermatt, F., Durance, I., Friberg, N., Dumbrell, A.J., Piggott, J.J., Tiegs, S.D., Tockner, K., Krug, C.B., Leadley, P.W., Woodward, G., 2016. Recommendations for the next generation of global freshwater biological monitoring tools. Adv. Ecol. Res. 55, 615–636.

Jaouen, K., Beasley, M., Schoeninger, M., Hublin, J.-J., Richards, M.P., 2016. Zinc isotope ratios of bones and teeth as new dietary indicators: results from a modern food web (Koobi Fora, Kenya). Sci. Rep. 6, 26281. http://dx.doi.org/10.1038/srep26281.

Jeliazkov, A., Bas, Y., Kerbiriou, C., Julien, J.-F., Penone, C., Le Viol, I., 2016. Large-scale semi-automated acoustic monitoring allows to detect temporal decline of bush-crickets. Glob. Ecol. Conserv. 6, 208–218. http://dx.doi.org/10.1016/j.gecco.2016.02.008.

Jenkins, B., Kitching, R.L., Pimm, S.L., 1992. Productivity, disturbance and food web structure at a local spatial scale in experimental container habitats. Oikos 65, 249. http://dx.doi.org/10.2307/3545016.

Jerde, C.L., Mahon, A.R., Chadderton, W.L., Lodge, D.M., 2011. "Sight-unseen" detection of rare aquatic species using environmental DNA: eDNA surveillance of rare aquatic species. Conserv. Lett. 4, 150–157. http://dx.doi.org/10.1111/j.1755-263X.2010.00158.x.

Jeschke, J.M., Bacher, S., Blackburn, T.M., Dick, J.T.A., Essl, F., Evans, T., Gaertner, M., Hulme, P.E., KüHn, I., MrugałA, A., Pergl, J., PyšEk, P., Rabitsch, W., Ricciardi, A., Richardson, D.M., Sendek, A., Vilà, M., Winter, M., Kumschick, S., 2014. Defining the impact of non-native species: impact of non-native species. Conserv. Biol. 28, 1188–1194. http://dx.doi.org/10.1111/cobi.12299.

Ji, Y., Ashton, L., Pedley, S.M., Edwards, D.P., Tang, Y., Nakamura, A., Kitching, R., Dolman, P.M., Woodcock, P., Edwards, F.A., Larsen, T.H., Hsu, W.W., Benedick, S., Hamer, K.C., Wilcove, D.S., Bruce, C., Wang, X., Levi, T., Lott, M., Emerson, B.C., Yu, D.W., 2013. Reliable, verifiable and efficient monitoring of biodiversity via metabarcoding. Ecol. Lett. 16, 1245–1257. http://dx.doi.org/10.1111/ele.12162.

Jones, J.B., Campana, S.E., 2009. Stable oxygen isotope reconstruction of ambient temperature during the collapse of a cod (Gadus morhua) fishery. Ecol. Appl. 19, 1500–1514.

Juen, A., Traugott, M., 2005. Detecting predation and scavenging by DNA gut-content analysis: a case study using a soil insect predator-prey system. Oecologia 142, 344–352. http://dx.doi.org/10.1007/s00442-004-1736-7.

Kaartinen, R., Stone, G.N., Hearn, J., Lohse, K., Roslin, T., 2010. Revealing secret liaisons: DNA barcoding changes our understanding of food webs. Ecol. Entomol. 35, 623–638. http://dx.doi.org/10.1111/j.1365-2311.2010.01224.x.

Kadoya, T., Ishii, H.S., Kikuchi, R., Suda, S., Washitani, I., 2009. Using monitoring data gathered by volunteers to predict the potential distribution of the invasive alien bumblebee Bombus terrestris. Biol. Conserv. 142, 1011–1017. http://dx.doi.org/10.1016/j.biocon.2009.01.012.

Kadoya, T., Osada, Y., Takimoto, G., 2012. IsoWeb: a Bayesian isotope mixing model for diet analysis of the whole food web. PLoS One 7. e41057. http://dx.doi.org/10.1371/journal.pone.0041057.

Kaiser-Bunbury, C.N., Blüthgen, N., 2015. Integrating network ecology with applied conservation: a synthesis and guide to implementation. AoB Plants 7, plv076.

Karsenti, E., Acinas, S.G., Bork, P., Bowler, C., De Vargas, C., Raes, J., Sullivan, M., Arendt, D., Benzoni, F., Claverie, J.-M., Follows, M., Gorsky, G., Hingamp, P., Iudicone, D., Jaillon, O., Kandels-Lewis, S., Krzic, U., Not, F., Ogata, H., Pesant, S., Reynaud, E.G., Sardet, C., Sieracki, M.E., Speich, S., Velayoudon, D., Weissenbach, J., Wincker, P., The Tara Oceans Consortium, 2011. A holistic approach to marine eco-systems biology. PLoS Biol. 9. e1001177. http://dx.doi.org/10.1371/journal.pbio.1001177.

Kartzinel, T.R., Chen, P.A., Coverdale, T.C., Erickson, D.L., Kress, W.J., Kuzmina, M.L., Rubenstein, D.I., Wang, W., Pringle, R.M., 2015. DNA metabarcoding illuminates dietary niche partitioning by African large herbivores. Proc. Natl. Acad. Sci. U.S.A. 26, 819–824.

Kasper, M.L., Reeson, A.F., Cooper, S.J.B., Perry, K.D., Austin, A.D., 2004. Assessment of prey overlap between a native (Polistes humilis) and an introduced (Vespula germanica) social wasp using morphology and phylogenetic analyses of 16S rDNA: prey overlap in a native and an exotic wasp. Mol. Ecol. 13, 2037–2048. http://dx.doi.org/10.1111/j.1365-294X.2004.02193.x.

Kays, R., Costello, R., Forrester, T., Baker, M.C., Parsons, A.W., Kalies, E.L., Hess, G., Millspaugh, J.J., McShea, W., 2015. Cats are rare where coyotes roam. J. Mammal. 96, 981–987. http://dx.doi.org/10.1093/jmammal/gyv100.

Kéfi, S., Miele, V., Wieters, E.A., Navarrete, S.A., Berlow, E.L., 2016. How structured is the entangled bank? The surprisingly simple organization of multiplex ecological networks leads to increased persistence and resilience. PLoS Biol. 14 (8), e1002527.

Kelling, S., Hochachka, W.M., Fink, D., Riedewald, M., Caruana, R., Ballard, G., Hooker, G., 2009. Data-intensive science: a new paradigm for biodiversity studies. Bioscience 59, 613–620. http://dx.doi.org/10.1525/bio.2009.59.7.12.

Kelly, B., Dempson, J.B., Power, M., 2006. The effects of preservation on fish tissue stable isotope signatures. J. Fish Biol. 69, 1595–1611. http://dx.doi.org/10.1111/j.1095-8649.2006.01226.x.

Kelly, R.P., Port, J.A., Yamahara, K.M., Martone, R.G., Lowell, N., Thomsen, P.F., Mach, M.E., Bennett, M., Prahler, E., Caldwell, M.R., Crowder, L.B., 2014. Harnessing DNA to improve environmental management. Science 344, 1455–1456. http://dx.doi.org/10.1126/science.1251156.

Kerfoot, W.C., Weider, L.J., 2004. Experimental paleoecology (resurrection ecology): chasing Van Valen's Red Queen hypothesis. Limnol. Oceanogr. 49, 1300–1316.

Kilianski, A., Haas, J.L., Corriveau, E.J., Liem, A.T., Willis, K.L., Kadavy, D.R., Rosenzweig, C., Minot, S.S., 2015. Bacterial and viral identification and differentiation by amplicon sequencing on the MinION nanopore sequencer. GigaScience 4, 12. http://dx.doi.org/10.1186/s13742-015-0051-z.

King, R.A., Read, D.S., Traugott, M., Symondson, W.O.C., 2008. Molecular analysis of predation: a review of best practice for DNA-based approaches. Mol. Ecol. 17, 947–963. http://dx.doi.org/10.1111/j.1365-294X.2007.03613.x.

Koenig, W.D., Liebhold, A.M., Bonter, D.N., Hochachka, W.M., Dickinson, J.L., 2013. Effects of the emerald ash borer invasion on four species of birds. Biol. Invasions 15, 2095–2103. http://dx.doi.org/10.1007/s10530-013-0435-x.

Kokkoris, G.D., Jansen, V.A.A., Loreau, M., Troumbis, A.Y., 2002. Variability in interaction strength and implications for biodiversity. J. Anim. Ecol. 71, 362–371.

Kõljalg, U., Larsson, K.-H., Abarenkov, K., Nilsson, R.H., Alexander, I.J., Eberhardt, U., Erland, S., Høiland, K., Kjøller, R., Larsson, E., Pennanen, T., Sen, R., Taylor, A.F.S., Tedersoo, L., Vrålstad, T., 2005. UNITE: a database providing web-based methods for the molecular identification of ectomycorrhizal fungi: methods. New Phytol. 166, 1063–1068. http://dx.doi.org/10.1111/j.1469-8137.2005.01376.x.

Kraaijeveld, K., de Weger, L.A., Ventayol García, M., Buermans, H., Frank, J., Hiemstra, P.S., den Dunnen, J.T., 2015. Efficient and sensitive identification and quantification of airborne pollen using next-generation DNA sequencing. Mol. Ecol. Resour. 15, 8–16. http://dx.doi.org/10.1111/1755-0998.12288.

Krause, S., Lüke, C., Frenzel, P., 2010. Succession of methanotrophs in oxygen-methane counter-gradients of flooded rice paddies. ISME J. 4, 1603–1607.

Kremen, C., Ullman, K.S., Thorp, R.W., 2011. Evaluating the quality of citizen-scientist data on pollinator communities: citizen-scientist pollinator monitoring. Conserv. Biol. 25, 607–617. http://dx.doi.org/10.1111/j.1523-1739.2011.01657.x.

Kurata, A., Fujiwara, A., Haruyama, N., Tsuchida, T., 2016. Multiplex PCR method for rapid identification of genetic group and symbiont infection status in Bemisia tabaci (Hemiptera: Aleyrodidae). Appl. Entomol. Zool. 51, 167–172. http://dx.doi.org/10.1007/s13355-015-0378-z.

Kurtz, Z.D., Müller, C.L., Miraldi, E.R., Littman, D.R., Blaser, M.J., Bonneau, R.A., 2015. Sparse and compositionally robust inference of microbial ecological networks. PLoS Comput. Biol. 11, e1004226. http://dx.doi.org/10.1371/journal.pcbi.1004226.

Kyle, M., Haande, S., Ostermaier, V., Rohrlack, T., 2015. The Red Queen race between parasitic chytrids and their host, Planktothrix: a test using a time series reconstructed from sediment DNA. PLoS One 10, e0118738. http://dx.doi.org/10.1371/journal.pone.0118738.

Lach, L., Tillberg, C.V., Suarez, A.V., 2010. Contrasting effects of an invasive ant on a native and an invasive plant. Biol. Invasions 12, 3123–3133. http://dx.doi.org/10.1007/s10530-010-9703-1.

Lafferty, K.D., Dobson, A.P., Kuris, A.M., 2006. Parasites dominate food web links. Proc. Natl. Acad. Sci. U.S.A. 103, 11211–11216. http://dx.doi.org/10.1073/pnas.0604755103.

Lafferty, K.D., Allesina, S., Arim, M., Briggs, C.J., De Leo, G., Dobson, A.P., Dunne, J.A., Johnson, P.T.J., Kuris, A.M., Marcogliese, D.J., Martinez, N.D., Memmott, J., Marquet, P.A., McLaughlin, J.P., Mordecai, E.A., Pascual, M., Poulin, R., Thieltges, D.W., 2008. Parasites in food webs: the ultimate missing links: parasites in food webs. Ecol. Lett. 11, 533–546. http://dx.doi.org/10.1111/j.1461-0248.2008.01174.x.

Lahoz-Monfort, J.J., Guillera-Arroita, G., Tingley, R., 2016. Statistical approaches to account for false-positive errors in environmental DNA samples. Mol. Ecol. Resour. 16, 673–685. http://dx.doi.org/10.1111/1755-0998.12486.

Lam, K.N., Cheng, J., Engel, K., Neufeld, J.D., Charles, T.C., 2015. Current and future resources for functional metagenomics. Front. Microbiol. 6, 1196. http://dx.doi.org/10.3389/fmicb.2015.01196.

Lamarche, J., Potvin, A., Pelletier, G., Stewart, D., Feau, N., Alayon, D.I.O., et al., 2015. Molecular detection of 10 of the most unwanted alien forest pathogens in Canada using real-time PCR. PLoS One 10. e0134265. http://dx.doi.org/10.1371/journal.pone.0134265.

Lansdown, K., McKew, B.A., Whitby, C., Heppell, C.M., Dumbrell, A.J., Binley, A., Olde, L., Trimmer, M., 2016. Importance and controls of anaerobic ammonium oxidation influenced by riverbed geology. Nat. Geosci. 9, 357–360.

Láruson, Á.J., Craig, S.F., Messer, K.J., Mackie, J.A., 2012. Rapid and reliable inference of mitochondrial phylogroups among Watersipora species, an invasive group of ship-fouling species (Bryozoa, Cheilostomata). Conserv. Genet. Resour. 4, 617–619. http://dx.doi.org/10.1007/s12686-012-9606-9.

Lavoie, C., Saint-Louis, A., Guay, G., Groeneveld, E., Villeneuve, P., 2012. Naturalization of exotic plant species in north-eastern North America: trends and detection capacity: naturalization and detection capacity of exotic plants. Divers. Distrib. 18, 180–190. http://dx.doi.org/10.1111/j.1472-4642.2011.00826.x.

Layman, C.A., Arrington, D.A., Montaña, C.G., Post, D.M., 2007. Can stable isotope ratios provide for community-wide measures of trophic structure? Ecology 88, 42–48.

Lehman, C.L., Tilman, D., 2000. Biodiversity, stability, and productivity in competitive communities. Am. Nat. 156, 534–552.

Lemmon, E.M., Lemmon, A.R., 2013. High-throughput genomic data in systematics and phylogenetics. Annu. Rev. Ecol. Evol. Syst. 44, 99–121. http://dx.doi.org/10.1146/annurev-ecolsys-110512-135822.

Leonardi, M., Librado, P., Der Sarkissian, C., Schubert, M., Alfarhan, A.H., Alquraishi, S.A., Al-Rasheid, K.A.S., Gamba, C., Willerslev, E., Orlando, L., 2016. Evolutionary patterns and processes: lessons from ancient DNA. Syst. Biol. 65. http://dx.doi.org/10.1093/sysbio/syw059.

Leray, M., Meyer, C.P., Mills, S.C., 2015. Metabarcoding dietary analysis of coral dwelling predatory fish demonstrates the minor contribution of coral mutualists to their highly partitioned, generalist diet. PeerJ 3. e1047. http://dx.doi.org/10.7717/peerj.1047.

Lester, P.J., Bosch, P.J., Gruber, M.A.M., Kapp, E.A., Peng, L., Brenton-Rule, E.C., et al., 2015. No evidence of enemy release in pathogen and microbial communities of common wasps (Vespula vulgaris) in their native and introduced range. PLoS One 10. e0121358. http://dx.doi.org/10.1371/journal.pone.0121358.

Levine, J.M., D'Antonio, C.M., 2003. Forecasting biological invasions with increasing international trade. Conserv. Biol. 17, 322–326.

Li, C., Chng, K.R., Boey, J.H.E., Ng, H.Q.A., Wilm, A., Nagarajan, N., 2016. INC-Seq: accurate single molecule reads using nanopore sequencing. GigaScience 5 (1), 34. http://dx.doi.org/10.1186/s13742-016-0140-7.

Liang, G.H., Jang, E.B., Heller, W.P., Chang, C.L., Chen, J.H., Zhang, F.P., Geib, S.M., 2015. A qPCR-based method for detecting parasitism of Fopius arisanus (Sonan) in oriental fruit flies, Bactrocera dorsalis (Hendel): qPCR-based detection of parasitoid F. arisanus in B. dorsalis. Pest Manag. Sci. 71, 1666–1674. http://dx.doi.org/10.1002/ps.3976.

Liu, L., Li, Y., Li, S., Hu, N., He, Y., Pong, R., Lin, D., Lu, L., Law, M., 2012. Comparison of next-generation sequencing systems. J. Biomed. Biotechnol. 2012, 1–11. http://dx.doi.org/10.1155/2012/251364.

Lodge, D.M., 1993. Biological invasions: lessons for ecology. Trends Ecol. Evol. 8, 133–137.

Loeuille, N., Leibold, M.A., 2008. Ecological consequences of evolution in plant defenses in a metacommunity. Theor. Popul. Biol. 74, 34–45. http://dx.doi.org/10.1016/j.tpb.2008.04.004.

Loman, N.J., Quinlan, A.R., 2014. Poretools: a toolkit for analyzing nanopore sequence data. Bioinformatics 30, 3399–3401. http://dx.doi.org/10.1093/bioinformatics/btu555.

Loman, N.J., Quick, J., Simpson, J.T., 2015. A complete bacterial genome assembled de novo using only nanopore sequencing data. Nat. Methods 12, 733–735. http://dx.doi.org/10.1038/nmeth.3444.

Lopezaraiza-Mikel, M.E., Hayes, R.B., Whalley, M.R., Memmott, J., 2007. The impact of an alien plant on a native plant–pollinator network: an experimental approach. Ecol. Lett. 10, 539–550.

López-Duarte, P.C., Carson, H.S., Cook, G.S., Fodrie, F.J., Becker, B.J., DiBacco, C., Levin, L.A., 2012. What controls connectivity? An empirical, multi-species approach. Integr. Comp. Biol. 52, 511–524. http://dx.doi.org/10.1093/icb/ics104.

Loreau, M., 2010. Linking biodiversity and ecosystems: towards a unifying ecological theory. Philos. Trans. R. Soc. B Biol. Sci. 365, 49–60. http://dx.doi.org/10.1098/rstb.2009.0155.

Lotz, A., Allen, C.R., 2007. Observer bias in anuran call surveys. J. Wildl. Manag. 71, 675–679. http://dx.doi.org/10.2193/2005-759.

Lu, M., Wingfield, M.J., Gillette, N.E., Mori, S.R., Sun, J.-H., 2010. Complex interactions among host pines and fungi vectored by an invasive bark beetle. New Phytol. 187, 859–866. http://dx.doi.org/10.1111/j.1469-8137.2010.03316.x.

Lueders, T., Wagner, B., Claus, P., Friedrich, M.W., 2004. Stable isotope probing of rRNA and DNA reveals a dynamic methylotroph community and trophic interactions with fungi and protozoa in oxic rice field soil. Environmental Microbiology 6, 60–72.

Lundgren, J.G., Fergen, J.K., 2011. Enhancing predation of a subterranean insect pest: a conservation benefit of winter vegetation in agroecosystems. Appl. Soil Ecol. 51, 9–16. http://dx.doi.org/10.1016/j.apsoil.2011.08.005.

Lundgren, J.G., Ellsbury, M.E., Prischmann, D.A., 2009. Analysis of the predator community of a subterranean herbivorous insect based on polymerase chain reaction. Ecol. Appl. 19, 2157–2166.

Lundgren, J.G., Saska, P., Honěk, A., 2013. Molecular approach to describing a seed-based food web: the post-dispersal granivore community of an invasive plant. Ecol. Evol. 3, 1642–1652. http://dx.doi.org/10.1002/ece3.580.

MacDougall, A.S., Turkington, R., 2005. Are invasive species the drivers or passengers of change in degraded ecosystems? Ecology 86, 42–55.

Mackie, J.A., Darling, J.A., Geller, J.B., 2012. Ecology of cryptic invasions: latitudinal segregation among Watersipora (Bryozoa) species. Sci. Rep. 2, 871. http://dx.doi.org/10.1038/srep00871.

Madoui, M.-A., Engelen, S., Cruaud, C., Belser, C., Bertrand, L., Alberti, A., Lemainque, A., Wincker, P., Aury, J.-M., 2015. Genome assembly using nanopore-guided long and error-free DNA reads. BMC Genomics. 16. http://dx.doi.org/10.1186/s12864-015-1519-z.

Maloy, A.P., Culloty, S.C., Slater, J.W., 2013. Dietary analysis of small planktonic consumers: a case study with marine bivalve larvae. J. Plankton Res. 35, 866–876. http://dx.doi.org/10.1093/plankt/fbt027.

Manefield, M., Whiteley, A.S., Griffiths, R.I., Bailey, M.J., 2002. RNA stable isotope probing, a novel means of linking microbial community function to phylogeny. Appl. Environ. Microbiol. 68, 5367–5373.

Massol, F., Dubart, M., Calcagno, V., Cazelles, K., Jacquet, C., Kéfi, S., Gravel, D., 2017. Island biogeography of food webs. Adv. Ecol. Res. 56, 183–262.

Mata, V.A., Amorim, F., Corley, M.F.V., McCracken, G.F., Rebelo, H., Beja, P., 2016. Female dietary bias towards large migratory moths in the European free-tailed bat (*Tadarida teniotis*). Biol. Lett. 12, 20150988. http://dx.doi.org/10.1098/rsbl.2015.0988.

Matsuzaki, S.S., Mabuchi, K., Takamura, N., Hicks, B.J., Nishida, M., Washitani, I., 2010. Stable isotope and molecular analyses indicate that hybridization with non-native domesticated common carp influence habitat use of native carp. Oikos 119, 964–971. http://dx.doi.org/10.1111/j.1600-0706.2009.18076.x.

May, R.M., 1973. Stability and complexity in model ecosystems. Monogr. Popul. Biol. 6, 1–235.

McCann, K., 2007. Protecting biostructure. Nature 446, 29.

McClenaghan, B., Gibson, J.F., Shokralla, S., Hajibabaei, M., 2015. Discrimination of grasshopper (Orthoptera: Acrididae) diet and niche overlap using next-generation sequencing of gut contents. Ecol. Evol. 5, 3046–3055. http://dx.doi.org/10.1002/ece3.1585.

McShea, W.J., Forrester, T., Costello, R., He, Z., Kays, R., 2016. Volunteer-run cameras as distributed sensors for macrosystem mammal research. Landsc. Ecol. 31, 55–66. http://dx.doi.org/10.1007/s10980-015-0262-9.

Médoc, V., Firmat, C., Sheath, D.J., Pegg, J., Andreou, D., Britton, J.R., 2017. Parasites and biological invasions: predicting ecological alterations at levels from individual hosts to whole networks. Adv. Ecol. Res. 57, 1–54.

Meehan, R.R., 2003. DNA methylation in animal development. Semin. Cell Dev. Biol. 14, 53–65.

Memmott, J., 2009. Food webs: a ladder for picking strawberries or a practical tool for practical problems? Philos. Trans. R. Soc. B Biol. Sci. 364, 1693–1699. http://dx.doi.org/10.1098/rstb.2008.0255.

Memmott, J., Waser, N.M., 2002. Integration of alien plants into a native flower–pollinator visitation web. Proc. R. Soc. Lond. B Biol. Sci. 269, 2395–2399.

Mercier, L., Darnaude, A.M., Bruguier, O., Vasconcelos, R.P., Cabral, H.N., Costa, M.J., Lara, M., Jones, D.L., Mouillot, D., 2011. Selecting statistical models and variable combinations for optimal classification using otolith microchemistry. Ecol. Appl. 21, 1352–1364.

Mergeay, J., Verschuren, D., Meester, L.D., 2006. Invasion of an asexual American water flea clone throughout Africa and rapid displacement of a native sibling species. Proc. R. Soc. B Biol. Sci. 273, 2839–2844. http://dx.doi.org/10.1098/rspb.2006.3661.

Miller, N., Estoup, A., Toepfer, S., Bourguet, D., Lapchin, L., Derridj, S., Kim, K.S., Reynaud, P., Furlan, L., Guillemaud, T., 2005. Multiple transatlantic introductions of the western corn rootworm. Science 310, 992. http://dx.doi.org/10.1126/science.1115871.

Mollot, G., Duyck, P.-F., Lefeuvre, P., Lescourret, F., Martin, J.-F., Piry, S., Canard, E., Tixier, P., 2014. Cover cropping alters the diet of arthropods in a banana plantation: a metabarcoding approach. PLoS One 9. e93740. http://dx.doi.org/10.1371/journal.pone.0093740.

Montoya, J.M., 2007. Evolution within food webs: the possible and the actual. Heredity 99, 477–478.

Mora, C.A., Paunescu, D., Grass, R.N., Stark, W.J., 2015. Silica particles with encapsulated DNA as trophic tracers. Mol. Ecol. Resour. 15, 231–241. http://dx.doi.org/10.1111/1755-0998.12299.

Mouquet, N., Lagadeuc, Y., Devictor, V., Doyen, L., Duputié, A., Eveillard, D., Faure, D., Garnier, E., Gimenez, O., Huneman, P., Jabot, F., Jarne, P., Joly, D., Julliard, R., Kéfi, S., Kergoat, G.J., Lavorel, S., Le Gall, L., Meslin, L., Morand, S., Morin, X., Morlon, H., Pinay, G., Pradel, R., Schurr, F.M., Thuiller, W., Loreau, M., 2015.

REVIEW: predictive ecology in a changing world. J. Appl. Ecol. 52, 1293–1310. http://dx.doi.org/10.1111/1365-2664.12482.

Munch, K., Boomsma, W., Huelsenbeck, J., Willerslev, E., Nielsen, R., 2008a. Statistical assignment of DNA sequences using Bayesian phylogenetics. Syst. Biol. 57, 750–757. http://dx.doi.org/10.1080/10635150802422316.

Munch, K., Boomsma, W., Willerslev, E., Nielsen, R., 2008b. Fast phylogenetic DNA barcoding. Philos. Trans. R. Soc. B Biol. Sci. 363, 3997–4002. http://dx.doi.org/10.1098/rstb.2008.0169.

Murray, D.C., Bunce, M., Cannell, B.L., Oliver, R., Houston, J., White, N.E., Barrero, R.A., Bellgard, M.I., Haile, J., 2011. DNA-based faecal dietary analysis: a comparison of qPCR and high throughput sequencing approaches. PLoS One 6, e25776. http://dx.doi.org/10.1371/journal.pone.0025776.

Naaum, A.M., Foottit, R.G., Maw, H.E.L., Hanner, R., 2014. Real-time PCR for identification of the soybean aphid, *Aphis glycines* Matsumura. J. Appl. Entomol. 138, 485–489. http://dx.doi.org/10.1111/jen.12114.

Nakamura, S., Masuda, T., Mochizuki, A., Konishi, K., Tokumaru, S., Ueno, K., Yamaguchi, T., 2013. Primer design for identifying economically important *Liriomyza* species (Diptera: Agromyzidae) by multiplex PCR. Mol. Ecol. Resour. 13, 96–102. http://dx.doi.org/10.1111/1755-0998.12025.

Nedwell, D.B., Underwood, G.J.C., McGenity, T.J., Whitby, C., Dumbrell, A.J., 2016. The Colne estuary: a long-term microbial ecology observatory. Adv. Ecol. Res. 55, 227–281.

Nei, M., 1987. Molecular Evolutionary Genetics. Columbia University Press, New York, 512 pp.

Neufeld, J.D., Wagner, M., Murrell, J.C., 2007. Who eats what, where and when? Isotope-labelling experiments are coming of age. The ISME Journal 1, 103–110.

Ngai, J.T., Srivastava, D.S., 2006. Predators accelerate nutrient cycling in a bromeliad ecosystem. Science 314, 963. http://dx.doi.org/10.1126/science.1132598.

Nolf, D., 1994. Studies on fish otoliths—the state of the art. In: Secor, D.H., Dean, J.M., Miller, A.B., Baruch, B.W. (Eds.), Recent Developments in Fish Otolith Research. University of South Carolina Press, Columbia, SC, pp. 513–544.

O'Brien, T.G., Baillie, J.E.M., Krueger, L., Cuke, M., 2010. The wildlife picture index: monitoring top trophic levels: the wildlife picture index. Anim. Conserv. 13, 335–343. http://dx.doi.org/10.1111/j.1469-1795.2010.00357.x.

Oikonomopoulos, S., Wang, Y.C., Djambazian, H., Badescu, D., Ragoussis, J., 2016. Benchmarking of the Oxford nanopore MinION sequencing for quantitative and qualitative assessment of cDNA populations. Sci. Rep. 6, Article 31602.

Olsson, K., Stenroth, P., Nyström, P., GranéLi, W., 2009. Invasions and niche width: does niche width of an introduced crayfish differ from a native crayfish? Freshw. Biol. 54, 1731–1740. http://dx.doi.org/10.1111/j.1365-2427.2009.02221.x.

O'Rorke, R., Lavery, S., Jeffs, A., 2012. PCR enrichment techniques to identify the diet of predators. Mol. Ecol. Resour. 12, 5–17. http://dx.doi.org/10.1111/j.1755-0998.2011.03091.x.

Ovaskainen, O., Abrego, N., Halme, P., Dunson, D., 2016. Using latent variable models to identify large networks of species-to-species associations at different spatial scales. Meth. Ecol. Evol. 7, 549–555. http://dx.doi.org/10.1111/2041-210X.12501.

Pace, M.L., Cole, J.J., Carpenter, S.R., Kitchell, J.F., 1999. Trophic cascades revealed in diverse ecosystems. Trends Ecol. Evol. 14, 483–488.

Paine, R.T., 2002. Trophic control of production in a rocky intertidal community. Science 296, 736–739. http://dx.doi.org/10.1126/science.1069811.

Palmer, M.E., Yan, N.D., 2013. Decadal-scale regional changes in Canadian freshwater zooplankton: the likely consequence of complex interactions among multiple anthropogenic stressors. Freshw. Biol. 58, 1366–1378. http://dx.doi.org/10.1111/fwb.12133.

Pansu, J., Giguet-Covex, C., Ficetola, G.F., Gielly, L., Boyer, F., Zinger, L., Arnaud, F., Poulenard, J., Taberlet, P., Choler, P., 2015a. Reconstructing long-term human impacts on plant communities: an ecological approach based on lake sediment DNA. Mol. Ecol. 24, 1485–1498. http://dx.doi.org/10.1111/mec.13136.

Pansu, J., Winkworth, R.C., Hennion, F., Gielly, L., Taberlet, P., Choler, P., 2015b. Long-lasting modification of soil fungal diversity associated with the introduction of rabbits to a remote sub-Antarctic archipelago. Biol. Lett. 11, 20150408. http://dx.doi.org/10.1098/rsbl.2015.0408.

Pantel, J.H., Bohan, D.A., Calcagno, V., David, P., Duyck, P.-F., Kamenova, S., Loeuille, N., Mollot, G., Romanuk, T.N., Thébault, E., Tixier, P., Massol, F., 2017. 14 Questions for invasion in ecological networks. Adv. Ecol. Res. 56, 293–340.

Papadopoulou, A., Taberlet, P., Zinger, L., 2015. Metagenome skimming for phylogenetic community ecology: a new era in biodiversity research. Mol. Ecol. 24, 3515–3517. http://dx.doi.org/10.1111/mec.13263.

Parnell, A.C., Phillips, D.L., Bearhop, S., Semmens, B.X., Ward, E.J., Moore, J.W., Jackson, A.L., Grey, J., Kelly, D.J., Inger, R., 2013. Bayesian stable isotope mixing models. Environmetrics 24, 387–399. http://dx.doi.org/10.1002/env.2221.

Paterson, G., Rush, S.A., Arts, M.T., Drouillard, K.G., Haffner, G.D., Johnson, T.B., Lantry, B.F., Hebert, C.E., McGoldrick, D.J., Backus, S.M., Fisk, A.T., 2014. Ecological tracers reveal resource convergence among prey fish species in a large lake ecosystem. Freshw. Biol. 59, 2150–2161. http://dx.doi.org/10.1111/fwb.12418.

Paula, D.P., Linard, B., Andow, D.A., Sujii, E.R., Pires, C.S.S., Vogler, A.P., 2015. Detection and decay rates of prey and prey symbionts in the gut of a predator through metagenomics. Mol. Ecol. Resour. 15, 880–892. http://dx.doi.org/10.1111/1755-0998.12364.

Pearse, I.S., Altermatt, F., 2013. Predicting novel trophic interactions in a non-native world. Ecol. Lett. 16, 1088–1094. http://dx.doi.org/10.1111/ele.12143.

Pedersen, M.W., Overballe-Petersen, S., Ermini, L., Sarkissian, C.D., Haile, J., Hellstrom, M., Spens, J., Thomsen, P.F., Bohmann, K., Cappellini, E., Schnell, I.B., Wales, N.A., Caroe, C., Campos, P.F., Schmidt, A.M.Z., Gilbert, M.T.P., Hansen, A.J., Orlando, L., Willerslev, E., 2014. Ancient and modern environmental DNA. Philos. Trans. R. Soc. B Biol. Sci. 370, 20130383. http://dx.doi.org/10.1098/rstb.2013.0383.

Pell, J.K., Baverstock, J., Roy, H.E., Ware, R.L., Majerus, M.E.N., 2008. Intraguild predation involving Harmonia axyridis: a review of current knowledge and future perspectives. BioControl 53, 147–168. http://dx.doi.org/10.1007/s10526-007-9125-x.

Peralta, G., Frost, C.M., Rand, T.A., Didham, R.K., Tylianakis, J.M., 2014. Complementarity and redundancy of interactions enhance attack rates and spatial stability in host–parasitoid food webs. Ecology 95, 1888–1896.

Pérez-Sayas, C., Pina, T., Gómez-Martínez, M.A., Camañes, G., Ibáñez-Gual, M.V., Jaques, J.A., Hurtado, M.A., 2015. Disentangling mite predator-prey relationships by multiplex PCR. Mol. Ecol. Resour. 15, 1330–1345. http://dx.doi.org/10.1111/1755-0998.12409.

Perga, M.E., Gerdeaux, D., 2005. "Are fish what they eat" all year round? Oecologia 144, 598–606. http://dx.doi.org/10.1007/s00442-005-0069-5.

Perga, M.-E., Desmet, M., Enters, D., Reyss, J.-L., 2010. A century of bottom-up- and top-down-driven changes on a lake planktonic food web: a paleoecological and paleoisotopic study of Lake Annecy, France. Limnol. Oceanogr. 55, 803–816.

Petchey, O.L., Beckerman, A.P., Riede, J.O., Warren, P.H., 2008. Size, foraging, and food web structure. Proc. Natl. Acad. Sci. U.S.A. 105, 4191–4196. http://dx.doi.org/10.1073/pnas.0710672105.

Peterson, A.T., Soberón, J., Pearson, R.G., Anderson, R.P., Martínez-Meyer, E., Nakamura, M., Araújo, M.B., 2011. Ecological Niches and Geographic Distributions. Princeton University Press, New Jersey.

Phillips, D.L., Koch, P.L., 2002. Incorporating concentration dependence in stable isotope mixing models. Oecologia 130, 114–125. http://dx.doi.org/10.1007/s004420100786.

Phillips, D.L., Inger, R., Bearhop, S., Jackson, A.L., Moore, J.W., Parnell, A.C., Semmens, B.X., Ward, E.J., 2014. Best practices for use of stable isotope mixing models in food-web studies. Can. J. Zool. 92, 823–835. http://dx.doi.org/10.1139/cjz-2014-0127.

Piaggio, A.J., Engeman, R.M., Hopken, M.W., Humphrey, J.S., Keacher, K.L., Bruce, W.E., Avery, M.L., 2014. Detecting an elusive invasive species: a diagnostic PCR to detect Burmese python in Florida waters and an assessment of persistence of environmental DNA. Mol. Ecol. Resour. 14, 374–380. http://dx.doi.org/10.1111/1755-0998.12180.

Pianezzola, E., Roth, S., Hatteland, B.A., 2013. Predation by carabid beetles on the invasive slug Arion vulgaris in an agricultural semi-field experiment. Bull. Entomol. Res. 103, 225–232. http://dx.doi.org/10.1017/S0007485312000569.

Pilliod, D.S., Goldberg, C.S., Arkle, R.S., Waits, L.P., Richardson, J., 2013. Estimating occupancy and abundance of stream amphibians using environmental DNA from filtered water samples. Can. J. Fish. Aquat. Sci. 70, 1123–1130. http://dx.doi.org/10.1139/cjfas-2013-0047.

Pocock, M.J.O., Evans, D.M., Memmott, J., 2012. The robustness and restoration of a network of ecological networks. Science 335, 973–977.

Pointier, J.P., Jourdane, J., 2000. Biological control of the snail hosts of schistosomiasis in areas of low transmission: the example of the Caribbean area. Acta Trop. 77, 53–60.

Poisot, T., Canard, E., Mouillot, D., Mouquet, N., Gravel, D., 2012. The dissimilarity of species interaction networks. Ecol. Lett. 15, 1353–1361. http://dx.doi.org/10.1111/ele.12002.

Poisot, T., Mouquet, N., Gravel, D., 2013. Trophic complementarity drives the biodiversity-ecosystem functioning relationship in food webs. Ecol. Lett. 16, 853–861. http://dx.doi.org/10.1111/ele.12118.

Polis, G.A., Sears, A.L.W., Huxel, G.R., Strong, D.R., Maron, J., 2000. When is a trophic cascade a trophic cascade? Trends Ecol. Evol. 15, 473–475.

Port, J.A., O'Donnell, J.L., Romero-Maraccini, O.C., Leary, P.R., Litvin, S.Y., Nickols, K.J., Yamahara, K.M., Kelly, R.P., 2016. Assessing vertebrate biodiversity in a kelp forest ecosystem using environmental DNA. Mol. Ecol. 25, 527–541. http://dx.doi.org/10.1111/mec.13481.

Post, D.M., 2002. Using stable isotopes to estimate trophic position: models, methods, and assumptions. Ecology 83, 703–718.

Post, D.M., Layman, C.A., Arrington, D.A., Takimoto, G., Quattrochi, J., Montaña, C.G., 2007. Getting to the fat of the matter: models, methods and assumptions for dealing with lipids in stable isotope analyses. Oecologia 152, 179–189. http://dx.doi.org/10.1007/s00442-006-0630-x.

Prinsloo, G., Chen, Y., Giles, K.L., Greenstone, M.H., 2002. Release and recovery in South Africa of the exotic aphid parasitoid Aphelinus hordei verified by the polymerase chain reaction. BioControl 47, 127–136.

Purvis, A., Hector, A., 2000. Getting the measure of biodiversity. Nature 405, 212–219.

Pyšek, P., Jarošík, V., Hulme, P.E., Pergl, J., Hejda, M., Schaffner, U., Vilà, M., 2012. A global assessment of invasive plant impacts on resident species, communities and ecosystems: the interaction of impact measures, invading species' traits and environment. Glob. Chang. Biol. 18, 1725–1737.

Quail, M., Smith, M.E., Coupland, P., Otto, T.D., Harris, S.R., Connor, T.R., Bertoni, A., Swerdlow, H.P., Gu, Y., 2012. A tale of three next generation sequencing platforms: comparison of Ion torrent, Pacific biosciences and Illumina MiSeq sequencers. BMC Genomics 13, 341.

Quast, C., Pruesse, E., Yilmaz, P., Gerken, J., Schweer, T., Yarza, P., Peplies, J., Glockner, F.O., 2013. The SILVA ribosomal RNA gene database project: improved data processing and web-based tools. Nucleic Acids Res. 41, D590–D596. http://dx.doi.org/10.1093/nar/gks1219.

Quéméré, E., Hibert, F., Miquel, C., Lhuillier, E., Rasolondraibe, E., Champeau, J., Rabarivola, C., Nusbaumer, L., Chatelain, C., Gautier, L., Ranirison, P., Crouau-Roy, B., Taberlet, P., Chikhi, L., 2013. A DNA metabarcoding study of a primate dietary diversity and plasticity across its entire fragmented range. PLoS One 8, e58971. http://dx.doi.org/10.1371/journal.pone.0058971.

Quick, J., Loman, N.J., Duraffour, S., Simpson, J.T., Severi, E., Cowley, L., Bore, J.A., Koundouno, R., Dudas, G., Mikhail, A., Ouédraogo, N., Afrough, B., Bah, A., Baum, J.H.J., Becker-Ziaja, B., Boettcher, J.P., Cabeza-Cabrerizo, M., Camino-Sánchez, Á., Carter, L.L., Doerrbecker, J., Enkirch, T., Dorival, I.G., Hetzelt, N., Hinzmann, J., Holm, T., Kafetzopoulou, L.E., Koropogui, M., Kosgey, A., Kuisma, E., Logue, C.H., Mazzarelli, A., Meisel, S., Mertens, M., Michel, J., Ngabo, D., Nitzsche, K., Pallasch, E., Patrono, L.V., Portmann, J., Repits, J.G., Rickett, N.Y., Sachse, A., Singethan, K., Vitoriano, I., Yemanaberhan, R.L., Zekeng, E.G., Racine, T., Bello, A., Sall, A.A., Faye, O., Faye, O., Magassouba, N., Williams, C.V., Amburgey, V., Winona, L., Davis, E., Gerlach, J., Washington, F., Monteil, V., Jourdain, M., Bererd, M., Camara, A., Somlare, H., Camara, A., Gerard, M., Bado, G., Baillet, B., Delaune, D., Nebie, K.Y., Diarra, A., Savane, Y., Pallawo, R.B., Gutierrez, G.J., Milhano, N., Roger, I., Williams, C.J., Yattara, F., Lewandowski, K., Taylor, J., Rachwal, P., Turner, D.J., Pollakis, G., Hiscox, J.A., Matthews, D.A., Shea, M.K.O., Johnston, A.M., Wilson, D., Hutley, E., Smit, E., Di Caro, A., Wölfel, R., Stoecker, K., Fleischmann, E., Gabriel, M., Weller, S.A., Koivogui, L., Diallo, B., Keïta, S., Rambaut, A., Formenty, P., Günther, S., Carroll, M.W., 2016. Real-time, portable genome sequencing for Ebola surveillance. Nature 530, 228–232. http://dx.doi.org/10.1038/nature16996.

Radajewski, S., Ineson, P., Parekh, N., Murrell, J., 2000. Stable-isotope probing as a tool in microbial ecology. Nature 403, 646–649.

Ramsey, D.S.L., MacDonald, A.J., Quasim, S., Barclay, C., Sarre, S.D., 2015. An examination of the accuracy of a sequential PCR and sequencing test used to detect the incursion of an invasive species: the case of the red fox in Tasmania. J. Appl. Ecol. 52, 562–570. http://dx.doi.org/10.1111/1365-2664.12407.

Rand, A.C., Jain, M., Eizenga, J., Musselman-Brown, A., Olsen, H.E., Akeson, M., Paten, B., 2016. Cytosine variant calling with high-throughput nanopore sequencing (No. biorxiv; 047134v1), http://dx.doi.org/10.1101/047134.

Ratnasingham, S., Hebert, P.D.N., 2007. BOLD: the barcode of life data system. Mol. Ecol. Notes 7, 355–364.

Rees, H.C., Bishop, K., Middleditch, D.J., Patmore, J.R.M., Maddison, B.C., Gough, K.C., 2014. The application of eDNA for monitoring of the great crested newt in the UK. Ecol. Evol. 4, 4023–4032. http://dx.doi.org/10.1002/ece3.1272.

Rennie, M.D., Sprules, W.G., Johnson, T.B., 2009. Resource switching in Wsh following a major food web disruption. Oecologia 159, 789–802.

Ribeiro, F.J., Przybylski, D., Yin, S., Sharpe, T., Gnerre, S., Abouelleil, A., Berlin, A.M., Montmayeur, A., Shea, T.P., Walker, B.J., Young, S.K., Russ, C., Nusbaum, C., MacCallum, I., Jaffe, D.B., 2012. Finished bacterial genomes from shotgun sequence data. Genome Res. 22, 2270–2277. http://dx.doi.org/10.1101/gr.141515.112.

Richardson, A.J., Walne, A.W., John, A.W.G., Jonas, T.D., Lindley, J.A., Sims, D.W., Stevens, D., Witt, M., 2006. Using continuous plankton recorder data. Prog. Oceanogr. 68, 27–74. http://dx.doi.org/10.1016/j.pocean.2005.09.011.

Richardson, R.T., Lin, C.-H., Quijia, J.O., Riusech, N.S., Goodell, K., Johnson, R.M., 2015a. Rank-based characterization of pollen assemblages collected by honey bees using a multi-locus metabarcoding approach. Appl. Plant Sci. 3, 1500043. http://dx.doi.org/10.3732/apps.1500043.

Richardson, R.T., Lin, C.-H., Sponsler, D.B., Quijia, J.O., Goodell, K., Johnson, R.M., 2015b. Application of ITS2 metabarcoding to determine the provenance of pollen collected by honey bees in an agroecosystem. Appl. Plant Sci. 3, 1400066. http://dx.doi.org/10.3732/apps.1400066.

Riemann, L., Alfredsson, H., Hansen, M.M., Als, T.D., Nielsen, T.G., Munk, P., Aarestrup, K., Maes, G.E., Sparholt, H., Petersen, M.I., Bachler, M., Castonguay, M., 2010. Qualitative assessment of the diet of European eel larvae in the Sargasso Sea resolved by DNA barcoding. Biol. Lett. 6, 819–822. http://dx.doi.org/10.1098/rsbl.2010.0411.

Risse, J., Thomson, M., Blakely, G., Koutsovoulos, G., Blaxter, M., Watson, M., 2015. A single chromosome assembly of Bacteroides fragilis strain BE1 from Illumina and MinION nanopore sequencing data. GigaScience 4, 60. http://dx.doi.org/10.1186/s13742-015-0101-6.

Ristaino, J.B., 2002. Tracking historic migrations of the Irish potato famine pathogen, Phytophthora infestans. Microbes Infect. 4, 1369–1377.

Rizzi, E., Lari, M., Gigli, E., De Bellis, G., Caramelli, D., 2012. Ancient DNA studies: new perspectives on old samples. Genet. Sel. Evol. 44, 1–21.

Roberts, R.L., Donald, P.F., Green, R.E., 2007. Using simple species lists to monitor trends in animal populations: new methods and a comparison with independent data. Anim. Conserv. 10, 332–339. http://dx.doi.org/10.1111/j.1469-1795.2007.00117.x.

Rohr, R.P., Scherer, H., Kehrli, P., Mazza, C., Bersier, L., 2010. Modeling food webs: exploring unexplained structure using latent traits. Am. Nat. 176, 170–177. http://dx.doi.org/10.1086/653667.

Rohr, R.P., Naisbit, R.E., Mazza, C., Bersier, L.-F., 2016. Matching–centrality decomposition and the forecasting of new links in networks. Proc. R. Soc. B Biol. Sci. 283, 20152702. http://dx.doi.org/10.1098/rspb.2015.2702.

Romanuk, T.N., Zhou, Y., Brose, U., Berlow, E.L., Williams, R.J., Martinez, N.D., 2009. Predicting invasion success in complex ecological networks. Philos. Trans. R. Soc. B Biol. Sci. 364, 1743–1754. http://dx.doi.org/10.1098/rstb.2008.0286.

Romanuk, T.N., Zhou, Y., Valdovinos, F.S., Martinez, N.D., 2017. Robustness trade-offs in model food webs: invasion probability decreases while invasion consequences increase with connectance. Adv. Ecol. Res. 56, 263–291.

Rougerie, R., Smith, M.A., Fernandez-Triana, J., Lopez-Vaamonde, C., Ratnasingham, S., Hebert, P.D.N., 2011. Molecular analysis of parasitoid linkages (MAPL): gut contents of adult parasitoid wasps reveal larval host: molecular analysis of parasitoid linkages. Mol. Ecol. 20, 179–186. http://dx.doi.org/10.1111/j.1365-294X.2010.04918.x.

Roussel, J.-M., Paillisson, J.-M., Tréguier, A., Petit, E., 2015. The downside of eDNA as a survey tool in water bodies. J. Appl. Ecol. 52, 823–826. http://dx.doi.org/10.1111/1365-2664.12428.

Roy, H., Wajnberg, E., 2008. From biological control to invasion: the ladybird Harmonia axyridis as a model species. In: Roy, H.E., Wajnberg, E. (Eds.), From Biological Control to Invasion: The Ladybird Harmonia axyridis as a Model Species. Springer Netherlands, Dordrecht, pp. 1–4.

Roy, H.E., Adriaens, T., Isaac, N.J.B., Kenis, M., Onkelinx, T., Martin, G.S., Brown, P.M.J., Hautier, L., Poland, R., Roy, D.B., Comont, R., Eschen, R., Frost, R., Zindel, R., Van Vlaenderen, J., Nedvěd, O., Ravn, H.P., Grégoire, J.-C., de Biseau, J.-C., Maes, D., 2012. Invasive alien predator causes rapid declines of native European ladybirds: alien predator causes declines of native ladybirds. Divers. Distrib. 18, 717–725. http://dx.doi.org/10.1111/j.1472-4642.2012.00883.x.

Rubenstein, D.R., Hobson, K.A., 2004. From birds to butterflies: animal movement patterns and stable isotopes. Trends Ecol. Evol. 19, 256–263. http://dx.doi.org/10.1016/j.tree.2004.03.017.

Rush, S.A., Paterson, G., Johnson, T.B., Drouillard, K.G., Haffner, G.D., Hebert, C.E., Arts, M.T., McGoldrick, D.J., Backus, S.M., Lantry, B.F., Lantry, J.R., Schaner, T., Fisk, A.T., 2012. Long-term impacts of invasive species on a native top predator in a large lake system: *lake trout diet shift*. Freshw. Biol. 57, 2342–2355. http://dx.doi.org/10.1111/fwb.12014.

Russo, L., Memmott, J., Montoya, D., Shea, K., Buckley, Y.M., 2014. Patterns of introduced species interactions affect multiple aspects of network structure in plant–pollinator communities. Ecology 95, 2953–2963.

Sagouis, A., Cucherousset, J., Villéger, S., Santoul, F., Boulêtreau, S., 2015. Non-native species modify the isotopic structure of freshwater fish communities across the globe. Ecography 38, 979–985. http://dx.doi.org/10.1111/ecog.01348.

Sakai, A.K., Allendorf, F.W., Holt, J.S., Lodge, D.M., Molofsky, J., With, K.A., Baughman, S., Cabin, R.J., Cohen, J.E., Ellstrand, N.C., et al., 2001. The population biology of invasive species. Annu. Rev. Ecol. Syst. 32, 305–332.

Sala, O.E., et al., 2000. Global biodiversity scenarios for the year 2100. Science 287, 1770–1774. http://dx.doi.org/10.1126/science.287.5459.1770.

Salathé, M., Kouyos, R., Bonhoeffer, S., 2008. The state of affairs in the kingdom of the Red Queen. Trends Ecol. Evol. 23, 439–445. http://dx.doi.org/10.1016/j.tree.2008.04.010.

Saltré, F., Duputié, A., Gaucherel, C., Chuine, I., 2015. How climate, migration ability and habitat fragmentation affect the projected future distribution of European beech. Glob. Chang. Biol. 21, 897–910. http://dx.doi.org/10.1111/gcb.12771.

Santamaría, S., Galeano, J., Pastor, J.M., Méndez, M., 2016. Removing interactions, rather than species, casts doubt on the high robustness of pollination networks. Oikos 125, 526–534. http://dx.doi.org/10.1111/oik.02921.

Sato, T., Watanabe, K., Kanaiwa, M., Niizuma, Y., Harada, Y., Lafferty, K.D., 2011. Nematomorph parasites drive energy flow through a riparian ecosystem. Ecology 92, 201–207. http://dx.doi.org/10.1890/09-1565.1.

Sato, T., Egusa, T., Fukushima, K., Oda, T., Ohte, N., Tokuchi, N., Watanabe, K., Kanaiwa, M., Murakami, I., Lafferty, K.D., 2012. Nematomorph parasites indirectly alter the food web and ecosystem function of streams through behavioural manipulation of their cricket hosts. Ecol. Lett. 15, 786–793. http://dx.doi.org/10.1111/j.1461-0248.2012.01798.x.

Savage, J., Vellend, M., 2015. Elevational shifts, biotic homogenization and time lags in vegetation change during 40 years of climate warming. Ecography 38, 546–555. http://dx.doi.org/10.1111/ecog.01131.

Savichtcheva, O., Debroas, D., Kurmayer, R., Villar, C., Jenny, J.P., Arnaud, F., Perga, M.E., Domaizon, I., 2011. Quantitative PCR enumeration of total/toxic Planktothrix rubescens and total cyanobacteria in preserved DNA isolated from lake sediments. Appl. Environ. Microbiol. 77, 8744–8753. http://dx.doi.org/10.1128/AEM.06106-11.

Sax, D.F., Gaines, S.D., 2003. Species diversity: from global decreases to local increases. Trends Ecol. Evol. 18, 561–566. http://dx.doi.org/10.1016/S0169-5347(03)00224-6.

Schindler, D.W., 1998. Replication versus realism: the need for ecosystem-scale experiments. Ecosystems 1, 323–334.

Schlaepfer, M.A., Sherman, P.W., Blossey, B., Runge, M.C., 2005. Introduced species as evolutionary traps: introduced species as evolutionary traps. Ecol. Lett. 8, 241–246. http://dx.doi.org/10.1111/j.1461-0248.2005.00730.x.

Schloss, P.D., Gevers, D., Westcott, S.L., 2011. Reducing the effects of PCR amplification and sequencing artifacts on 16S rRNA-based studies. PLoS One 6, e27310. http://dx.doi.org/10.1371/journal.pone.0027310.

Schmidt, B.R., Kéry, M., Ursenbacher, S., Hyman, O.J., Collins, J.P., 2013a. Site occupancy models in the analysis of environmental DNA presence/absence surveys: a case study of an emerging amphibian pathogen. Meth. Ecol. Evol. 4, 646–653. http://dx.doi.org/10.1111/2041-210X.12052.

Schmidt, P.-A., Bálint, M., Greshake, B., Bandow, C., Römbke, J., Schmitt, I., 2013b. Illumina metabarcoding of a soil fungal community. Soil Biol. Biochem. 65, 128–132. http://dx.doi.org/10.1016/j.soilbio.2013.05.014.

Scriver, M., Marinich, A., Wilson, C., Freeland, J., 2015. Development of species-specific environmental DNA (eDNA) markers for invasive aquatic plants. Aquat. Bot. 122, 27–31. http://dx.doi.org/10.1016/j.aquabot.2015.01.003.

Scully, E.D., Geib, S.M., Hoover, K., Tien, M., Tringe, S.G., Barry, K.W., Glavina del Rio, T., Chovatia, M., Herr, J.R., Carlson, J.E., 2013. Metagenomic profiling reveals lignocellulose degrading system in a microbial community associated with a wood-feeding beetle. PLoS One 8. e73827. http://dx.doi.org/10.1371/journal.pone.0073827.

Secondi, J., Dejean, T., Valentini, A., Audebaud, B., Miaud, C., 2016. Detection of a global aquatic invasive amphibian, Xenopus laevis, using environmental DNA. Amphibia-Reptilia 37, 131–136.

Shehzad, W., Riaz, T., Nawaz, M.A., Miquel, C., Poillot, C., Shah, S.A., Pompanon, F., Coissac, E., Taberlet, P., 2012. Carnivore diet analysis based on next-generation sequencing: application to the leopard cat (Prionailurus bengalensis) in Pakistan: leopard cat diet. Mol. Ecol. 21, 1951–1965. http://dx.doi.org/10.1111/j.1365-294X.2011.05424.x.

Sheppard, S.K., Bell, J., Sunderland, K.D., Fenlon, J., Skervin, D., Symondson, W.O.C., 2005. Detection of secondary predation by PCR analyses of the gut contents of invertebrate generalist predators: secondary predation and PCR. Mol. Ecol. 14, 4461–4468. http://dx.doi.org/10.1111/j.1365-294X.2005.02742.x.

Shokralla, S., Porter, T.M., Gibson, J.F., Dobosz, R., Janzen, D.H., Hallwachs, W., Golding, G.B., Hajibabaei, M., 2015. Massively parallel multiplex DNA sequencing for specimen identification using an Illumina MiSeq platform. Sci. Rep. 5, 9687. http://dx.doi.org/10.1038/srep09687.

Simpson, J.T., Workman, R., Zuzarte, P.C., David, M., Duris, L.J., Timp, W., 2016. Detecting DNA methylation using the Oxford Nanopore Technologies MinION sequencer. BioRxiv. http://dx.doi.org/10.1101/047142.

Sint, D., Traugott, M., 2015. Food web designer: a flexible tool to visualize interaction networks. J. Pest Sci. 89, 1–5. http://dx.doi.org/10.1007/s10340-015-0686-7.

Sint, D., Raso, L., Traugott, M., 2012. Advances in multiplex PCR: balancing primer efficiencies and improving detection success. Meth. Ecol. Evol. 3, 898–905.

Sloggett, J.J., Obrycki, J.J., Haynes, K.F., 2009. Identification and quantification of predation: novel use of gas chromatography-mass spectrometric analysis of prey alkaloid markers. Funct. Ecol. 23, 416–426. http://dx.doi.org/10.1111/j.1365-2435.2008.01492.x.

Smith, C.A., Gardiner, M.M., 2013. Oviposition habitat influences egg predation of native and exotic coccinellids by generalist predators. Biol. Control 67, 235–245. http://dx.doi.org/10.1016/j.biocontrol.2013.07.019.

Smith, M.A., Eveleigh, E.S., McCann, K.S., Merilo, M.T., McCarthy, P.C., Van Rooyen, K.I., 2011. Barcoding a quantified food web: crypsis, concepts, ecology and hypotheses. PLoS One 6, e14424. http://dx.doi.org/10.1371/journal.pone.0014424.

Snäll, T., Kindvall, O., Nilsson, J., Pärt, T., 2011. Evaluating citizen-based presence data for bird monitoring. Biol. Conserv. 144, 804–810. http://dx.doi.org/10.1016/j.biocon.2010.11.010.

Soininen, E.M., Ehrich, D., Lecomte, N., Yoccoz, N.G., Tarroux, A., Berteaux, D., Gauthier, G., Gielly, L., Brochmann, C., Gussarova, G., Ims, R.A., 2014. Sources of variation in small rodent trophic niche: new insights from DNA metabarcoding and stable isotope analysis. Isot. Environ. Health Stud. 50, 361–381. http://dx.doi.org/10.1080/10256016.2014.915824.

Sotton, B., Guillard, J., Anneville, O., Maréchal, M., Savichtcheva, O., Domaizon, I., 2014. Trophic transfer of microcystins through the lake pelagic food web: evidence for the role of zooplankton as a vector in fish contamination. Sci. Total Environ. 466–467, 152–163.

Sović, I., Šikic, M., Wilm, A., Fenlon, S.N., Chen, S., Nagarajan, N., 2016. Fast and sensitive mapping of nanopore sequencing reads with GraphMap. Nat. Commun. 7, 11307. http://dx.doi.org/10.1038/ncomms11307.

Spence, K.O., Rosenheim, J.A., 2005. Enrichment in herbivorous insects: a comparative field-based study of variation. Oecologia 146, 89–97.

Spikmans, F., van Tongeren, T., van Alen, T., van der Velde, G., Op den Camp, H., 2013. High prevalence of the parasite Sphaerothecum destruens in the invasive topmouth gudgeon Pseudorasbora parva in the Netherlands, a potential threat to native freshwater fish. Aquat. Invasions 8, 355–360. http://dx.doi.org/10.3391/ai.2013.8.3.12.

Spotswood, E.N., Meyer, J.-Y., Bartolome, J.W., 2012. An invasive tree alters the structure of seed dispersal networks between birds and plants in French Polynesia. J. Biogeogr. 39, 2007–2020.

Srivathsan, A., Sha, J.C.M., Vogler, A.P., Meier, R., 2015. Comparing the effectiveness of metagenomics and metabarcoding for diet analysis of a leaf-feeding monkey (*Pygathrix nemaeus*). Mol. Ecol. Resour. 15, 250–261. http://dx.doi.org/10.1111/1755-0998.12302.

Srivathsan, A., Ang, A., Vogler, A.P., Meier, R., 2016. Fecal metagenomics for the simultaneous assessment of diet, parasites, and population genetics of an understudied primate. Front. Zool. 13, 17. http://dx.doi.org/10.1186/s12983-016-0150-4.

Stachowicz, J.J., Fried, H., Osman, R.W., Whitlatch, R.B., 2002. Biodiversity, invasion resistance, and marine ecosystem function: reconciling pattern and process. Ecology 83, 2575–2590.

Stafford, R., Hart, A.G., Collins, L., Kirkhope, C.L., Williams, R.L., et al., 2010. Eu-social science: the role of internet social networks in the collection of bee biodiversity data. PLoS One 5. e14381. http://dx.doi.org/10.1371/journal.pone.0014381.

Staudacher, K., Jonsson, M., Traugott, M., 2016. Diagnostic PCR assays to unravel food web interactions in cereal crops with focus on biological control of aphids. J. Pest Sci. 89, 281–293. http://dx.doi.org/10.1007/s10340-015-0685-8.

Stock, B.C., Semmens, B.X., 2016. Unifying error structures in commonly used biotracer mixing models. Ecology 97, 2562–2569.

Stoof-Leichsenring, K.R., Epp, L.S., Trauth, M.H., Tiedemann, R., 2012. Hidden diversity in diatoms of Kenyan Lake Naivasha: a genetic approach detects temporal variation. Mol. Ecol. 21, 1918–1930. http://dx.doi.org/10.1111/j.1365-294X.2011.05412.x.

Storkey, J., Macdonald, A.J., Bell, J.R., Clark, I.M., Gregory, A.S., Hawkins, N.J., Todman, L.C., Whitmore, A.P., 2016. The unique contribution of Rothamsted to ecological research at large temporal scales. Adv. Ecol. Res. 55, 3–42.

Strand, G.-H., 1996. Detection of observer bias in ongoing forest health monitoring programmes. Can. J. For. Res. 26, 1692–1696.

Strayer, D.L., Eviner, V.T., Jeschke, J.M., Pace, M.L., 2006. Understanding the long-term effects of species invasions. Trends Ecol. Evol. 21, 645–651. http://dx.doi.org/10.1016/j.tree.2006.07.007.

Stuart, M.K., Greenstone, M.H., 1990. Beyond ELISA: a rapid, sensitive, specific immunodot assay for identification of predator stomach contents. Ann. Entomol. Soc. Am. 83, 1101–1107.

Stull, G.W., Moore, M.J., Mandala, V.S., Douglas, N.A., Kates, H.-R., Qi, X., Brockington, S.F., Soltis, P.S., Soltis, D.E., Gitzendanner, M.A., 2013. A targeted enrichment strategy for massively parallel sequencing of angiosperm plastid genomes. Appl. Plant Sci. 1, 1200497. http://dx.doi.org/10.3732/apps.1200497.

Sturrock, A.M., Trueman, C.N., Darnaude, A.M., Hunter, E., 2012. Can otolith elemental chemistry retrospectively track migrations in fully marine fishes? J. Fish Biol. 81, 766–795. http://dx.doi.org/10.1111/j.1095-8649.2012.03372.x.

Sunnucks, P., Wilson, A.C.C., Beheregaray, L.B., Zenger, K., French, J., Taylor, A.C., 2000. SSCP is not so difficult: the application and utility of single-stranded conformation polymorphism in evolutionary biology and molecular ecology. Mol. Ecol. 9, 1699–1710.

Symondson, W.O.C., 2002. Molecular identification of prey in predator diets. Mol. Ecol. 11, 627–641.

Szalay, T., Golovchenko, J.A., 2015. De novo sequencing and variant calling with nanopores using PoreSeq. Nat. Biotechnol. 33, 1087–1091. http://dx.doi.org/10.1038/nbt.3360.

Taberlet, P., Coissac, E., Hajibabaei, M., Rieseberg, L.H., 2012a. Environmental DNA. Mol. Ecol. 21, 1789–1793.

Taberlet, P., Coissac, E., Pompanon, F., Brochmann, C., Willerslev, E., 2012b. Towards next-generation biodiversity assessment using DNA metabarcoding. Mol. Ecol. 21, 2045–2050.

Takahara, T., Minamoto, T., Yamanaka, H., Doi, H., Kawabata, Z., 2012. Estimation of fish biomass using environmental DNA. PLoS One 7. e35868. http://dx.doi.org/10.1371/journal.pone.0035868.

Tang, S., Allesina, S., 2014. Reactivity and stability of large ecosystems. Front. Ecol. Evol. 2, 1–8. http://dx.doi.org/10.3389/fevo.2014.00021.

Tang, S., Pawar, S., Allesina, S., 2014. Correlation between interaction strengths drives stability in large ecological networks. Ecol. Lett. 17, 1094–1100. http://dx.doi.org/10.1111/ele.12312.

Tang, M., Hardman, C.J., Ji, Y., Meng, G., Liu, S., Tan, M., Yang, S., Moss, E.D., Wang, J., Yang, C., Bruce, C., Nevard, T., Potts, S.G., Zhou, X., Yu, D.W., 2015. High-throughput monitoring of wild bee diversity and abundance via mitogenomics. Meth. Ecol. Evol. 6, 1034–1043. http://dx.doi.org/10.1111/2041-210X.12416.

Tanner, S.E., Vasconcelos, R.P., Reis-Santos, P., Cabral, H.N., Thorrold, S.R., 2011. Spatial and ontogenetic variability in the chemical composition of juvenile common sole (Solea solea) otoliths. Estuar. Coast. Shelf Sci. 91, 150–157. http://dx.doi.org/10.1016/j.ecss.2010.10.008.

Tayeh, A., Hufbauer, R.A., Estoup, A., Ravigné, V., Frachon, L., Facon, B., 2015. Biological invasion and biological control select for different life histories. Nat. Commun. 6, 7268. http://dx.doi.org/10.1038/ncomms8268.

Thalinger, B., Oehm, J., Mayr, H., Obwexer, A., Zeisler, C., Traugott, M., 2016. Molecular prey identification in Central European piscivores. Mol. Ecol. Resour. 16, 123–137. http://dx.doi.org/10.1111/1755-0998.12436.

Thieltges, D.W., Amundsen, P.-A., Hechinger, R.F., Johnson, P.T.J., Lafferty, K.D., Mouritsen, K.N., Preston, D.L., Reise, K., Zander, C.D., Poulin, R., 2013. Parasites as prey in aquatic food webs: implications for predator infection and parasite transmission. Oikos 122, 1473–1482. http://dx.doi.org/10.1111/j.1600-0706.2013.00243.x.

Thierry, M., Bile, A., Grondin, M., Reynaud, B., Becker, N., Delatte, H., 2015. Mitochondrial, nuclear, and endosymbiotic diversity of two recently introduced populations of the invasive Bemisia tabaci MED species in La Réunion. Insect Conserv. Divers. 8, 71–80. http://dx.doi.org/10.1111/icad.12083.

Thomas, A.C., Jarman, S.N., Haman, K.H., Trites, A.W., Deagle, B.E., 2014. Improving accuracy of DNA diet estimates using food tissue control materials and an evaluation of proxies for digestion bias. Mol. Ecol. 23, 3706–3718. http://dx.doi.org/10.1111/mec.12523.

Thomas, A.C., Deagle, B.E., Eveson, J.P., Harsch, C.H., Trites, A.W., 2016a. Quantitative DNA metabarcoding: improved estimates of species proportional biomass using correction factors derived from control material. Mol. Ecol. Resour. 16, 714–726. http://dx.doi.org/10.1111/1755-0998.12490.

Thomas, S.M., Kiljunen, M., Malinen, T., Eloranta, A.P., Amundsen, P.-A., Lodenius, M., Kahilainen, K.K., 2016b. Food-web structure and mercury dynamics in a large subarctic lake following multiple species introductions. Freshw. Biol. 61, 500–517. http://dx.doi.org/10.1111/fwb.12723.

Thompson, A.A., Mapstone, B.D., 1997. Observer effects and training in underwater visual surveys of reef fishes. Mar. Ecol. Prog. Ser. 154, 53–63.

Thompson, R.M., Brose, U., Dunne, J.A., Hall, R.O., Hladyz, S., Kitching, R.L., Martinez, N.D., Rantala, H., Romanuk, T.N., Stouffer, D.B., Tylianakis, J.M., 2012. Food webs: reconciling the structure and function of biodiversity. Trends Ecol. Evol. 27, 689–697. http://dx.doi.org/10.1016/j.tree.2012.08.005.

Thompson, M.S.A., Bankier, C., Bell, T., Dumbrell, A.J., Gray, C., Ledger, M.E., Lehmann, K., McKew, B.A., Sayer, C.D., Shelley, F., Trimmer, M., Warren, S.L., Woodward, G., 2015. Gene-to-ecosystem impacts of a catastrophic pesticide spill: testing a multilevel bioassessment approach in a river ecosystem. Freshw. Biol 61, 2037–2050. http://dx.doi.org/10.1111/fwb.12676.

Thomsen, P.F., Willerslev, E., 2015. Environmental DNA—an emerging tool in conservation for monitoring past and present biodiversity. Biol. Conserv. 183, 4–18. http://dx.doi.org/10.1016/j.biocon.2014.11.019.

Thomsen, P.F., Kielgast, J., Iversen, L.L., Møller, P.R., Rasmussen, M., Willerslev, E., 2012. Detection of a diverse marine fish fauna using environmental DNA from seawater samples. PLoS One 7. e41732. http://dx.doi.org/10.1371/journal.pone.0041732.

Thorrold, S.R., Jones, C.M., Campana, S.E., 1997. Response of otolith microchemistry to environmental variations experienced by larval and juvenile Atlantic croaker (Micropogonias undulatus). Limnol. Oceanogr. 42, 102–111.

Thuiller, W., Albert, C., Araújo, M.B., Berry, P.M., Cabeza, M., Guisan, A., Hickler, T., Midgley, G.F., Paterson, J., Schurr, F.M., Sykes, M.T., Zimmermann, N.E., 2008. Predicting global change impacts on plant species' distributions: future challenges. Perspect. Plant Ecol. Evol. Syst. 9, 137–152. http://dx.doi.org/10.1016/j.ppees.2007.09.004.

Thuiller, W., Münkemüller, T., Lavergne, S., Mouillot, D., Mouquet, N., Schiffers, K., Gravel, D., 2013. A road map for integrating eco-evolutionary processes into biodiversity models. Ecol. Lett. 16, 94–105. http://dx.doi.org/10.1111/ele.12104.

Tiede, J., Wemheuer, B., Traugott, M., Daniel, R., Tscharntke, T., Ebeling, A., Scherber, C., 2016. Trophic and non-trophic interactions in a biodiversity experiment assessed by next-generation sequencing. PLoS One 11. e0148781. http://dx.doi.org/10.1371/journal.pone.0148781.

Tiedeken, E.J., Stout, J.C., 2015. Insect-flower interaction network structure is resilient to a temporary pulse of floral resources from invasive Rhododendron ponticum. PLoS One 10. e0119733. http://dx.doi.org/10.1371/journal.pone.0119733.

Tillberg, C.V., Holway, D.A., LeBrun, E.G., Suarez, A.V., 2007. Trophic ecology of invasive Argentine ants in their native and introduced ranges. Proc. Natl. Acad. Sci. U.S.A. 104, 20856–20861. http://dx.doi.org/10.1073/pnas.0706903105.

Torti, A., Lever, M.A., Jørgensen, B.B., 2015. Origin, dynamics, and implications of extracellular DNA pools in marine sediments. Mar. Genomics 24, 185–196. http://dx.doi.org/10.1016/j.margen.2015.08.007.

Townsend, D.W., Radtke, R.L., Malone, D.P., Wallinga, J.P., 1995. Use of otolith strontium: calcium ratios for hind- casting larval cod Gadus morhua distributions relative to water masses on Georges Bank. Mar. Ecol. Prog. Ser. 119, 37–44.

Traugott, M., Symondson, W.O.C., 2008. Molecular analysis of predation on parasitized hosts. Bull. Entomol. Res. 98, 223–231. http://dx.doi.org/10.1017/S0007485308005968.

Traugott, M., Zangerl, P., Juen, A., Schallhart, N., Pfiffner, L., 2006. Detecting key parasitoids of lepidopteran pests by multiplex PCR. Biol. Control 39, 39–46. http://dx.doi.org/10.1016/j.biocontrol.2006.03.001.

Traugott, M., Bell, J.R., Broad, G.R., Powell, W., Van Veen, F.J.F., Vollhardt, I.M.G., Symondson, W.O.C., 2008. Endoparasitism in cereal aphids: molecular analysis of a whole parasitoid community. Mol. Ecol. 17, 3928–3938. http://dx.doi.org/10.1111/j.1365-294X.2008.03878.x.

Traugott, M., Kamenova, S., Ruess, L., 2013. Empirically characterising trophic networks: what emerging DNA-based methods, stable isotope and fatty acid analyses can offer. Adv. Ecol. Res. 49, 177–224.

Traveset, A., Richardson, D., 2006. Biological invasions as disruptors of plant reproductive mutualisms. Trends Ecol. Evol. 21, 208–216. http://dx.doi.org/10.1016/j.tree.2006.01.006.

Tréguier, A., Paillisson, J.-M., Dejean, T., Valentini, A., Schlaepfer, M.A., Roussel, J.-M., 2014. Environmental DNA surveillance for invertebrate species: advantages and technical limitations to detect invasive crayfish Procambarus clarkii in freshwater ponds. J. Appl. Ecol. 51, 871–879. http://dx.doi.org/10.1111/1365-2664.12262.

Troedsson, C., Simonelli, P., Nägele, V., Nejstgaard, J.C., Frischer, M.E., 2009. Quantification of copepod gut content by differential length amplification quantitative PCR (dla-qPCR). Mar. Biol. 156, 253–259. http://dx.doi.org/10.1007/s00227-008-1079-8.

Turner, C.R., Uy, K.L., Everhart, R.C., 2015. Fish environmental DNA is more concentrated in aquatic sediments than surface water. Biol. Conserv. 183, 93–102. http://dx.doi.org/10.1016/j.biocon.2014.11.017.

Uchii, K., Doi, H., Minamoto, T., 2016. A novel environmental DNA approach to quantify the cryptic invasion of non-native genotypes. Mol. Ecol. Resour. 16, 415–422. http://dx.doi.org/10.1111/1755-0998.12460.

Vacher, C., Daudin, J.-J., Piou, D., Desprez-Loustau, M.-L., 2010. Ecological integration of alien species into a tree-parasitic fungus network. Biol. Invasions 12, 3249–3259. http://dx.doi.org/10.1007/s10530-010-9719-6.

Vacher, C., Tamaddoni-Nezhad, A., Kamenova, S., Peyrard, N., Moalic, Y., Sabbadin, R., Schwaller, L., Chiquet, J., Smith, M.A., Vallance, J., Fievet, V., Jakuschkin, B., Bohan, D.A., 2016. Learning ecological networks from next-generation sequencing data. In: Woodward, G., Bohan, D. (Eds.), Advances in Ecological Research. Elsevier Academic Press, Cambridge, MA, pp. 1–39.

Václavík, T., Meentemeyer, R.K., 2012. Equilibrium or not? Modelling potential distribution of invasive species in different stages of invasion: equilibrium and invasive species distribution models. Divers. Distrib. 18, 73–83. http://dx.doi.org/10.1111/j.1472-4642.2011.00854.x.

Valentini, A., Miquel, C., Nawaz, M.A., Bellemain, E., Coissac, E., Pompanon, F., Gielly, L., Cruaud, C., Nascetti, G., Wincker, P., Swenson, J.E., Taberlet, P., 2009a. New perspectives in diet analysis based on DNA barcoding and parallel pyrosequencing: the trn L approach. Mol. Ecol. Resour. 9, 51–60. http://dx.doi.org/10.1111/j.1755-0998.2008.02352.x.

Valentini, A., Pompanon, F., Taberlet, P., 2009b. DNA barcoding for ecologists. Trends Ecol. Evol. 24, 110–117. http://dx.doi.org/10.1016/j.tree.2008.09.011.

Valentini, A., Taberlet, P., Miaud, C., Civade, R., Herder, J., Thomsen, P.F., Bellemain, E., Besnard, A., Coissac, E., Boyer, F., Gaboriaud, C., Jean, P., Poulet, N., Roset, N., Copp, G.H., Geniez, P., Pont, D., Argillier, C., Baudoin, J.-M., Peroux, T., Crivelli, A.J., Olivier, A., Acqueberge, M., Le Brun, M., Møller, P.R., Willerslev, E., Dejean, T., 2016. Next-generation monitoring of aquatic biodiversity using environmental DNA metabarcoding. Mol. Ecol. 25, 929–942. http://dx.doi.org/10.1111/mec.13428.

Valles, S.M., Oi, D.H., Yu, F., Tan, X.-X., Buss, E.A., 2012. Metatranscriptomics and pyrosequencing facilitate discovery of potential viral natural enemies of the invasive Caribbean crazy ant, Nylanderia pubens. PLoS One 7. e31828. http://dx.doi.org/10.1371/journal.pone.0031828.

Valles, S.M., Shoemaker, D., Wurm, Y., Strong, C.A., Varone, L., Becnel, J.J., Shirk, P.D., 2013. Discovery and molecular characterization of an ambisense densovirus from South American populations of Solenopsis invicta. Biol. Control 67, 431–439. http://dx.doi.org/10.1016/j.biocontrol.2013.09.015.

Vander Zanden, M.J., Vadeboncoeur, Y., 2002. Fishes as integrators of benthic and pelagic food webs in lakes. Ecology 83, 2152–2161.

Vander Zanden, M.J., Shuter, B.J., Lester, N., Rasmussen, J.B., 1999. Patterns of food chain length in lakes: a stable isotope study. Am. Nat. 154, 406–416.

Vander Zanden, H.B., Soto, D.X., Bowen, G.J., Hobson, K.A., 2016. Expanding the isotopic toolbox: applications of hydrogen and oxygen stable isotope ratios to food web studies. Front. Ecol. Evol. 4, 1–19. http://dx.doi.org/10.3389/fevo.2016.00020.

Vanderklift, M.A., Ponsard, S., 2003. Sources of variation in consumer-diet? 15N enrichment: a meta-analysis. Oecologia 136, 169–182. http://dx.doi.org/10.1007/s00442-003-1270-z.

van Leeuwen, J.F.N., Schäfer, H., van der Knapp, W.O., Rittenour, T.M., Björck, S., Ammann, B., 2005. Native or introduced? Fossil pollen and spores may say. An example from the Azores Islands. NEOBIOTA 6, 27–34.

Varennes, Y.-D., Boyer, S., Wratten, S.D., 2014. Un-nesting DNA Russian dolls—the potential for constructing food webs using residual DNA in empty aphid mummies. Mol. Ecol. 23, 3925–3933. http://dx.doi.org/10.1111/mec.12633.

Vasanthakumar, A., Handelsman, J.O., Schloss, P.D., Bauer, L.S., Raffa, K.F., 2008. Gut microbiota of an invasive subcortical beetle, Agrilus planipennis Fairmaire, across various life stages. Environ. Entomol. 37, 1344–1353.

Vernière, C., Bui Thi Ngoc, L., Jarne, P., Ravigné, V., Guérin, F., Gagnevin, L., Le Mai, N., Chau, N.M., Pruvost, O., 2014. Highly polymorphic markers reveal the establishment of an invasive lineage of the citrus bacterial pathogen Xanthomonas citri pv. citri in its area of origin. Environ. Microbiol. 16, 2226–2237. http://dx.doi.org/10.1111/1462-2920.12369.

Vesterinen, E.J., Ruokolainen, L., Wahlberg, N., Peña, C., Roslin, T., Laine, V.N., Vasko, V., Sääksjärvi, I.E., Norrdahl, K., Lilley, T.M., 2016. What you need is what you eat? Prey selection by the bat Myotis daubentonii. Mol. Ecol. 25, 1581–1594. http://dx.doi.org/10.1111/mec.13564.

Vestheim, H., Jarman, S.N., 2008. Blocking primers to enhance PCR amplification of rare sequences in mixed samples—a case study on prey DNA in Antarctic krill stomachs. Front. Zool. 5, 12. http://dx.doi.org/10.1186/1742-9994-5-12.

Vilà, M., Bartomeus, I., Dietzsch, A.C., Petanidou, T., Steffan-Dewenter, I., Stout, J.C., Tscheulin, T., 2009. Invasive plant integration into native plant-pollinator networks across Europe. Proc. R. Soc. B Biol. Sci. 276, 3887–3893. http://dx.doi.org/10.1098/rspb.2009.1076.

Vilcinskas, A., 2015. Pathogens as biological weapons of invasive species. PLoS Pathog. 11. e1004714. http://dx.doi.org/10.1371/journal.ppat.1004714.

Vilcinskas, A., Stoecker, K., Schmidtberg, H., Röhrich, C.R., Vogel, H., 2013. Invasive harlequin ladybird carries biological weapons against native competitors. Science 340, 862–863.

Vitousek, P.M., Mooney, H.A., Lubchenco, J., Melillo, J.M., 1997. Human domination of earth's ecosystems. Science 277, 494–499.

Von Berg, K., Traugott, M., Scheu, S., 2012. Scavenging and active predation in generalist predators: a mesocosm study employing DNA-based gut content analysis. Pedobiologia 55, 1–5. http://dx.doi.org/10.1016/j.pedobi.2011.07.001.

Ward, E.J., Semmens, B.X., Phillips, D.L., Moore, J.W., Bouwes, N., 2011. A quantitative approach to combine sources in stable isotope mixing models. Ecosphere 2, art19. http://dx.doi.org/10.1890/ES10-00190.1.

Watson, M., Thomson, M., Risse, J., Talbot, R., Santoyo-Lopez, J., Gharbi, K., Blaxter, M., 2015. poRe: an R package for the visualization and analysis of nanopore sequencing data. Bioinformatics 31, 114–115.

Webb, D.A., 1985. What are the criteria for presuming native status? Watsonia 15, 231–236.

Weir, L.A., Royle, J.A., Nanjappa, P., Jung, R.E., 2005. Modeling anuran detection and site occupancy on North American Amphibian Monitoring Program (NAAMP) routes in Maryland. J. Herpetol. 39, 627–639. http://dx.doi.org/10.1670/0022-1511(2005)039 [0627:MADASO]2.0.CO;2.

Willerslev, E., Cappellini, E., Boomsma, W., Nielsen, R., Hebsgaard, M.B., Brand, T.B., Hofreiter, M., Bunce, M., Poinar, H.N., Dahl-Jensen, D., Johnsen, S., Steffensen, J.P., Bennike, O., Schwenninger, J.-L., Nathan, R., Armitage, S., de Hoog, C.-J., Alfimov, V., Christl, M., Beer, J., Muscheler, R., Barker, J., Sharp, M., Penkman, K.E.H., Haile, J., Taberlet, P., Gilbert, M.T.P., Casoli, A., Campani, E., Collins, M.J., 2007. Ancient biomolecules from deep ice cores reveal a forested Southern Greenland. Science 317, 111–114. http://dx.doi.org/10.1126/science.1141758.

Willerslev, E., Davison, J., Moora, M., Zobel, M., Coissac, E., Edwards, M.E., Lorenzen, E.D., Vestergård, M., Gussarova, G., Haile, J., Craine, J., Gielly, L., Boessenkool, S., Epp, L.S., Pearman, P.B., Cheddadi, R., Murray, D., Bråthen, K.A., Yoccoz, N., Binney, H., Cruaud, C., Wincker, P., Goslar, T., Alsos, I.G., Bellemain, E., Brysting, A.K., Elven, R., Sønstebø, J.H., Murton, J., Sher, A., Rasmussen, M., Rønn, R., Mourier, T., Cooper, A., Austin, J., Möller, P., Froese, D., Zazula, G., Pompanon, F., Rioux, D., Niderkorn, V., Tikhonov, A., Savvinov, G., Roberts, R.G., MacPhee, R.D.E., Gilbert, M.T.P., Kjær, K.H., Orlando, L., Brochmann, C., Taberlet, P., 2014. Fifty thousand years of Arctic vegetation and megafaunal diet. Nature 506, 47–51. http://dx.doi.org/10.1038/nature12921.

Williams, R.J., Martinez, N.D., 2000. Simple rules yield complex food webs. Nature 404, 180–183.

Williams, R.J., Anandanadesan, A., Purves, D., 2010. The probabilistic niche model reveals the niche structure and role of body size in a complex food web. PLoS One 5. e12092. http://dx.doi.org/10.1371/journal.pone.0012092.

Willis, K.J., Birks, H.J.B., 2006. What is natural? The need for a long-term perspective in biodiversity conservation. Science 314, 1261–1265.

Wilson, E.E., Wolkovich, E.M., 2011. Scavenging: how carnivores and carrion structure communities. Trends Ecol. Evol. 26, 129–135. http://dx.doi.org/10.1016/j.tree.2010.12.011.

Wilson, E.E., Mullen, L.M., Holway, D.A., 2009. Life history plasticity magnifies the ecological effects of a social wasp invasion. Proc. Natl. Acad. Sci. U.S.A. 106, 12809–12813. http://dx.doi.org/10.1073/pnas.0902979106.

Wilson, E.E., Sidhu, C.S., LeVan, K.E., Holway, D.A., 2010a. Pollen foraging behaviour of solitary Hawaiian bees revealed through molecular pollen analysis: pollen analysis reveals bee foraging patterns. Mol. Ecol. 19, 4823–4829. http://dx.doi.org/10.1111/j.1365-294X.2010.04849.x.

Wilson, E.E., Young, C.V., Holway, D.A., 2010b. Predation or scavenging? Thoracic muscle pH and rates of water loss reveal cause of death in arthropods. J. Exp. Biol. 213, 2640–2646. http://dx.doi.org/10.1242/jeb.043117.

Wilson-Rankin, E.E., 2015. Level of experience modulates individual foraging strategies of an invasive predatory wasp. Behav. Ecol. Sociobiol. 69, 491–499. http://dx.doi.org/10.1007/s00265-014-1861-1.

Yachi, S., Loreau, M., 2007. Does complementary resource use enhance ecosystem functioning? A model of light competition in plant communities: light-use complementarity and complementarity index. Ecol. Lett. 10, 54–62. http://dx.doi.org/10.1111/j.1461-0248.2006.00994.x.

Yan, N.D., Girard, R., Boudreau, S., 2002. An introduced invertebrate predator (Bythotrephes) reduces zooplankton species richness. Ecol. Lett. 5, 481–485.

Yang, C.-C.S., Shoemaker, D.D., Wu, J.-C., Lin, Y.-K., Lin, C.-C., Wu, W.-J., Shih, C.-J., 2009. Successful establishment of the invasive fire ant *Solenopsis invicta* in Taiwan: insights into interactions of alternate social forms. Divers. Distrib. 15, 709–719. http://dx.doi.org/10.1111/j.1472-4642.2009.00577.x.

Yeakel, J.D., Novak, M., Guimarães, P.R., Dominy, N.J., Koch, P.L., Ward, E.J., Moore, J.W., Semmens, B.X., 2011. Merging resource availability with isotope mixing models: the role of neutral interaction assumptions. PLoS One 6. e22015. http://dx.doi.org/10.1371/journal.pone.0022015.

Yeakel, J.D., Bhat, U., Elliott Smith, E.A., Newsome, SD., 2016. Exploring the isotopic niche: isotopic variance, physiological incorporation, and the temporal dynamics of foraging. Front. Ecol. Evol. 4, 1. http://dx.doi.org/10.3389/fevo.2016.00001.

Yoccoz, N.G., 2012. The future of environmental DNA in ecology. Mol. Ecol. 21, 2031–2038.

Yoccoz, N.G., Bråthen, K.A., Gielly, L., Haile, J., Edwards, M.E., Goslar, T., Von Stedingk, H., Brysting, A.K., Coissac, E., Pompanon, F., SøNstebø, J.H., Miquel, C., Valentini, A., De Bello, F., Chave, J., Thuiller, W., Wincker, P., Cruaud, C., Gavory, F., Rasmussen, M., Gilbert, M.T.P., Orlando, L., Brochmann, C., Willerslev, E., Taberlet, P., 2012. DNA from soil mirrors plant taxonomic and growth form diversity: DNA from soil mirrors plant diversity. Mol. Ecol. 21, 3647–3655. http://dx.doi.org/10.1111/j.1365-294X.2012.05545.x.

Yoshida, K., Burbano, H.A., Krause, J., Thines, M., Weigel, D., Kamoun, S., 2014. Mining herbaria for plant pathogen genomes: back to the future. PLoS Pathog. 10. e1004028. http://dx.doi.org/10.1371/journal.ppat.1004028.

Yoshida, K., Sasaki, E., Kamoun, S., 2015. Computational analyses of ancient pathogen DNA from herbarium samples: challenges and prospects. Front. Plant Sci. 6, 771. http://dx.doi.org/10.3389/fpls.2015.00771.

Yu, J., Wong, W.-K., Hutchinson, R.A., 2010. Modeling experts and novices in citizen science data for species distribution modeling. In: IEEE, pp. 1157–1162. http://dx.doi.org/10.1109/ICDM.2010.103.

Zaiko, A., Martinez, J.L., Schmidt-Petersen, J., Ribicic, D., Samuiloviene, A., Garcia-Vazquez, E., 2015a. Metabarcoding approach for the ballast water surveillance—an advantageous solution or an awkward challenge? Mar. Pollut. Bull. 92, 25–34. http://dx.doi.org/10.1016/j.marpolbul.2015.01.008.

Zaiko, A., Samuiloviene, A., Ardura, A., Garcia-Vazquez, E., 2015b. Metabarcoding approach for nonindigenous species surveillance in marine coastal waters. Mar. Pollut. Bull. 100, 53–59. http://dx.doi.org/10.1016/j.marpolbul.2015.09.030.

Zarzoso-Lacoste, D., Bonnaud, E., Corse, E., Gilles, A., Meglecz, E., Costedoat, C., Gouni, A., Vidal, E., 2016. Improving morphological diet studies with molecular ecology: an application for invasive mammal predation on island birds. Biol. Conserv. 193, 134–142. http://dx.doi.org/10.1016/j.biocon.2015.11.018.

Zavaleta, E.S., Hobbs, R.J., Mooney, H.A., 2001. Viewing invasive species removal in a whole-ecosystem context. Trends Ecol. Evol. 16, 454–459.

Zazzo, A., Smith, G.R., Patterson, W.P., Dufour, E., 2006. Life history reconstruction of modern and fossil sockeye salmon (Oncorhynchus nerka) by oxygen isotopic analysis of otoliths, vertebrae, and teeth: implication for paleoenvironmental reconstructions. Earth Planet. Sci. Lett. 249, 200–215. http://dx.doi.org/10.1016/j.epsl.2006.07.003.

Zenetos, A., Koutsogiannopoulos, D., Ovalis, P., Poursanidis, D., et al., 2013. The role played by citizen scientists in monitoring marine alien species in Greece. Cah. Biol. Mar. 54, 419–426.

Zhang, A.-B., Muster, H.-B., Liang, C.-D., Crozier, R., Wan, P., Feng, J., Ward, R.D., 2012. A fuzzy-set-theory-based approach to analyse species membership in DNA barcoding. Mol. Ecol. 21, 1848–1863.

Zhao, L., Lu, M., Niu, H., Fang, G., Zhang, S., Sun, J., 2013. A native fungal symbiont facilitates the prevalence and development of an invasive pathogen–native vector symbiosis. Ecology 94, 2817–2826.

Zhou, X., Li, Y., Liu, S., Yang, Q., Su, X., Zhou, L., Tang, M., Fu, R., Li, J., Huang, Q., 2013. Ultra-deep sequencing enables high-fidelity recovery of biodiversity for bulk arthropod samples without PCR amplification. GigaScience 2, 4.

CHAPTER FOUR

Island Biogeography of Food Webs

F. Massol*,1, M. Dubart†,‡, V. Calcagno§, K. Cazelles¶,||,#, C. Jacquet¶,||,#, S. Kéfi‡, D. Gravel¶,||,**

*CNRS, Université de Lille, UMR 8198 Evo-Eco-Paleo, SPICI Group, Lille, France
†CNRS, Université de Lille-Sciences et Technologies, UMR 8198 Evo-Eco-Paleo, SPICI Group, Villeneuve d'Ascq, France
‡Institut des Sciences de l'Évolution, Université de Montpellier, CNRS, IRD, EPHE, CC065, Montpellier, France
§Université Côte d'Azur, CNRS, INRA, ISA, France
¶Université du Québec à Rimouski, Rimouski, QC, Canada
||Quebec Center for Biodiversity Science, Montréal, QC, Canada
#UMR MARBEC (MARine Biodiversity, Exploitation and Conservation), Université de Montpellier, Montpellier, France
**Faculté des Sciences, Université de Sherbrooke, Sherbrooke, QC, Canada
1Corresponding author: e-mail address: francois.massol@univ-lille1.fr

Contents

Advances in Ecological Research, Volume 56
ISSN 0065-2504
http://dx.doi.org/10.1016/bs.aecr.2016.10.004

Abstract

To understand why and how species invade ecosystems, ecologists have made heavy use of observations of species colonization on islands. The theory of island biogeography, developed in the 1960s by R.H. MacArthur and E.O. Wilson, has had a tremendous impact on how ecologists understand the link between species diversity and characteristics of the habitat such as isolation and size. Recent developments have described how the inclusion of information on trophic interactions can further inform our understanding of island biogeography dynamics. Here, we extend the trophic theory of island biogeography to assess whether certain food web properties on the mainland affect colonization/extinction dynamics of species on islands. Our results highlight that both food web connectance and size on the mainland increase species diversity on islands. We also highlight that more heavily tailed degree distributions in the mainland food web correlate with less frequent but potentially more important extinction cascades on islands. The average shortest path to a basal species on islands follows a hump-shaped curve as a function of realized species richness, with food chains slightly longer than on the mainland at intermediate species richness. More modular mainland webs are also less persistent on islands. We discuss our results in the context of global changes and from the viewpoint of community assembly rules, aiming at pinpointing further theoretical developments needed to make the trophic theory of island biogeography even more useful for fundamental and applied ecology.

1. INTRODUCTION

1.1 Island Biogeography

Islands have always fascinated ecologists. Since the earliest stages of ecology as a scientific discipline, the fauna and flora of islands have been considered as objects worthy of study because they capture the essence of colonization-extinction and ecoevolutionary dynamics shaping natural systems (Whittaker and Fernández-Palacios, 2007). Among the iconic rules of ecology, the "island rule" (Lomolino, 1985) suggests that the span of body sizes found on islands is much narrower than on continents, reflecting the remarkable examples of gigantism in small herbivores/granivores and of dwarfism in predators observed. Simple, general theoretical models of faunal build-up on islands were proposed some 50 years ago (Levins and Heatwole, 1963; MacArthur and Wilson, 1963). More recently, extensions and applications of these models have been made, constituting what has been dubbed "eco-evolutionary island biogeography" which incorporates trophic, functional and local adaptation information into the island biogeography framework (Farkas et al., 2015; Gravel et al., 2011b). Empirically, islands have provided ecologists with a playground to understand community assembly (Piechnik et al., 2008; Simberloff, 1976; Simberloff and Abele, 1976), species diversity (Condit et al., 2002; Ricklefs and Renner, 2012; Volkov et al., 2003) and metapopulation dynamics (Hanski, 1999; Ojanen et al., 2013).

The historical "theory of island biogeography" (TIB), presented independently by Levins and Heatwole (1963) and MacArthur and Wilson (1963), and summarized in the well-known book of the same name by MacArthur and Wilson (1967), is based on the idea that the ecological communities found on islands are a sample of those found on continents. The size of this "sample" results from two opposing processes: island colonization by external species and their local extinction on the island (see also Preston, 1962). The original formulation of the model by MacArthur and Wilson (1963) takes the form of a master equation linking the probability $P_S(t)$ that the focal island has exactly S species (from the original T total number of species found on the mainland) at time t to the rate of colonization by a new species (λ_S) and the rate of species extinction (μ_S):

$$\frac{dP_S}{dt} = \lambda_{S-1} P_{S-1} + \mu_{S+1} P_{S+1} - (\lambda_S + \mu_S) P_S \tag{1}$$

Adding the assumption of constant colonization and extinction rates (i.e. $\lambda_S = (T - S)c$ and $\mu_S = Se$, where c is the species colonization rate and e, their extinction rate), this naturally leads to the following equation on the expected species richness $\bar{S}$ at time t:

$$\frac{d\bar{S}}{dt} = \sum_S S \frac{dP_S}{dt} = c(T - \bar{S}) - e\bar{S} \tag{2}$$

Solved at equilibrium, Eq. (2) yields the expected number of species on the island at any given time:

$$\bar{S}^* = T \frac{c}{c + e} \tag{3}$$

Eq. (3) lends itself to the interpretation of species–area and species–distance curves (i.e. curves linking the number of species present on an island to the area of the island or its distance from the continent) as the log-derivative of Eq. (3) with respect to e/c yields:

$$\frac{d \log \bar{S}^*}{d (e/c)} = -\frac{1}{1 + e/c} \tag{4}$$

For large e/c, Eq. (4) reads approximately as:

$$\frac{d \log \bar{S}^*}{d (e/c)} \approx -1 \tag{5}$$

As long as e and c are supposed to be monotonic functions of area and distance, with the intuitive slopes (i.e. c increases with area and decreases with distance, and the opposite relationships hold for e) from the continent respectively, Eq. (5) yields species–area and species–distance curves compatible with observed patterns (MacArthur and Wilson, 1963, 1967).

The TIB is a cornerstone of invasion biology for several reasons. First, it is very likely that colonization by external species is a strong component of the forces shaping community composition on long time scales for islands, whereas for mainland communities random species extinction and larger population sizes make coevolution more likely to be the dominant factor structuring communities on the same time scales. In a sense, remote islands can be seen as complete population sinks for all species (i.e. black–hole sinks or nearly so) and, as such, have very little to no impact on the overall coevolutionary patterns observed in species (Holt et al., 2003; Kawecki, 2004; Massol and Cheptou, 2011; Rousset, 1999), except for island endemics. From an other point of view, islands can also be considered as natural experiments for understanding biological invasions. The dynamics of successive species colonization from the mainland to the island constitutes an "accelerated" version of what could happen in other less extinction-prone habitats. Understanding how waves of colonization events can take place on an island, depending on island and mainland community characteristics, will give information on the conditions favouring the invasibility of habitats by exotic species—be they habitat characteristics or species traits. In the case of the TIB, colonizing species are not assumed to be fundamentally different in terms of their ability to invade an island; thus, the variability in diversity on islands predicted by the TIB boils down to variability in colonization and extinction parameters among islands, which in turn is assumed to only depend on island remoteness and area (MacArthur and Wilson, 1963, 1967).

Predictions from the TIB can be tested in different ways. First, the species–area curves can be fitted to infer underlying extinction-to-colonization ratios and/or to test whether such ratios are indeed constant across different areas, during a given period or among taxa (Cameron et al., 2013; Guilhaumon et al., 2008; Triantis et al., 2012). While this approach has produced some success in the past, it is not a strong test per se as different theories can produce the same curves. Second, the TIB also predicts the value of the variance-to-mean ratio of species richness to be about one in species-poor island communities (MacArthur and Wilson, 1963). Indeed, taking the original master Eq. (1) and plugging $\lambda_S = (T - S)c$ and $\mu_S = Se$, one finds that:

$$\frac{d\overline{S^2}}{dt} = 2c(\overline{S}T - \overline{S^2}) - 2e\overline{S^2} + c(T - \overline{S}) + e\overline{S} \tag{6}$$

which, combined with Eq. (2), yields the differential equation for the variance $V[S]$ of S:

$$\frac{dV[S]}{dt} = \frac{d\overline{S^2}}{dt} - 2\overline{S}\frac{d\overline{S}}{dt} = c(T - \overline{S}) + e\overline{S} - 2(c + e)V[S] \tag{7}$$

Taken at equilibrium, Eq. (7) entails the following relation between mean and variance of S:

$$V[S] = \frac{c(T - \overline{S}) + e\overline{S}}{2(c + e)} \tag{8}$$

Plugging Eq. (3) into Eq. (8), one finally finds:

$$\frac{V[S]}{\overline{S}} = 1 - \frac{\overline{S}}{T} \tag{9}$$

which is approximately equal to 1 when $\overline{S} \ll T$.

A third prediction of the TIB concerns the distribution of species richness in equally small and remote islands: by taking the equilibrium solution of Eq. (1), iterating from the equation of P_0 and getting up to P_T, one finds:

$$P_S = \frac{c(T - S + 1)P_{S-1}}{eS} = \ldots = \binom{T}{S}\left(\frac{c}{e}\right)^S P_0 \tag{10}$$

where $\binom{T}{S}$ is the binomial coefficient $\dfrac{T!}{S!(T - S)!}$. With the sum of all P_S being equal to one, Eq. (10) leads to:

$$P_S = \binom{T}{S}\left(\frac{c}{c + e}\right)^S \left(\frac{e}{c + e}\right)^{T-S} \tag{11}$$

which is simply a binomial distribution with parameter $c/c + e$. Eq. (11) can also be found by considering that, at equilibrium, each species presence/ absence is a Bernoulli-distributed random variable, thus summing over all species ends up having species richness follow a binomially distributed random variable. Eq. (11) can be used to estimate the ratio $\alpha = c/e$ using different observations of species richness on the same island over time, with sufficiently temporally distant observations to remove the effect of temporal autocorrelation, or comparing the species richness of several equally small and remote islands.

Finally, a fourth prediction of the TIB comes from looking at the dynamics of P_S in vector format:

$$\frac{d\mathbf{P}}{dt} = \mathbf{M} \cdot \mathbf{P} \tag{12}$$

where $\mathbf{P}$ is the vector $(P_0, P_1,\ldots, P_T)$ and $\mathbf{M}$ is the matrix given by:

$$\mathbf{M} = \begin{pmatrix} -cT & e & & 0 \\ cT & -c(T-1)-e & & \\ & c(T-1) & \ddots & Te \\ 0 & & c & -Te \end{pmatrix} \tag{13}$$

Following Eq. (12), the solutions for $\mathbf{P}(t)$ beginning in state $\mathbf{P}_0$ at $t=0$ are given by:

$$\mathbf{P(t)} = e^{t\mathbf{M}} \cdot \mathbf{P}_0 \tag{14}$$

It is easy to check that the vector $\mathbf{P}^*$ given by Eq. (11) is in the kernel of $\mathbf{M}$, i.e. that $\mathbf{M} \cdot \mathbf{P}^* = \mathbf{0}$. As such, $e^{t\mathbf{M}} \cdot \mathbf{P}^* = \mathbf{P}^*$ at all time and thus one recovers that $\mathbf{P}^*$ is indeed the equilibrium distribution of species richness on the island. The nonzero eigenvalues of $\mathbf{M}$ are exactly equal to the sequence $\{-(c+e), -2(c+e),\ldots, -T(c+e)\}$, which entails that the typical time of convergence to steady state is $\tau_S = 1/(c+e)$; meanwhile, each species, once present on the island, has a typical time to local extinction equal to $\tau_X = 1/e$. Thus, based on Eq. (3), the TIB predicts the following relation between species richness at equilibrium and the ratio τ_S/τ_X:

$$\bar{S}^* = T\left(1 - \frac{\tau_S}{\tau_X}\right) \tag{15}$$

Such a prediction, which is equivalent to Eq. (3), can be tested based on time series of species colonization and extinction on islands, such as obtained from experimental defaunation data (Wilson and Simberloff, 1969) or long-term surveys.

Thus, as initially envisaged by its first proponents (Levins and Heatwole, 1963; MacArthur and Wilson, 1963, 1967), the TIB tries to explain species–area and species–distance curves on islands using a simple theory that assumes species accumulate due to colonization from the mainland and die out on islands at a constant rate. Although precisely formalized in mathematical terms, the TIB is not a mechanistic model per se as it assumes that extinction decreases with island area and that colonization decreases with distance between island and mainland. While these assumptions seem justified in general, the underlying mechanisms generating these relationships are omitted from the TIB. In particular, the links between island area, its underlying habitat heterogeneity (and hence its diversity in terms of potential species niches) and species carrying capacities (and thus extinction probabilities) have opposite effects on extinction rate, because habitat heterogeneity trades off with habitat size in an island of limited size (Allouche et al., 2012). Moreover, the TIB does not take species interactions into account in the sense that colonization and extinction rates are independent from current species composition on the island. Species are considered equivalent, but the model differs from neutral theory in that there is no competitive interaction (Hubbell's model assumes very strong preemptive competition). As a consequence, the TIB cannot predict any aspect of community structure based on species–specific attributes, such as the dominance by generalists, functional composition or successional sequence. Gravel et al. (2011b) therefore introduced an extension to the classic TIB called the trophic theory of island biogeography (TTIB) with the aim of predicting the variation in food web structure with area and isolation. When trying to model empirical data akin to mainland–islands datasets such as Havens' (1992), they found that the TTIB performs significantly better (in terms of statistical goodness–of–fit indicators) without introducing any new parameter into the TIB.

1.2 Spatial Food Webs

To understand how species coexist, it is necessary to understand how they interact. At the simplest level, species can be considered as competing under limiting factors (resources, predators, etc.). Competition need not be as simple as scramble resource competition. Indirect interactions through shared natural enemies (Holt, 1977; Leibold, 1996) or more generally any other species' abundances or occurrences (e.g. species recycling nutrients from detritus; Daufresne and Hedin, 2005) can also be experienced as limiting factors and result in "apparent" competition sensu lato. Competition is the only kind of interaction taken into account in most metacommunity models (Hubbell, 2001; Leibold et al., 2004; Massol et al., 2011). In general, however, species interactions are not only competitive. Food webs, i.e. networks of species that feed on one another, represent antagonistic trophic interactions among species and, as such, have to be taken into account as a structuring force behind coexistence patterns within communities. While other types of ecological interaction networks do exist (e.g. mutualistic trophic networks between plants, fungus and soil microbes, or antagonistic nontrophic networks among antibiotic-producing bacteria), food webs are ubiquitous and their complexity seems to be tightly associated with species richness, abundances, functioning and dynamics of communities (Chase et al., 2000; Cohen and Briand, 1984; Cohen and Łuczak, 1992; Downing and Leibold, 2002; Dunne et al., 2004; Kéfi et al., 2012; Pimm et al., 1991; Post et al., 2000).

Inspired by the work of Huffaker, recent work on the subject of food web dynamics has emphasized the necessity to consider food webs as spatialized entities (Duggins et al., 1989; Estes and Duggins, 1995; Estes et al., 1998; Huffaker, 1958; Huffaker et al., 1963; Polis and Hurd, 1995; Polis et al., 2004) in order to understand patterns such as species turnover, species richness, food chain length or nutrient recycling dynamics (Calcagno et al., 2011; Gravel et al., 2011a; Massol et al., 2011; McCann et al., 2005; Pillai et al., 2010, 2011; Takimoto et al., 2012). Considering food webs only as local, spatially disconnected entities may lead to misunderstanding, e.g. not recognizing population sinks because populations are maintained by the dispersal of detritus among habitat patches (Gravel et al., 2010). However, all species do not perceive space with the same grain, with habitat patches for one species not being the same as for another species at a different trophic level (Massol et al., 2011; McCann et al., 2005). Taking into account the spatial aspect of food webs is thus a necessity that requires understanding the complexity of spatial scales of species interactions. Acknowledging this spatial component of food webs is also a prerequisite to fully grasp the concept of limiting factors (Gravel et al., 2010; Haegeman and Loreau, 2014; Massol et al., 2011): when species and abiotic nutrients move from habitat patch to habitat patch, the effective "limitation" of species growth by an environmental factor depends not only on the local conditions but also on conditions in nearby patches contributing to influxes of organisms and nutrients into the focal patch. For these reasons, considering the spatial aspect of food webs is an important issue worthy of both theoretical and empirical development.

When considering food webs as spatialized entities, the fluxes of organisms and abiotic material between different locations present a duality of perspective of ecological fluxes (Massol and Petit, 2013; Massol et al., 2011): on the one hand, these fluxes contribute to the demographics of all species present in the different locations under study through immigration and emigration; more generally, such fluxes participate in the shaping of species and abiotic material stocks through source–sink dynamics (Loreau et al., 2013), while, on the other hand, the movement of plants, animals or simply abiotic material can be translated in fluxes of energy and nutrients that, together with other energy/nutrient fluxes due to local species interactions, represent the dynamics of nutrients and energy at large scales, flowing through trophic levels and between spatially distinct locations (in a manner rather reminiscent of Ulanowicz' ascendant perspective; Ulanowicz, 1997). Under this second perspective, there are common currencies (energy, carbon, nitrogen, phosphorus, etc.) behind each and every flux of organisms across the spatial food web that must comply with basic conservation laws (mass balance), thus constraining the possibilities of source–sink patterns among trophic levels (Loreau and Holt, 2004). Coupled with a consideration of species stoichiometric needs in terms of C:N:P ratios, such a perspective can potentially help us understand the spatial dynamics of nutrient enrichment through, for example death, reproduction, excretion and the foraging of organisms in different habitats (Hannan et al., 2007; Helfield and Naiman, 2002; Jefferies et al., 2004; Nakano and Murakami, 2001).

In the case of the TIB, the necessity of taking into account the spatial aspects of food webs has been construed as mandating the use of food web information in TIB models. The first model bridging food web theory and TIB was named the TTIB and makes use of trophic information for which species preys on what to correct effective colonization and extinction rates (Cazelles et al., 2015b; Cirtwill and Stouffer, 2016; Gravel et al., 2011b). In this first model, this correction takes a very simple form: species cannot colonize an island when they have no prey on the island, and the extinction of the prey species of a colonizer will also extirpate it from the island (Holt, 2002). More generally, this framework can be extended to account for arbitrary changes in colonization and extinction probabilities that depend on the presence or absence of other species on the island (Cazelles et al., 2015b). Thus, the TTIB can represent a complex picture of rates of transitions from one community state (i.e. the set of species occurring on the island) to another, through species colonization and extinction (Morton and Law, 1997; Slatkin, 1974). As illustrated in Fig. 1, the TIB and

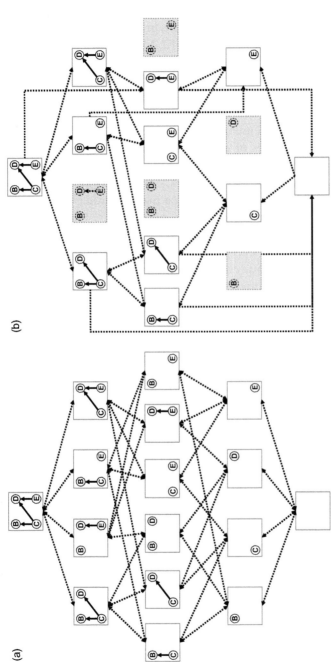

Fig. 1 Representations of (a) the TIB and (b) the TTIB in terms of possible transitions among communities, following the framework of Morton and Law (1997). The total community (with species B, C, D and E) is represented at the *top of both panels*, while the empty community is at the *bottom*. *Arrows* represent either colonization (upwards) or extinction (downwards) events. In the case of the TIB (a), all transitions are reversible and species richness does not "leap backwards" as species are lost one by one. In the case of the TTIB (b), certain communities (in grey with dotted borders) are impossible because they violate the principle of at least one prey species sustaining a predator species; moreover, the extinction of a single species can lead to losing more than one species, e.g. species C getting extinct entails losing both species C and B, and thus there is a possible transition between the total community towards the (D, E) community. While this scheme highlights the conceptual differences arising from taking into account the principle of sequential, bottom-up, food web assembly, more flexible approaches can be developed, e.g. by allowing predator species to sustain themselves in the absence of preys (albeit with more difficulty) or allowing predator species to invade the community "in advance" of their prey (Cazelles et al, 2015b).

TTIB are extremes of a continuum of approaches that weight differently transitions among possible community states.

1.3 Invasions in Food Webs, Eco-evolutionary Perspectives

Community assembly, broadly envisioned as the accumulation of species diversity over time in a novel habitat, has been modeled, classically, as an ecological process driven by sequential colonizations of species from a regional pool. This approach effectively considers species as having fixed characteristics and traits (e.g. prey range, feeding preferences, antipredator defences, etc.). This form of community assembly produces what are often called "invasion-structured" food webs (Rummel and Roughgarden, 1985). Invasion-structured food webs may be an adequate concept for ecosystems that are well connected to mainland, in which immigration events and invasions occur at a frequent pace. However, when considering more isolated areas, such as remote islands characterized by a very low colonization rate (low c value in the TIB framework), species may undergo a significant amount of evolutionary change in the interval of time between successive species arrivals. In such conditions, it is no longer legitimate to consider community assembly as a purely ecological process, and one needs to consider the effect of species evolution as well. The process of community assembly should thus be considered as involving three different timescales:

(i)   a rapid ecological timescale corresponding to community dynamics following species invasion;

(ii) a slow immigration timescale corresponding to the arrival of individuals from the main source of species;

(iii) a slow timescale corresponding to species (co)evolution in-between immigration events.

The relative speed of the last two timescales will be contingent on the level of isolation of the focal community directly through the rate of species colonization and indirectly through the amount of gene flow (slowing down local adaptation in the colonized community) and local selective pressures.

We can consider two extreme cases of community assembly: (i) where immigration events are common enough to completely outpace the effects of in situ coevolution, we recover the classic "invasion-structured" food webs; and, (ii) where evolution has plenty of time to proceed in-between invasion events, we may obtain "adaptive radiations", i.e. the in situ formation of new species by evolution (see also Vanoverbeke et al., 2016). This second extreme case is sometimes called "evolutionary community assembly" (Bonsall et al., 2004; Brännström et al., 2012; Doebeli and Dieckmann, 2000; HilleRisLambers et al., 2012; Loeuille and Leibold, 2014; Pillai and Guichard, 2012; Tokita and Yasutomi, 2003). Evolutionary community assembly has been mostly applied to competitive communities, but some studies have considered evolutionary diversification in the context of food webs. Doebeli and Dieckmann (2000) have shown how coevolution could yield diversification and greater specialization in predator–prey interactions. In the same vein, Loeuille and Leibold (2008) explored how the evolution of specific vs nonspecific (but costly) plant defences affects the topology of plant-herbivore food web modules. While these approaches treated trophic positions as fixed, and evolution occurred within trophic levels, more general approaches have also been employed. Rossberg et al. (2006) considered the coevolution of foraging and vulnerability traits in an abstract set of species and could reproduce interaction matrices similar to that of natural food webs (see also Drossel et al., 2004 for similar approaches). Loeuille and Loreau (2005) and Allhoff et al. (2015), using dynamical models of trophic interactions based on body mass, have studied how the level of trophic structuring and a number of trophic levels emerge from single ancestor species. In the case of mutualistic networks, Nuismer et al. (2013) assessed how the coevolution of mutualistic partners along a phenotypic trait continuum that governs both local adaptation and partner match/mismatch affected the topology of the mutualistic network.

Although studying the two extreme cases of community assembly is useful, it should be kept in mind that most natural food webs are likely not assembled by only invasions or only evolution, but probably by a combination of the two forces. Hence most ecosystems should harbour "coevolution-structured" food webs, using the terminology coined by Rummel and Roughgarden (1985). Unfortunately, comparatively few studies have examined the simultaneous action of sequential invasions and in situ coevolution on the dynamics of community assembly. Moreover, these studies have either only considered generalized Lotka–Volterra (or replicator) equations (Tokita and Yasutomi, 2003) or competitive interactions (Rummel and Roughgarden, 1985) and not trophic interactions. To some extent, predictions from models of asymmetric competitive interactions might be extrapolated to trophic interactions, but more work is needed to explore the interaction of invasion and coevolution (both directional evolution and diversification) in assembling food webs.

A few general conclusions can nevertheless be drawn. First, invasion- and coevolution-assembled communities can possess very different characteristics (Rummel and Roughgarden, 1985; Tokita and Yasutomi, 2003). It is therefore crucial to understand how the type of assembly history affects the resistance/resilience of food webs, in particular their susceptibility to invaders and their ability to sustain functioning following invasions (Romanuk et al., 2017). Second, the impact of evolution in-between invasion events is threefold:

(i) the evolution of species traits (e.g. body mass) can result in a rewiring of the network of ecological interactions (e.g. greater specialization/ clustering, less generalism and omnivory);

(ii) these changes of interactions might, ultimately, cause (delayed) species extinctions throughout the food web;

(iii) on longer timescales, coevolution might cause species diversification, and thus entail (delayed) species additions throughout the network.

Finally, a general finding is that coevolutionary trajectories will feature more "loops", i.e. would more often engage in cyclic successions of community states (for a theoretical example, see Loeuille and Leibold, 2014). This tendency of coevolution to generate sustained change in species traits and composition is a general property of asymmetric competition models ("taxon-cycle"; Nordbotten and Stenseth, 2016; Rummel and Roughgarden, 1985) that seems to extend to trophic interactions (Allhoff et al., 2015; Dieckmann et al., 1995) and that relates to the well-known Red Queen dynamics between hosts and parasites/pathogens (Boots et al., 2014; Gandon et al., 2008; Salathé et al., 2008).

Returning to the assembly trajectories depicted in Fig. 1, this would mean that taking into account the effect of species coevolution could: (i) alter the number and identity of species that go extinct following an invasion; (ii) allow more than one species to appear following an invasion, thus introducing longer upwards links; and (iii) introduce more cycles in the assembly graphs. All these effects seemingly make the assembly-trajectory graphs more complex, adding to the challenge of analysing and predicting food web dynamics. However, evolutionary changes might also make certain pathways much more likely than others, thus making assembly trajectories in effect less variable, more predictable.

1.4 Invasions in Other Spatially Structured Networks

Despite focusing here on trophic interactions, we must acknowledge that other types of networks can result in particular colonization/extinction dynamics when merged with the TIB (MacArthur and Wilson, 1963) or with Levins' metapopulation model (Levins, 1969). For instance, Fortuna and Bascompte (2006) have argued that representing patch–occupancy dynamics of plant–animal mutualistic networks can help understand the effects of habitat loss on the observed structure of those networks (see also Astegiano et al., 2015; Fortuna et al., 2013). In practice, these effects stem from the asymmetry in the mutualistic network, in which one type of mutualistic partners mandatorily needs the other type to be able to colonize a patch (e.g. pollinators in need of plants as local trophic resources), while the other type of partners benefits from the interaction at a larger scale through an increase in its colonization ability (e.g. plant reproduction enhanced by animal-mediated pollination).

More generally, we can analyse the potential effects of species interaction on TIB-like dynamics through a translation of Nee and May's (1992) generalization of Levins' (1969) model to the case of island biogeography dynamics. We can describe the dynamics of the probability p_i that a given species i is present on the focal island as:

$$\frac{dp_i}{dt} = c_i(h_i - p_i) - e_i p_i \tag{16}$$

where c_i denotes species colonization rate, h_i denotes the probability that the island is hospitable to species i and e_i is species extinction rate once present on the island. Any type of species interaction can influence Eq. (16) for any species by making these three parameters depend on the expected state of the network on the island. For instance, in the case of the TTIB (Gravel et al., 2011b), h_i and e_i both change in time due to expected successive colonizations of the island by different species—successive arrivals of species increase h_i for species at all trophic levels (the more species are present on the island, the more likely it is that any species not yet present on the island will find a suitable prey species), while increase in species richness with time tends to decrease expected e_i for all species.

In the case of mutualistic networks, one must distinguish the effect of network structure on obligate vs facultative mutualistic partners (Fortuna and Bascompte, 2006). For obligate mutualistic partners, network structure is usually expressed through a strong dependency of h_i on current island network, i.e. an obligate mutualist cannot invade an island that does not contain at least one its partners. It also underlies a strong dependency of e_i on current island network structure because, as with predators in the TTIB, obligate mutualists are expected to go extinct with their last mutualistic partner on the island. In the case of facultative mutualists, the weak dependency of the species on the presence of mutualistic partners on the island could be translated in one of two ways, either in: an increase in c_i when mutualists are present on the island (e.g. plants more likely to invade an island on which the animal seed dispersers are already present); and/or a decrease of e_i in the presence of beneficial partners (e.g. hermaphroditic plants on an island with pollinators can maintain larger, more genetically diverse, and thus less extinction-prone, populations). It is remarkable that, in the case of Levins' metapopulation model describing the dynamics of plant–pollinator interaction networks (Astegiano et al., 2015; Fortuna and Bascompte, 2006), a natural choice for the effect of pollinators on plant parameters is to consider that patches with pollinators contribute more to the pool of propagules (i.e. increase their contribution to the overall colonization × occupancy rate).

In the particular case of networks of competitive interactions, TIB dynamics might be altered through four different mechanisms. First, the presence of competitors, especially dominant competitors, can drive other species out of the island, i.e. by increasing their extinction rate e_i. This would mimic competitive exclusion dynamics on the island. An early attempt at understanding how competition would affect species occurrence patterns in a TIB framework was made by Hastings (1987). A second possibility is that there is a priority effect, affecting those species colonizing the island, so that certain species might exclude other species from colonizing the island when they are present, i.e. effectively decreasing h_i for the other species (Shurin et al., 2004). A third possibility is to make habitat quality change with current community composition, as a consequence of ecosystem engineering (Wright et al., 2004). Finally, there might be a gradient of competitive and colonization abilities in the pool of species, so that species more likely to colonize the island first would also be more likely to be displaced by more competitive species afterwards, causing "ecological succession-like" assembly trajectories. Although the competition–colonization trade-off has been traditionally used in the context of Levins' (1969) metapopulation model (Calcagno et al., 2006; Hastings, 1980; Tilman, 1994), it can also be incorporated in an island biogeography context to describe community transitions before and after colonization by a superior competitor, e.g. borrowing from Slatkin's (1974) model and adapting it to the TIB.

For the remainder of this chapter, we cast the TTIB in terms of classic food webs, i.e. networks of species feeding on one another through predator–prey interactions. However, we would note that integrating host–parasite (or more generally, host-symbiont) interactions within this context is feasible, provided Eq. (16) accommodates the effect of the network structure on symbiont colonization and extinction dynamics. One modelling choice is to consider that obligate symbionts cannot colonize if their host is absent, so that the symbiont can: (i) colonize the island together with its host; or, (ii) colonize the island once its host is already there (through some immigration of infected hosts which is not accounted for in the dynamics of the host occupancy because it has already colonized the island). In the case of facultative symbionts, this modelling option can be combined with the possibility of a symbiont-only colonization event, possibly occurring at a different (lower) rate. Regarding extinction rates, obligate symbionts will inevitably die out when their host goes extinct on the island, while facultative symbionts could persist (but possibly with an increased extinction rate) after their last host species disappears from the island.

2. ISLAND BIOGEOGRAPHY OF FOOD WEBS

2.1 The Model

2.1.1 Explaining the TTIB With a Simple Example

The TTIB can be understood using a simple example consisting in five species (A, B, C, D, E), arranged in three trophic levels, with species C and E being basal species, species B and D primary consumers and species A the top predator (Fig. 2). The core idea of the TTIB is that colonization and extinction of species on the island obey the principle of "at least one prey species per predator". In practical terms, this principle can be translated as follows:

1. Colonization of a given species can only take place when at least one of its prey species is present (Fig. 3). This means that the colonization rate is now modulated by the probability that the island food web contains at least one prey species for the focal species;

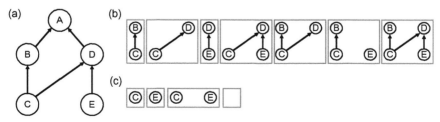

Fig. 2 Simple food web used to illustrate the Trophic Theory of Island Biogeography. (a) The complete food web (on the mainland) consists in five different species (A, B, C, D, E), (b) subfood webs that species A can colonize and (c) subfood webs that species A cannot colonize.

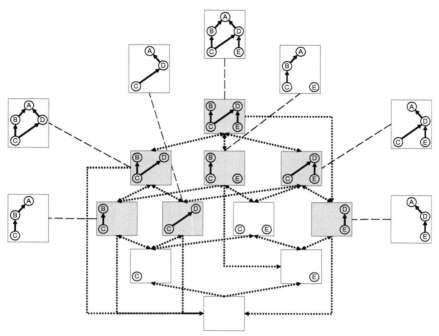

Fig. 3 Restriction of colonization under the TTIB. Transitions among community states that do not include species A are depicted by *dotted arrows*, as in Fig. 1. Communities in grey represent communities that can be colonized by species A (i.e. the fraction $h_A - P_{X_A=1}$ of the possible island communities). The associated species A-colonized communities (the "outer communities", i.e. the fraction $P_{X_A=1}$ of the possible island communities) are linked to them with a *thick dashed l*.

2. Extinction of a given species now depends not only on the single-species extinction rate of the focal species, but also on the rate at which the focal species gets caught in an "extinction cascade" (Fig. 4). Extinction cascades happen when a species "down", or lower, in the local food web supports a whole portion of the food web and goes extinct (e.g. species C in the (A, B, C, D) community in Fig. 4).

For a general idea of the TTIB dynamics, we can write that the probability of occurrence of a given species Z obeys the following differential equation (rewriting Eq. 16):

$$\frac{dP_{X_Z=1}}{dt} = c(h_Z - P_{X_Z=1}) - e_Z P_{X_Z=1} \qquad (17)$$

where h_Z is the probability that island community is hospitable to species Z (i.e. community contains at least one prey of species Z) and e_Z is the effective

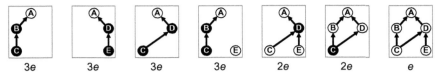

| 3e | 3e | 3e | 3e | 2e | 2e | e |

Fig. 4 Increase in the extinction rate of species A as a result of cascading extinctions. Each of the panels represents one the community that includes species A (the "outer communities" in Fig. 3). Species in *black dots* are species that contribute to the additional extinction rate of species A. For instance, in the (A, B, C) community, if species B or species C goes extinct, species A must also go extinct. The mention under each panel is the extinction rate of species A in the configuration depicted in the matching panel, e.g. in the above mentioned example, the rate of extinction of species A is 3e because it goes extinct at its own rate plus that of species B and that of species C. In this particular case, the total extinction rate of species A can be deduced from that of species B because if species C goes extinct, then species B must also go extinct. Thus, in this instance, the extinction rate of A is simply that of species B plus e. However, this additivity of extinction rate is not always in place. For instance, in the (A, B, C, D) community, species B and species D both experience a "2e" extinction rate, but species A only experiences a "2e" extinction rate as well because the single extinction of either species B or D has no effect on A, while the extinction of C makes the whole food web collapse.

extinction rate of species Z computed using e, the number of ways it can go extinct through a single-species extinction in the island food web and the weights associated with these extinction events (which are computed from the occurrence probabilities of species "down" the food web).

Let us start this example using the mainland food web given in Fig. 2a. Species A can only colonize a restricted set of island food webs (Figs 2b and 3). In other island food webs, it simply cannot settle because there is neither prey species B nor D (Fig. 2c). Once species A is on the island, the community must be in one of the "outer community" states of Fig. 3. The extinction rate of species A when established in one of these outer communities depends on how many species down the food web may provoke an extinction cascade affecting species A (Fig. 4). For instance, when only one species down the food web can provoke the extinction of species A through cascading extinction, then the extinction rate of species A is $2e$; if two species can provoke such an extinction (e.g. if the community is (A, B, C)), then the extinction rate of species A is $3e$.

Although the graphical representation of transitions among communities and colonization-prone communities for a focal species is useful to fully grasp the principles of the TTIB, a more quantitative approach can be obtained by focusing on the master equation behind food web dynamics.

The following explanation starts with the corresponding master equation in the TIB and then introduces an equivalent formulation in the case of the TTIB.

The TIB model, as expressed using Eqs (1) and (2), does not distinguish species based on any feature. However, the underlying random variable describing the number of species present on the island, S, can be decomposed as a sum of indicator variables X_i which describe the presence/absence of species i, so that at all times:

$$S(t) = \sum_i X_i(t) \tag{18}$$

Under the TIB, each of the $X_i(t)$ is a random variable the value of which changes from 1 to 0 with rate e and from 0 to 1 with rate c. The corresponding master equation for a single species is thus given by two coupled differential equations (indices i are omitted for the sake of clarity):

$$\frac{dP_{X=0}}{dt} = eP_{X=1} - cP_{X=0} \tag{19a}$$

$$\frac{dP_{X=1}}{dt} = cP_{X=0} - eP_{X=1} \tag{19b}$$

Noting $P_{X=1} = p$ and $P_{X=0} = 1 - p$, we obtain the well-known equation for the occupancy of islands by a single species under MacArthur and Wilson's framework:

$$\frac{dp}{dt} = c(1 - p) - ep \tag{20}$$

When compared with the framework set by Eqs (16) or (17), Eq. (20) means that: (i) there are no effects of network structure on any one of the three parameters of Eq. (16); and (ii) the probability that an island is hospitable to colonization is always 1 (i.e. there is no restriction to species colonization potential).

The stationary distribution of X following Eq. (19) is a Bernoulli distribution of parameter $c/(c + e)$, associated with the eigenvalue 0 of the matrix defining the process defined in Eq. (19). In other words, solving for the equilibrium of Eq. (19) is equivalent to finding the eigenvalues and eigenvectors of a 2×2 matrix, and the equilibrium is given by the eigenvector associated with the eigenvalue 0. The other eigenvalue, $-(c + e)$, is associated with the eigenvector $(-1, 1)$ (i.e. any discrepancy in the probability of occurrence of the species from the stationary distribution of X), so that an initially absent species at time $t=0$ has a probability of occurrence at time t equal to:

$$P_{X(t)=1|X(0)=0} = \frac{c}{c+e}\left(1 - e^{-(c+e)t}\right) \tag{21a}$$

while an initially present species at time $t=0$ has occurrence probability:

$$P_{X(t)=1|X(0)=1} = \frac{c}{c+e} + \frac{e}{c+e}e^{-(c+e)t} \tag{21b}$$

Now let us proceed with the simple four species community (B, C, D, E) given in Fig. 1 under the TIB, i.e. without using the principle of "at least one prey species per predator species" on the island. Each of the P_X, where X now represents a community, obeys a differential equation similar to Eq. (19b), with losses due to both extinction of local species and colonization by new species, and gains due to "upwards" transitions from species-poor communities and "downwards" transitions from species-rich ones. For instance, the equation for the community (D, E) is:

$$\frac{dP_{DE}}{dt} = c(P_D + P_E) + e(P_{BDE} + P_{CDE}) - 2(c+e)P_{DE} \tag{22}$$

More generally, noting **P** the vector of all P_X, the master equation can be written as:

$$\frac{d\mathbf{P}}{dt} = \mathbf{G} \cdot \mathbf{P} \tag{23}$$

where **G** is a matrix describing all the coefficients of the master equation. In the case of the TIB acting on the food web described in Fig. 1a, matrix **G** is given in Table 1.

Solving the equation $\mathbf{G} \cdot \mathbf{P} = 0$ (i.e. finding a vector of sum equal to 1 associated with the eigenvalue 0 of matrix **G**) yields the probability of finding the food web in the different states. In the case of the TIB, the result is somehow easy to find without having to resort to the study of matrix eigenvalues and eigenvectors; Eq. (11) already gives the probability of finding a community with exactly S species present. Dividing this expression by the number of combinations of S species among T yields the following general formula:

$$P_X = \left(\frac{c}{c+e}\right)^{|X|}\left(\frac{e}{c+e}\right)^{T-|X|} \tag{24}$$

Table 1 Matrix **G** of Eq. (23) in the Case of the TIB Acting on the Food Web Presented in Fig. 1

To	Empty	B	C	D	E	BC	BD	BE	CD	CE	DE	BCD	BCE	BDE	CDE	BCDE
										From						
Empty	$-4c$	e	e	e	e											
B	c	$-3c-e$				e	e	e								
C	c		$-3c-e$			e			e	e						
D	c			$-3c-e$			e		e		e					
E	c				$-3c-e$			e		e	e					
BC		c	c			$-2(c+e)$						e	e			
BD		c		c			$-2(c+e)$					e		e		
BE		c			c			$-2(c+e)$					e	e		
CD			c	c					$-2(c+e)$			e			e	
CE			c		c					$-2(c+e)$			e		e	
DE				c	c						$-2(c+e)$			e	e	
BCD						c	c		c			$-c-3e$				e
BCE						c		c		c			$-c-3e$			e
BDE							c	c			c			$-c-3e$		e
CDE									c	c	c				$-c-3e$	e
BCDE												c	c	c	c	$-4e$

Empty cells are equal to zero (omitted for clarity). Columns indicate "giver" community states, *rows* indicate "receiver" community states. The sum of each column equals zero because the sum of incoming and outgoing state transition rates must balance (the sum of the probability of all states must always be equal to 1). A single equation such as Eq. (22) can be retrieved by following the matching row of the matrix and adding/subtracting appropriate terms based on matrix coefficients on that row.

where $|X|$ is the cardinality of community X (i.e. its species richness). The probability that a single species, say C, is present on the island can be obtained by summing Eq. (24) over all communities that include species C:

$$P_{X_C=1} = \sum_{C \in X} P_X = \sum_{k=0}^{T} \sum_{\substack{C \in X \\ |X|=k}} \left(\frac{c}{c+e}\right)^k \left(\frac{e}{c+e}\right)^{T-k}$$

$$= \sum_{k=1}^{T} \binom{T-1}{k-1} \left(\frac{c}{c+e}\right)^k \left(\frac{e}{c+e}\right)^{T-k} = \frac{c}{c+e} \quad (25)$$

We now shift to the case of the TTIB, using the example given in Fig. 1b, i.e. the same food web as the one used above, but with an intrinsic dependency between species occurrences due to the underlying TTIB principle. In this context, certain communities cannot exist, i.e. (B), (D), (B, D), (B, E), (B, D, E) (grey communities in Fig. 1b). The associated $\mathbf{G}$ matrix is given in Table 2.

Solving $\mathbf{G} \cdot \mathbf{P} = \mathbf{0}$ using the $\mathbf{G}$ matrix given in Table 2 yields complicated expressions. However, the same type of computations as the ones used to go from Eq. (24) to Eq. (25) can be applied to this stable distribution of community states to obtain the stable distribution of each species occurrence probability. For instance, one obtains that the probability of observing species B is given by:

$$P_{X_B=1} = \frac{c^2}{(c+e)(c+2e)} \quad (26)$$

The food chain (B, C) being particularly simple (see also Section 2.2.1), this result is easy to interpret: the probability of occurrence of species B relies on species C being present (with probability $c/(c+e)$ as species C is a basal species), and species B having colonized (rate c), and the whole food chain not collapsing (rate of species B extinction $2e$).

2.1.2 A More General Presentation of the TTIB

A simple consequence of the rules governing the dynamics of the TTIB is that the two variables needed to compute the dynamics of a given species Z, h_Z and e_Z (Eq. (17)) can be obtained based only on the probability of occurrence of species "down" the food web. In other words, looking "up" the food web (i.e. at species that depend on Z for colonization) or "laterally" (i.e. at species that have no dependence relation with Z for their colonization or for species Z colonization) is not needed when assessing h_Z and e_Z. This

Table 2 Matrix **G** of Eq. (23) in the Case of the TTIB Acting on the Food Web Presented in Fig. 1

To		From									
	Empty	**C**	**E**	**BC**	**CD**	**CE**	**DE**	**BCD**	**BCE**	**CDE**	**BCDE**
Empty	$-2c$	e	e	e	e		e	e			
C	c	$-3c-e$		e	e	e					
E	c		$-2c-e$			e	e		e		
BC		c		$-2(c+e)$				e	e		
CD		c			$-2(c+e)$			e		e	
CE		c	c			$-2(c+e)$			e	e	
DE			c				$-c-2e$			e	e
BCD				c	c			$-c-3e$			e
BCE				c		c			$-c-3e$		e
CDE					c	c	c			$-c-3e$	e
BCDE								c	c	c	$-4e$

Empty cells are equal to zero (omitted for clarity). Interpretation of this table follows the same legend as Table 1.

means that, when focusing on species Z, we can focus on species Z and the species it depends on to colonize (the ones "down" the food web), and thus describe the dynamics of community states forgetting about the occurrence of all the other species in the mainland food web. For instance, following the example given above (Fig. 3), if we were to focus on species B, the only community states to focus on would be the empty community, the one with species C and the one with species (B, C). When forgetting about the rest of the food web, the "empty state" refers to any state in which both species B and C are absent, the "C community" refers to all states in which species C is present but not species B, and the "(B, C) community" refers to all states in which both species are present. In this instance, the (B, C, D) community would "count" under the (B, C) modeled community state. In the following, we present a new analytical derivation of the expressions for h_Z and e_Z under the assumptions of the TTIB, which was not provided by Gravel et al. (2011b).

This can be formalized more rigorously using notations that will help us navigate the set of possible communities:

- Species Y is a **foundation species** for species Z if Y is part of at least one path linking a basal species to species Z. The set of all foundation species for species Z is noted F_Z.
- Among foundation species for species Z, we note G_Z the set of prey species of species Z. For convenience, we also note $H_Z = F_Z \cup \{Z\}$, i.e. the set of foundation species and the focal species itself. The intuitive notion of species being "up" or "down" the food web can be understood through the following statement: species Y is in H_Z if and only if H_Y is included in H_Z.
- We will call a community, K, **TTIB-compatible** when all species in the community are connected to at least one basal species in the community by at least one path of species present in the community.
- The trimmed community, $\lfloor K \rfloor$, is obtained by removing the minimal number of species from community K so that the community obtained is TTIB-compatible.
- We note Ω the **set of all TTIB-compatible communities** containing from 0 to T species of the mainland food web.
- We note Ω_Z the set all TTIB-compatible communities that include species Z.
- For TTIB-compatible community K, we note Φ_K its **TTIB-compatible combination set**, i.e. the set of all TTIB-compatible communities consisting only of combinations of species in K. Naturally, $K \in \Phi_K$.

- **The expansion set** of TTIB-compatible community K, noted $\langle K \rangle$, is the set of all TTIB-compatible communities that naturally expand community K with the addition of any number of species not found in K. K is always in its expansion set and it is the smallest community in this set.
- By extension, we will note the expansion set of species Z, $\langle Z \rangle$, the set consisting in the union of all expansions of TTIB-compatible communities containing Z. Quite intuitively $\langle Z \rangle = \Omega_Z$ as any TTIB-compatible community containing species Z is in its own expansion set.
- The expansion set of K constrained by community D (not necessarily TTIB-compatible), noted $\langle K \rangle_D$, corresponds to natural expansions of K that do not include species that are part of D. As a special case, $\langle \varnothing \rangle_D$ refers to the D-constrained expansion of the empty community, i.e. the set of all communities that do not include species in community D.
- Extending the probability measure introduced on communities, the **measure P_U of the set of communities** U is equal to the sum of the P_C of all communities K in the set U.
- Finally, we introduce the indicator function for species Z in community K, noted $\mathbf{1}_Z(K)$, which is equal to one if and only if species Z is part of community K.

A consequence of the bottom–up control of species presence-absence under the TTIB framework is that the probability of species Z presence, also equal to $P_{\langle Z \rangle}$, can be computed based on the knowledge of all the $P_{\langle K \rangle_{H_Z}}$ for all communities K in Φ_{F_Z}. Taking again the example given above (Figs 2–4), assessing the probability that species A occurs on the island can be done by acknowledging that (B, C, D, E) is the set of foundation species of A, F_A (and thus $(A, B, C, D, E) = H_A$), and working on the measures (probabilities) of $\langle \varnothing \rangle_{H_A}$, $\langle (C) \rangle_{H_A}$, $\langle (B, C) \rangle_{H_A}$, etc., to find the measures of $\langle (A, B, C) \rangle_{H_A}$, $\langle (A, D, E) \rangle_{H_A}$, etc., which, together, yield $P_{\langle A \rangle}$, also equal to $P_{X_A=1}$.

Using the above notations, the probability that an island contains at least one prey of species Z, h_Z, is given by:

$$h_Z = P_{\underset{Y \in G_Z}{\cup} \Omega_Y} = P_{\underset{Y \in G_Z}{\cup} \langle Y \rangle} \qquad (27)$$

Indeed, $\underset{Y \in G_Z}{\cup} \Omega_Y$ corresponds to the union of all sets of TTIB-compatible communities that include at least one prey species of species Z. It is also equal to the union of expansion sets of all species in G_Z, $\underset{Y \in G_Z}{\cup} \langle Y \rangle$. Eq. (27) can also be rewritten by separating all communities K that allow colonization by species Z, i.e. all communities, possibly expanded (with a constraint on species in F_Z to avoid counting a community more than once), which are in Φ_{F_Z} but not in $\Phi_{F_Z \backslash G_Z}$ (consisting in Z-foundation species but with at least one prey species of species Z):

$$h_Z = \sum_{K \in \Phi_{F_Z} \backslash \Phi_{F_Z \backslash G_Z}} P_{\langle K \rangle_{F_Z}} \tag{28}$$

In other words, to find a community that is hospitable to species Z, one must first find a TTIB-compatible community made only of species that are foundation species for species Z and which include at least one prey of species Z, and then expand this community with species that are not foundation species for species Z (i.e. "decorative species" with respect to species Z colonization capacity).

In the same way, the extinction rate of species Z, e_Z, is given by:

$$e_Z = e + e \sum_{K \in \Omega_Z} \frac{P_C}{P_{\langle Z \rangle}} \sum_{Y \in K \cap F_Z} (1 - \mathbf{1}_Z[\lfloor (K \cap H_Z) \backslash \{Y\} \rfloor]) \tag{29}$$

In Eq. (29), species Z can go extinct by itself (the first e term) and also by losing a species that is necessary for its maintenance (the sum term). Enumerating these cases is made on all communities that contain Z (i.e. on $C \in \Omega_Z$); the probability to be in community state K, given that species Z is present, is $P_K / P_{\Omega_Z} = P_K / P_{\langle Z \rangle}$. Once we know that the community state is K, then we have to enumerate all the ways in which species Z can go extinct through an extinction cascade; this can only happen when one of the foundation species of species Z, i.e. a species $Y \in F_Z$ which is also present in community K (hence, $Y \in K \cap F_Z$), is such that removing it would also remove species Z from $K \cap H_Z$ after trimming "dead branches" in the ensuing community. Working on $K \cap H_Z$ (and not $K \cap F_Z$) is necessary. Indeed, one can think of situations in which removing one species from $K \cap F_Z$ might make community K TTIB-incompatible, but not by severing the path between Z and basal species. For instance, if we take species B in the last panel of Fig. 4 and imagine that a chain of species are linked to species B as their "support" species; removing species B would not affect species A in this instance, but it would sever the link between this chain of species and species C, and thus make it TTIB-incompatible. However, this should not increase the effective extinction rate of species A. As we deduced with Eq. (28) from (27), we can simplify the writing of Eq. (29) by restricting the sum to communities in $\Phi_{H_Z}\backslash\Phi_{F_Z}$, i.e. TTIB-compatible communities consisting of species in F_Z and always including species Z:

$$e_Z = e + e \sum_{K \in \Phi_{H_Z}\backslash\Phi_{F_Z}} \frac{P_{\langle K \rangle_{H_Z}}}{P_{\langle Z \rangle}} \sum_{Y \in C} (1 - \mathbf{1}_Z[\lfloor K\backslash\{Y\}\rfloor]) \tag{30}$$

By acknowledging that:

- the sets $\Phi_{H_Z}\backslash\Phi_{F_Z}$ and $\Phi_{F_Z}\backslash\Phi_{F_Z\backslash G_Z}$ are isomorphic (a community in the first set becomes one in the second by removing species Z, a community in the second set becomes on in the first by adding species Z);
- the set $\langle Z \rangle$ can be exactly decomposed as the disjoint union of the H_Z-constrained expansions of communities in $\Phi_{H_Z}\backslash\Phi_{F_Z}$, i.e. any TTIB-compatible community including Z must be an expansion of a community in $\Phi_{H_Z}\backslash\Phi_{F_Z}$ (the constraint on expansion makes it impossible to count a community twice in the union of expansions, and hence make them disjoint);
- the difference between h_Z and $P_{\langle Z \rangle}$ is a variation of Eq. (28) in which the expansions of communities are constrained by H_Z rather than by F_Z, i.e.:

$$h_Z - P_{\langle Z \rangle} = \sum_{K \in \Phi_{F_Z}\backslash\Phi_{F_Z\backslash G_Z}} P_{\langle K \rangle_{H_Z}} \tag{31}$$

and expressing the dynamics of $P_{\langle Z \rangle}$ using Eqs (17), (28), (30) and (31), we get:

$$\frac{dP_{\langle Z \rangle}}{dt} = \sum_{K \in \Phi_{H_Z}\backslash\Phi_{F_Z}} \left[cP_{\langle K\backslash\{Z\}\rangle_{H_Z}} - e \left(1 + \sum_{Y \in K} (1 - \mathbf{1}_Z[\lfloor K\backslash\{Y\}\rfloor]) \right) P_{\langle K \rangle_{H_Z}} \right] \tag{32}$$

Although more compact than Eqs (28) and (30), Eq. (32) is still no closer to an analytical approximation of the TTIB. In the supplementary information of Gravel et al. (2011b), an approximation of the TTIB was derived to analytically compute $P_{\langle Z \rangle}$ for any species Z. This approximation, which we do not reiterate here, is based on many assumptions. First, it is assumed that the presence of any prey within the diet of a predator on the island is independent from the presence of other prey species from the predator diet on the same island. Second, it is assumed that the absence of a predator on the island induces no statistical distortion of the probability that its prey species are present on the island. Third, the case of cascading extinction is limited to the extinction of species within the immediate diet of the focal predator, i.e. the extra extinction rate incurred by species A in the case of community (A, B, C, D) in Fig. 4 is ignored. Overall, all these assumptions can be translated as follows: for any species Z, the identity $F_Z = \bigcup_{Y \in G_Z} H_Y$ is considered as a union over disjoint sets, with empty intersections between any two H_Y's among the predator's prey species.

2.2 Simple Insights

In this section, we apply the rationale of the TTIB to very simple food web topologies in order to gain insight into the interaction of colonization/extinction dynamics with the position of species within food webs on species occurrence.

2.2.1 A Linear Food Chain

Let us first focus on a simple food chain consisting in N different species, noted Z_1, Z_2, ... so that Z_1 is a basal species, Z_2 preys on Z_1, etc. If we note P_i the probability of occurrence of species i, the following equations define the TTIB dynamics for this food chain:

$$\frac{dP_1}{dt} = c(1 - P_1) - eP_1 \tag{33a}$$

$$\frac{dP_2}{dt} = c(P_1 - P_2) - 2eP_2 \tag{33b}$$

$$...$$

$$\frac{dP_k}{dt} = c(P_{k-1} - P_k) - keP_k \tag{33c}$$

The equilibrium solution to the system (33) is given by

$$P_i = \frac{c^i}{e^i \left(\frac{c}{e} + i\right)_i} = \frac{\alpha^i}{(\alpha + i)_i} \tag{34}$$

where $(x)_i = x(x-1)(x-2)...(x-i+1)$ is the falling factorial or Pochhammer symbol and $\alpha = c/e$. As in the case of food chains in the metacommunity context (Calcagno et al., 2011), the take-home message of Eq. (34) is that constraining predator occurrence by the occurrence of its preys limits food chain length on islands (Fig. 5). This is also visible in

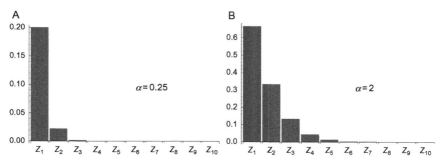

Fig. 5 Probabilities P_i to find species Z_1, Z_2, ... from the mainland food chain on the island. Probabilities never go to zero according to Eq. (34), but are negligible at trophic levels higher than three when $\alpha = 0.25$ (A) or higher than five when $\alpha = 2$ (B).

the measure of the average trophic level on the island, $\widehat{TL}$, which is given by summing Eq. (34) over all trophic levels with weights equal to i:

$$\widehat{TL} = \sum_i iP_i = \alpha \tag{35}$$

In other words, in a food chain, the expectation for the average trophic level of species that have colonized the island is simply the colonization-to-extinction ratio. To close on the topic of food chain length, the expected species richness of a food chain on an island is given as:

$$S = \sum_i i(P_i - P_{i+1}) = \frac{\alpha}{1+\alpha} + \frac{e^\alpha \alpha^{-\alpha}}{1+\alpha} \gamma(2+\alpha, \alpha) \tag{36}$$

where γ is the lower incomplete gamma function and the size of the mainland food chain is assumed infinite. In contrast with the TIB (Eq. 3), the expectation of species richness is not proportional to the number of species on mainland in the case of a food chain.

2.2.2 One Predator, Several Prey

We now turn our attention towards the case of one predator preying on N different prey species. As the different prey species are not identifiable (no difference in colonization or extinction rates), the TTIB can be understood as "the presence/absence of the predator" × "the number of prey species on the island". The probability of occurrence of the different communities will be noted as:

- P_i for the probability of finding i prey species but not the predator (i can be equal to 0);

• Q_i for the probability of finding i prey species and the predator $(i \geq 1)$. The equations defining the TTIB in such a case are:

$$\frac{dP_0}{dt} = e(P_1 + Q_1) - cNP_0 \tag{37a}$$

$$\frac{dP_i}{dt} = c(N - i + 1)P_{i-1} + e(i + 1)P_{i+1} - [c(N - i) + ei]P_i - cP_i + eQ_i \tag{37b}$$

$$\frac{dQ_i}{dt} = c(N - i + 1)Q_{i-1} + e(i + 1)Q_{i+1} - [c(N - i) + ei]Q_i + cP_i - eQ_i \tag{37c}$$

Solving system (37) in the general case is quite complicated. The path to a complete (but difficult to express) solution lies in rewriting system (37) for quantities $F_i = P_i + Q_i$ and $D_i = P_i - Q_i$, identifying the dynamics of F_i as those of the TIB (and hence its probabilities are known and given by Eq. 11), simplifying the recursions on D_i and finally expressing D_i in terms of the F_i's. We found no general form for this solution, but an interesting relationship between the expected number of prey species present to support a predator and the probability of predator occurrence emerges at equilibrium:

$$\sum_{k \geq 1} kQ_k = \frac{N\alpha^2}{2} + \frac{N\alpha}{4}\sum_{k \geq 1} Q_k \tag{38}$$

where $\alpha = c/e$. Eq. (38) always holds exactly for this system (i.e. it is not an approximation).

To obtain a general formula for the probability of predator occurrence, i.e. $\sum_{k \geq 1} Q_k$, we can assume that $\alpha \ll 1$. As the colonization by the predator species entails a colonization event on top of the ones already needed for the prey species to occur, we assume that the quantities $F_k = P_k + Q_k$ are asymmetrically separated, i.e. we assume that the Q_k's can be written as:

$$Q_k = a_k \alpha F_k \tag{39}$$

Developing Eq. (37c) in powers of α, we find that:

$$a_1 = \frac{N}{2} \tag{40a}$$

$$a_i = \frac{1 + i a_{i-1}}{1 + i} \tag{40b}$$

Solving recursion (40b), we finally have the following expression for the a_i's:

$$a_i = \frac{i-1+N}{1+i} \tag{41}$$

Plugged into Eq. (39) and summed over all k's, Eq. (41), take at the first available order in α, yields:

$$\sum_{k \geq 1} Q_k \approx \frac{N^2 \alpha^2}{2} \tag{42}$$

Plugged into Eq. (38), Eq. (42) implies:

$$\frac{\sum_{k \geq 1} k Q_k}{\sum_{k \geq 1} Q_k} \approx \frac{1}{N} + \frac{N\alpha}{4} \tag{43}$$

What Eqs (42) and (43) mean is that: (i) the probability of occurrence of the predator at low α is proportional to the square of its degree (i.e. the number of preys it can feed on); and, (ii) the expected number of prey species occurring with the predator, conditionally on its occurrence, is proportional to its degree when it is not too low. Eq. (42) can also be reinterpreted given that, when α is low, $\sum_{k \geq 1} F_k \approx N\alpha$:

$$\sum_{k \geq 1} Q_k \approx \frac{N\alpha}{2} \sum_{k \geq 1} F_k \tag{44}$$

Hence, a predator only occupies a "fraction", $N\alpha/2$, of the probability space given by its preys.

2.2.3 Multipartite Network as a Food Web

We now extend Eqs (42)–(44), i.e. assuming that $\alpha = c/e$ is small and consider a multipartite food web in which each and every species can be assigned a precise trophic level and only prey on species at the immediate trophic level down the web. We will note D_k the random variable giving the degree (as a predator, i.e. the in–degree) of species at trophic level k in the mainland food web. We assume that there are many species at each trophic level and we note N_k the number of species at trophic level k in the mainland food web. We will also note S_k the number of species at trophic level k on the island.

Following the TIB, we know that the expected number of basal species on the island is:

$$E[S_1] \approx N_1 \alpha \tag{45}$$

Using relation (42), we find that the expected number of species of the second trophic level occurring on the island is given by:

$$E[S_2] \approx \frac{N_2 \alpha^2}{2} E[D_2^2] \tag{46}$$

Generalizing approximation (46) to the next trophic levels, we finally get:

$$E[S_k] \approx \frac{N_k \alpha^k}{2^{k-1}} \prod_{i=1}^{k} E[D_i^2] \tag{47}$$

When the distribution of degrees is the same across trophic levels (i.e. $E[D_i^2] = E[D^2]$ for all i), noting $T = \sum N_k$ the total number of species on the mainland and $S = \sum S_k$ the equivalent on the island, we can transform Eq. (47) using $\beta = \alpha E[D^2]/2$:

$$E[S] \approx \alpha T \sum_{k \geq 1} \beta^{k-1} \frac{N_k}{T} \tag{48}$$

Eq. (48) expresses the relationship between $E[S]$ and T as a function of α (as in the classic TIB) and the distribution of degrees across the food web (through β) and the distribution of trophic levels in the food web (through N_k/T). The approximation remains valid as long as $E[D^2] = Var[D] + E[D]^2$ remains small compared to $1/\alpha$, ensuring a relatively low degree of dependence between species occurrence within the same trophic level.

2.3 Interpretation in Terms of Food Web Transitions

2.3.1 Reformulation in Terms of Transitions Between Community States

As suggested in Section 2.1.2, an analytical alternative for the study of the TTIB is to derive a master equation for community states rather than individual species. The mathematical object is then the random process $C_{t>0} = \cup_{i=1}^{T} X_i$ which is a vector of 0 and 1 describing the presence and absence of all species at any time t. For T species, there are 2^T community states. Table 3 provides an example for species B, C, D and E presented in Fig. 2.

Table 3 Community States Associated to the (B, C, D, E) Network, i.e. All the Combinations of Presence and Absence of Species (Columns B, C, D, E)

Community States	B	C	D	E	TTIB-Compatible
S_0	0	0	0	0	Yes
S_1	0	0	0	1	No
S_2	0	0	1	0	Yes
S_3	0	0	1	1	Yes
S_4	0	1	0	0	Yes
S_5	0	1	0	1	Yes
S_6	0	1	1	0	Yes
S_7	0	1	1	1	No
S_8	1	0	0	0	No
S_9	1	0	0	1	No
S_{10}	1	0	1	0	No
S_{11}	1	0	1	1	Yes
S_{12}	1	1	0	0	Yes
S_{13}	1	1	0	1	Yes
S_{14}	1	1	1	0	Yes
S_{15}	1	1	1	1	Yes

At any time t, $C_{t>0}$ represents the species composition of the island and thus is in one of the 16 possible S_k states. Here all communities are considered regardless of whether they are TTIB-compatible as highlighted in the *rightmost* column.

Deriving the equation associated to a given community state S_k requires study of the transition probabilities between community states between t and $t + dt$ (dt is assumed to be short enough to permit only one transition). As all community states are included in the set $S_k, k \in \{1, 2, ..., 2^T\}$, following the law of total probability, for any community state:

$$P_{C_{t+dt}=S_k} = \sum_{l=0}^{2^T-1} P_{C_{t+dt}=S_k|C_t=S_l} P_{C_t=S_l} \qquad (49)$$

The transition probabilities $P_{C_{t+dt}=S_k|C_t=S_l}$ are assumed to be linear functions of dt, when dt is small enough. For $k \neq l$, the transition matrix $\mathbf{G}$, similar to the one presented in Eq. (23), is defined by:

$$P_{C_{t+dt}=S_k|C_t=S_l} = g_{kl}dt \tag{50}$$

and, as $\sum_l P_{C_{t+dt}=S_l|C_t=S_k} = 1$:

$$P_{C_{t+dt}=S_k|C_t=S_k} = 1 - \sum_{l\neq k} g_{lk}dt \tag{51}$$

g_{kl} reflects the rate at which community state switches from state S_l to state S_k, and it can be greater than one as long as $g_{kl}dt < 1$. The TTIB acknowledges the existence of trophic interactions by assuming:

1. $g_{kl} = 0$, when the transition from S_l to S_k involves the colonization of a predator without any prey,
2. $g_{kl} = e$, when removing a species from S_l and then successively removing all predators not sustained by any prey species transforms the community into state S_k.

Plugging Eqs (50) and (51) in Eq. (49), we obtain:

$$\frac{P_{C_{t+dt}=S_k} - P_{C_t=S_k}}{dt} = -\left(\sum_{l\neq k} g_{lk}\right)P_{C_t=S_k} + \sum_{l\neq k} g_{kl}P_{C_t=S_l} \tag{52}$$

When $dt \to 0$, this approach provides the master equation than can be written in vector format to integrate the dynamics of all community states $\mathbf{P} = \left(P_{S_1}, P_{S_2}, \ldots, P_{S_{2^T}}\right)$:

$$\frac{d\mathbf{P}}{dt} = \mathbf{G} \cdot \mathbf{P} \tag{53}$$

This equation is the same as Eq. (23); $\mathbf{P}$ includes all community states and the coefficients of the matrix $\mathbf{G}$ generally depend on the community. Nevertheless, the form of the solution remains equivalent:

$$\mathbf{P}(t) = e^{t\mathbf{G}}\mathbf{P_0} \tag{54}$$

This approach describes a continuous-time Markov chain. When all the community states communicate (i.e. the Markov chain is irreducible), their probabilities reach an equilibrium $\mathbf{P}^*$ given by the vector in the kernel of $\mathbf{G}$ the elements of which sum to one.

2.3.2 Reasonable Approximations

The formulation above allows the study of nonindependence between species occurrences, but suffers from its generality: for T species, the matrix

G must be filled with $\left(2^T - 1\right) \times \left(2^T - 1\right)$ coefficients (the "-1" acknowledges that at any time t the elements of $\mathbf{P}(t)$ sum to one). Even if the knowledge of a particular network may help find these coefficients, reasonable assumptions can be made to decrease the complexity of **G**.

- First, community compositions between t and $t + dt$ cannot differ in more than one species, i.e. $g_{kl} = 0$ when $||S_k| - |S_l|| > 1$. This turns **G** into a sparse matrix: at most $T \times \left(2^T - 1\right)$ of its coefficients are not equal to zero. This assumption is not possible under the TTIB as the extinction of a prey can lead to multiple extinctions. However, this issue is easy to circumvent by allowing the predator to survive a (very) short period alone on the island. This can be done using large value for g_{kl} that measures the transition of a community with a predator to the same community without it.

- A second assumption is that colonization processes may be independent of interactions. That is, a predator can actually colonize an island without any prey. This is reasonable if the extinction probability of this predator on such an island is high, as is recommended to comply with the first assumption. Therefore, we can assume $g_{kl} = c$ when $|S_k| - |S_l| = 1$. This assumption is also useful to integrate variability in dispersal capacities among species (Cazelles et al., 2015b).

Table 4 presents matrix **G** for species C, D and E of the example presented in Figs 2–4, i.e. community states from S_0 to S_8. Once the colonization probability is determined, the remaining $T \times \left(2^T - 2\right)$ coefficients can be found based on the biological knowledge of species studies, e.g. the nature and the strength of interactions.

2.3.3 Deriving Species Richness

The solution $\mathbf{P}^*$ contains the probabilities of all community states at equilibrium. This information is actually more than the knowledge of individual species presence and species richness. Indeed the probabilities of species occurrence are particular unions of community states; therefore, $\mathbf{P}^*$ is sufficient to derive them at equilibrium. The probability of a species being present at equilibrium is given by:

$$P_{X_i=1} = \sum_{k|S_{k,i}=1} \mathbf{P}_k^* \qquad (55)$$

where $S_{k,i} = 1$ means that species i is in community state S_k and $\mathbf{P}_k^*$ is the kth component of $\mathbf{P}^*$. Similarly, the species richness is given by the sum of $\mathbf{P}^*$ weighted by the cardinality of the community states it refers to:

Table 4 The Transition Matrix **G** of the Continuous-Time Markov Chain Associated with All Combinations of C, D and E Species

	S_0	S_1	S_2	S_3	S_4	S_5	S_6	S_7
S_0	$-3c$	g_{01}	g_{02}		g_{04}			
S_1	c	$-2c-g_{01}$		g_{13}		g_{15}		
S_2	c		$-2c-g_{02}$	g_{23}			g_{26}	
S_3		c	c	$-c-g_{13}-g_{23}$				g_{37}
S_4	c				$-2c-g_{04}$	g_{45}	g_{46}	
S_5		c			c	$-c-g_{15}-g_{45}$		g_{57}
S_6			c		c		$-c-g_{26}-g_{46}$	g_{67}
S_7				c		c	c	$-g_{37}-g_{57}-g_{67}$

See Figs 2–4 for details on this simplistic food web. The present table is of intermediate density between Table 1 (the classic TIB model, dense transition matrix) and Table 2 (the TTIB model, sparse matrix involving cascading extinctions but no colonization in the absence of prey species).

$$S = \sum_{k=0}^{2^T-1} |S_k| \mathbf{P}_k^*$$ (56)

In a similar fashion, many probabilities can be derived regarding either a particular set of species or a particular property of the community. For instance, this framework allows deriving the probability of finding a given set of predators but also the mean trophic level expected and even the probability of having a trophic chain of at least p levels. In all these situations, computation of probabilities requires the identification of the community states to be summed.

3. EFFECTS OF MAINLAND FOOD WEB PROPERTIES ON COMMUNITY ASSEMBLY

In this section, our aim is to explore how the structural properties of mainland food webs affect the colonization/extinction dynamics and the resulting structural properties of island food webs. We first assess the effects of the degree distribution, the network size, the network connectance and the proportion of primary producers on island food web structural properties. In a second stage, we assess the effects of modularity in the mainland food web on island food web properties.

The degree distribution (i.e. the distribution of the number of links per node) is known to be a fundamental property of real-world, complex networks (Barabási and Albert, 1999; Newman et al., 2001) and will have important implications for their stability (Allesina et al., 2015; May, 1972). Food web models classically used in ecology—the niche model (Williams and Martinez, 2000), the cascade model (Cohen and Newman, 1985), etc.—are able to reproduce some properties of real food webs. However, they do not constrain the degree distribution, and typically display distributions of links that follow an exponential law for which there is mixed evidence in empirical data (Camacho et al., 2002; Dunne et al., 2002a). Moreover, those models do not control for the connectance or the proportion of basal species, contrary to the model used by Thébault and Fontaine (2010), for example. In the same vein, classic network generation models used in network science—the Erdős-Rényi model (Erdős and Rényi, 1959), the preferential attachment algorithm (Barabási and Albert, 1999; De Solla Price, 1976), etc.—are able to generate different degree distributions but in a stochastic way. Furthermore, one version of the Erdős–Rényi model allows the number of links to be fixed, but this is not the case for the Barabási–Albert algorithm in which network topology arises through a growth process (i.e. preferential or random attachment) and in which each vertex is added with a certain number of links. Finally, the growth process of the Barabási–Albert algorithm could lead to particular features in the resulting networks other than just the degree distribution (Newman, 2005; Stumpf and Porter, 2012). We therefore use a network generation algorithm based on degree sequence (Bollobás et al., 2001; Chung et al., 2003; Kim et al., 2009; Miklós et al., 2013).

3.1 Simulating the Model

3.1.1 Generating Directed Networks With Given Structural Properties

Because we want to assess the effects of different structural properties of the mainland food web on the colonization/extinction dynamics of island food webs, we need a network generation model that is able to be constrained by different properties. In particular, we want to fix the size of the network, the number of edges, the proportion of primary producers and the degree distribution. The typical size of food webs and other ecological networks is at least several orders of magnitude below that of network datasets from physics or computer science. For this reason, we want to use a network-building algorithm that yields a food web with properties (size, connectance, degree distribution, etc.) set at exact values, in order to remove the noise that could result from working with food webs that only obey "expected" constraints.

The sequence of degrees, which lists the total number of links of each node, allows the simultaneous fixing of three different metrics: the degree distribution, the network connectance and the size of the network. Thus, we generated the sequences of degrees in a deterministic way, using the procedure of Astegiano et al. (2015), adapted for unipartite networks, which was based on the algorithm of Chung et al. (2003).

3.1.1.1 Generating the Sequence of Degrees for a Given Degree Distribution

We want a degree sequence $w = \{w_1, w_2, ..., w_n\}$ representing the number of undirected links per node in a decreasing order, i.e. such that $w_i \geq w_{i+1}$ for all i, with n the total number of vertices, and such that the sequence of degrees follows a given probability density function f. One way of deriving such a sequence can be found by solving for w_i:

$$\frac{i}{n} \approx 1 - F(w_i) \tag{57a}$$

or equivalently

$$w_i \approx F^{-1}\left(1 - \frac{i}{n}\right) \tag{57b}$$

where F is the cumulative distribution function associated with the probability density function f. Given a fixed connectance C, the parameters of the probability density function have to be selected such that:

$$2C \approx \frac{1}{n(n-1)} \sum_i w_i \tag{58a}$$

with C the desired directed connectance;

$$C = L/n(n-1) \tag{58b}$$

and L the number of directed links in the network. The denominator in Eq. (58) is $n(n-1)$, because we consider that there is no self-loop (i.e. cannibalism) to avoid species that would stay on an island without any prey except itself. One can adjust the degree distribution (i.e. through f) to obey this constraint by setting the average undirected degree to be exactly $2C(n-1)$. We target twice the desired connectance $(2C)$ because the algorithm for directed links leads to a loss of half of the links (due to the passage from undirected to directed links). Finally, this sequence has to be feasible (i.e. the sequence has to be graphic) for a simple connected graph (with no cannibalism, i.e. self-reference, or multiple edges, and with a unique component). In other words, at least one simple connected graph must exist which satisfies the degree sequence (Berger, 2014; Brualdi and Ryser, 1991; Ryser, 1957).

We use truncated degree distributions, and so do not allow node degrees to be below or over certain predetermined limits. Indeed, the maximal degree of a vertex within a community of n species is $n-1$, in the absence of cannibalism, and the minimal degree is one because we target simply connected graph (i.e. without disconnected species). When high values of network connectance are targeted with a Power law for the degree distribution, obtained degree sequences are generally not graphic. In those cases, distributions are left-truncated in order to obtain graphic sequences that maintain a heavy tail. We also round degrees obtained for exponential distribution, because this law is continuous. This is not needed for Poisson and Power law (Zipf law) distributions, which are discrete. It should be noted that Power law distributions have been included mostly for the sake of comparison, even though it is very difficult to evidence the existence of such distributions in data unless the distribution of degrees spans at least three orders of magnitude (e.g. with degrees spanning the whole 1–1000 interval; Clauset et al., 2009; Stumpf and Porter, 2012).

3.1.1.2 Generating a Graph With a Prescribed Degree Sequence

If the sequence is graphic, the second step is to generate random undirected connected simple graphs that satisfy the sequence of degrees (and therefore the other desired properties as well), drawn uniformly from all the possible sets of graphs. To do so, we use the procedure described in Viger and Latapy (2005), which improves the MCMC algorithm from Gkantsidis et al. (2003) and is implemented in the R package `igraph` (Csardi and Nepusz, 2006).

3.1.1.3 Directing the Links

Finally, the generated networks have to be directed in order to be construed as food webs. There is, to our knowledge, no method currently available to direct a graph such that it has a fixed number of primary producers and to ensure that all vertices can be reached from a basal species (there are of course methods to generate directed graphs from undirected ones obeying certain constraints, e.g. Stanley, 1973, but these algorithms were not designed to keep certain nodes as incoming or outgoing ends of the graph, as is needed in the case of food webs with controlled number of basal species).

To do this, we first randomly selected the desired number of basal vertices. These vertices have to be disconnected. Indeed, if two vertices are connected, necessarily one of them cannot be a primary producer. Because of constraints on connectance and degree distribution, some proportions of basal species seem to be unreachable. To confirm that there is no solution in those instances, we use a graph colouring algorithm (Blöchliger and Zufferey, 2008). This algorithm tries to find the minimal number of colours necessary to colour each vertex such that two adjacent vertices have a different colour. Thus, a set of disconnected vertices is associated to each colour. If this algorithm finds not a single colour-associated set containing at least the desired number of primary producers, then we consider that our basal species assignment problem has no solution. In fact, all parameter sets are not possible, especially for degree distributions following a Poisson law (see later), or when connectance reaches relatively high values ($\sim$0.1). Then, the network is directed by an algorithm inspired from breadth-first search on graphs (see Appendix). The basic idea is that, at each time step, one proceeds from the current "trophic level" to the next up the food chain by exploring the graph from each of the node of the current trophic level, taking all nodes that can be reached in one step from all the nodes in the current trophic level. The next trophic level consists of all nodes thus explored that were not explored earlier; all links between the current level and the next are directed "upwards" (i.e. from current to the next), while links between nodes in the next trophic level are randomly directed. The next step of the algorithm then takes place after taking the next trophic level as the current one.

3.1.2 Generating Modular Networks

Modularity, i.e. the propensity of nodes to form densely connected clusters with few links between them, has been suggested as a potential explanation for properties such as ecological network stability (Thébault and Fontaine, 2010), persistence (Stouffer and Bascompte, 2011) and feasibility (Rozdilsky and Stone, 2001). In the context of the TTIB, our objective is to assess whether the modular structure of a mainland food web affects expectation on islands and, if so, how?

Modular networks are generated with a stochastic block model (SBM; Govaert and Nadif, 2008). Vertices are grouped into groups, and edges are added following the Erdős–Rényi model (Erdős and Rényi, 1959), with a probability q_w of finding an edge between two vertices in the same group, and a probability q_b between two vertices in different groups. If all modules have the same size, the expected network connectance is:

$$E\left[2C\right] = \left[\frac{1}{m} - \frac{1}{n-1}\left(1 - \frac{1}{m}\right)\right](q_w - q_b) + q_b \qquad (59)$$

with m is the number of modules. Note that when n (number of vertices/ species) is large enough, the expected connectance can be approximated by the following equation:

$$E\left[2C\right] \approx \frac{q_w - q_b}{m} + q_b \qquad (60)$$

As described in Section 3.1.4.4, modularity is computed with the walktrap algorithm (Pons and Latapy, 2006), and with the flow-based infomap algorithm (Rosvall and Bergstrom, 2008). Modularity cannot be controlled exactly, but can be approximated through generations of networks with different q_w and q_b such that the connectance stays constant.

An approximate log-linear relation can be found between the logarithm of modularity and the ratio $\beta = q_b/q_w < 1$ (at a constant connectance C):

$$\log(Q) \approx a_C\beta + b_C \tag{61}$$

Using a linear regression, a and b values can be computed. Based on approximations (60) and (61), we can approximately control both the expectation of the connectance and the modularity of the network.

3.1.3 Defining a Null Model

When we compare the properties of local food webs on islands relative to their mainland counterpart, there can possibly be two confounding effects. On the one hand, observed differences could be due to the sequential dependence of colonization/extinction events. On the other hand, differences could arise just because local food webs are limited subsets of the mainland food webs. In particular, some network properties could vary in a systematic way with the size of local networks (Poisot and Gravel, 2014).

In order to disentangle these two confounding effects, we use a null model, which is simply MacArthur and Wilson's TIB model, i.e. without any interaction. For each of the 50 replicates of each parameter set, and for each value of the colonization-to-extinction ratio ($\alpha = c/e$), we randomly draw a sample of species on the mainland. It is noteworthy that to get a broad range of local web sizes, we had to use a wider range of α values for the null model than for the simulations with interactions. The size of the sample is a proportion of the total number of species on mainland, whose value is determined by the value predicted by MacArthur and Wilson's model (i.e. Eq. 3). In this case, all species have the same colonization and extinction rates and are thus considered sampled with the same probability. We do not add any constraint on this random sample, which can thus lead to unrealistic local food webs (e.g. an intermediate species in the mainland food web could appear as a "basal" species in the island network generated by the null model, or a large proportion of species could end up being completely disconnected from the island network, especially for small α, weak connectance and small size of the mainland food web).

3.1.4 Network Properties of Interest

3.1.4.1 Connectance

We consider the directed connectance, which is the number of directed links found in the network relative to the maximal number of directed links possible, given by Eq. (58b).

We have to note that connectance for the null model is computed considering all species on the island, be they connected or not. When α is small (i.e. $c \ll e$), the proportion of disconnected species (i.e. species with no link) increases and thus, the connectance rapidly decreases.

3.1.4.2 Trophic Levels
We use the notion of the shortest trophic level (Williams et al., 2002), which is, the shortest path from a species i to any basal species. This metric is computed using the adjacency matrix ($\mathbf{A}$, with $a_{ij} = 1$ when there is a link going from species j to species i) of the food web in the following way; for a species i, its trophic level is one plus the smallest τ such that:

$$\mathbf{x}_i^T \cdot \mathbf{A}^\tau \cdot \mathbf{b} \neq 0 \qquad (62)$$

where $\mathbf{b}$ is the vector of basal species (i.e. if species j is a basal species, then $b_j = 1$, else $b_j = 0$) and $\mathbf{x}_i$ is a vector of zeroes with a single "1" at position i. Noting τ_i the shortest path from a species i to basal species, the mean shortest trophic level over the whole island food web is:

$$\widehat{TL} = \frac{1}{n}\sum_i (1 + \tau_i) \qquad (63)$$

where n is the number of species in the food web. The maximal trophic level in the same food web is:

$$TL_{\max} = \max_i (1 + \tau_i) \qquad (64)$$

3.1.4.3 Degree Distribution
Identification of degree distribution is done by fitting several models (Uniform, Poisson, Exponential and Power law) and retaining the one with the lowest Akaike Information Criterion (AIC). Likelihood and parameter estimates are computed with the R package `fitdistrplus`. When this is done on mainland webs, in which the degree distribution is imposed, the best-fit distribution does not always correspond to the desired distribution. For example, relatively small networks with a weak connectance tend to be characterized by uniform or Poisson distributions. Similarly, Power law-constructed networks can appear as following an exponential distribution of degrees when there are not enough well-connected nodes (Newman, 2005; Stumpf and Porter, 2012). For results about degree distributions, we consider the distribution given by the best-fit model rather than the

distribution used to create the input networks so as to better relate our findings to the type of information empiricists have access to.

3.1.4.4 Modularity

As previously noted, modules can be defined as a set of nodes that interact significantly more with each other than with nodes in the rest of the network (Newman, 2004, 2006a,b; Newman and Girvan, 2004). The identification of modules relies on partitioning the nodes of a given network in different modules such that this partition maximizes a property named *modularity*. In ecology, the definition commonly used is the one of Newman and Girvan (2004). With this definition, directed links (i.e. arrow from one species to another) in ecological networks are generally considered as undirected (but see Arenas et al., 2007; Leicht and Newman, 2008, for ways to extend classic modularity to directed networks). As food webs present directed links that represent energy or matter fluxes, this information could be important for constructing modules. Thus, we also use a flow-based approach, the infomap algorithm (Rosvall and Bergstrom, 2008) to partition food webs. With this approach, the network is decomposed into modules by optimally compressing a description of a random walk (representing information, matter or energy) on the network. This method could be more capable of detecting modules associated with energy channels (as in Rooney et al., 2006) than the classic optimization of modularity. A value of modularity can be obtained with the partition of the food web obtained by the infomap algorithm. However, we instead use the compression rate, i.e. one minus the ratio of description length after/before compression, the description lengths being given in bits necessary to describe typical random walks on the network, as a proxy for modularity.

3.1.5 Running Simulations

As many properties are susceptible to scale with the complexity of the network (Wood et al., 2015), simulations are done with mainland food webs of different sizes (100, 250, 500 nodes/species) and connectance values (0.01, 0.025, 0.05 and 0.1) as well as for different proportions of basal species (5%, 10%, 20%, 40%), since this could alter the outcome of the process by forming a more or less solid foundation for food chain construction. Three different degree distributions are used: Zipf (Power law), Poisson and Exponential (geometric) distributions. Thirty-two combinations of parameter sets are not feasible because they either lead to food webs that do not consist of a single component (disconnected webs) or are unable to host the targeted proportion of basal species. Fifty networks are constructed for each of the 112 feasible parameter sets.

Two connectance values (0.05 and 0.1) and four approximate levels of modularity ($\sim$0.1, $\sim$0.25, $\sim$0.45, $\sim$0.6) are used for simulations of modular networks. Food web size ($n=200$), the proportion of primary producers (10%) and the degree distribution (detected as Poisson) are constant among these simulations. Fifty food webs are generated for each parameter set.

All simulations start with empty islands which are then colonized by species from the mainland species network, in which the size, connectance, degree distribution and number of basal species are controlled, as explained earlier.

Simulations are done in continuous time following the Gillespie algorithm for the simulation of Poisson processes (Gillespie, 1976, 1977). The colonization rate, c, is fixed to 0.1 per capita, and the extinction rate, e, varies such that $\alpha = c/e$ takes the following values: $\alpha = \{0.16, 0.25, 0.5, 0.75, 1, 1.5, 2, 4\}$. Given that consumers must have at least one prey present on the island to colonize and species present on the island go extinct with their last prey species, at each moment, the global colonization rate is:

$$\widetilde{c} = \sum_{i \in S_C \backslash S_L} cY_i \tag{65}$$

where $Y_i = 0$ if species i has no prey on the island and $Y_i = 1$ if species i has at least one prey on the island; S_C and S_L are, respectively, the set of species on the mainland and on the island. In the same vein, the global extinction rate is:

$$\widetilde{e} = \sum_{i \in S_L} e = |S_L|e \tag{66}$$

The waiting time between two consecutive events is drawn from a negative exponential law of parameter $\lambda = \widetilde{c} + \widetilde{e}$. This event is a colonization with probability $\widetilde{c}/\lambda$ or otherwise an extinction. If it is a colonization event, a species is randomly selected from the set of species which are not already on the island and which have at least one prey on the island. If it is an extinction, a species is randomly selected from the set of species on the island.

Each simulation runs for 5000 events. The local network structure is saved every 50 events. At each time step, the local diversity and the number of secondary extinctions is recorded. Only networks recorded after 2500 events are analysed (2500 events is large enough to attain the stationary state).

3.2 Results Obtained When Controlling Mainland Food Web Richness, Connectance, Degree Distribution and the Proportion of Primary Producers

3.2.1 Effects on Species Diversity on the Island

The incorporation of trophic interactions into the theory of island biogeography generally leads to a reduction in the species richness observed on islands. This effect is stronger for smaller and less connected mainland food webs and effectively vanishes for large and well-connected networks (Fig. 6).

The shape of the mainland degree distribution does not seem to have a significant effect on the mean diversity observed on islands (Fig. 6). The proportion of basal species has a positive effect on species richness for small and poorly connected mainland food webs (Fig. S1 (doi:10.1016/bs.aecr.2016.10.004)): the more primary producers in those networks, the closer the species richness is from the null expectation (i.e. from MacArthur and Wilson's prediction). These results agree with expectations from the analytic model of Gravel et al. (2011b).

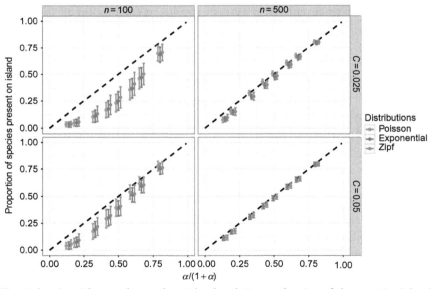

Fig. 6 Species richness observed on islands relative to the size of the mainland food web as a function of the colonization-to-extinction ratio (α, abscissas) compared to the species richness predicted by MacArthur and Wilson's model (the *dotted line*), for different sizes (*columns; n* = 100 or *n* = 500) and connectance values (*rows; C* = 0.025 or *C* = 0.05) of the mainland food web. *Different colours* correspond to the different degree distributions of the mainland network. Error bars represent the variance among replicates; points are slightly jittered for clarity. Here, proportion of primary producers is 10%.

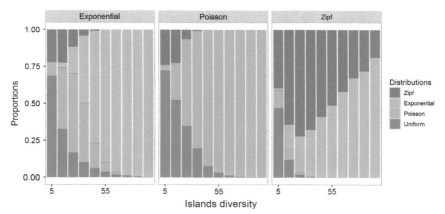

Fig. 7 Fitted degree distributions for island food webs as a function of diversity on islands (abscissas) and mainland degree distributions (from *left to right panel*: exponential, Poisson and Power law). Mainland food webs have 250 species, and a connectance of 0.025. All proportions of primary producers are merged. *Colours* correspond to fitted distributions (legend on the right of the three panels).

3.2.2 Effects on the Degree Distribution on the Island

Regardless of the parameter sets, all simulations show the same pattern when local diversity is low. Indeed, when the number of species on the island is less than a few dozen species, the local degree distribution tends to be statistically identified as uniform, Poisson or Power law, independently of the shape of the mainland degree distribution (Fig. 7). This pattern is stronger when mainland connectance and mainland diversity decrease. However, it can be attributed to a statistical power issue. Indeed, the proportion of cases in which the second best-fit degree distribution is less than 2 units of AIC away from the best model (Fig. 8) decreases with species diversity on the island and with mainland connectance and mainland diversity, but tends towards higher values for small local food webs. Thus, it is difficult to identify a degree distribution with certainty, especially when networks are small.

It is noteworthy that these situations correspond to colonization-to-extinction ratios near or inferior to 0.25, which is large enough when compared to empirical estimations, e.g. from species–area curves (Preston, 1962). For example, from the 50-lake dataset of Havens (1992), the mean local diversity is 38.8 ± 14.8, while the "mainland" food web has 210 species. This corresponds to a mean colonization-to-extinction ratio close to 0.23 following the original TIB (Eq. 3). Similarly, from the Simberloff and Wilson (1969) dataset on islands (after a defaunation experiment), mean local diversity is 26.8 ± 7.4 for a mainland diversity of 250. This corresponds

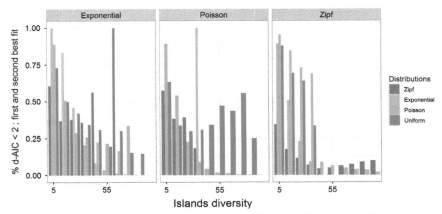

Fig. 8 Proportion of ΔAIC less than 2 units away from the best model. Abscissas represent species richness on the island. Bars represent the proportions of simulated island networks of a certain size for which the second best fit for degree distribution is less than two units of AIC away from the best-fit degree distribution and is given by the distribution associated with the colour of the bar. Parameter values are as in Fig. 7. Mainland degree distribution changes between panels (from *left to right*: exponential, Poisson and Zipf distributions).

to a colonization-to-extinction ratio close to 0.12. Thus, from our results with colonization-to-extinction lower than 0.25, mainland degree distribution does not seem to be predictable from a sample observed on a local food web (at least, from the comparison of AIC values).

When the number of species on the island increases (which is not necessarily a realistic situation, as mentioned earlier), the local degree distribution tends to become more similar to the mainland distribution (Figs 7 and 8). However, this similarity is not so high when the mainland food web is characterized by a heavy-tailed degree distribution (i.e. a Zipf distribution; see Fig. S2 (doi:10.1016/bs.aecr.2016.10.004)). In this case, a large proportion of local networks are identified to have exponential distributions. For intermediate values of connectance $(0.025 \leq C \leq 0.05)$, the majority of island networks follow an exponential degree distribution, especially for intermediate sizes of the local networks. When connectance is relatively strong $(C \approx 0.1)$, the larger the size of the local network, the higher the proportion of networks with a degree distribution detected as heavy tailed (Fig. S2).

3.2.3 Effects on Secondary Extinctions on the Island
The size and the frequency of cascading extinctions in the local networks following the extinction of a species are affected by the degree distribution

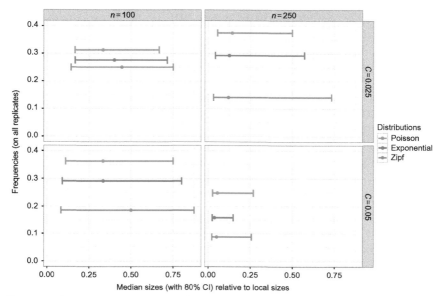

Fig. 9 Median size (with 80% CI) relative to local sizes and frequencies of secondary extinctions, as function of mainland fixed degree distribution and for different size (*columns*) and connectance (*rows*) of the mainland food web. Here, proportion of primary producers is 5%, and alpha ratio is 0.25.

of the mainland network. Indeed, secondary extinctions are less frequent, but generally more variable in size, on islands that are being colonized from mainland food webs with more heavily tailed degree distributions (Fig. 9). This pattern is reinforced in small and poorly connected mainland webs. These results are consistent with previous work on "virtual removal experiments" in networks (Albert et al., 2000; Cohen et al., 2000, 2001; Dunne et al., 2002b), although the processes involved are slightly different (removal of vertex/species vs colonization/extinction process, see also Curtsdotter et al., 2011).

A higher proportion of primary producers on mainland is positively linked to a more robust base for network construction: communities with a high proportion of basal species are less prone to undergo secondary extinctions, and when secondary extinctions do occur, their results are less catastrophic (Fig. S3 (doi:10.1016/bs.aecr.2016.10.004)).

It is interesting to note that, even if some different mainland degree distributions are statistically detected as the same distribution (e.g. Zipf and Exponential laws could both be identified statistically as Exponential), the median size and frequencies of secondary extinctions depend on the "ideal" distribution (Fig. S4 (doi:10.1016/bs.aecr.2016.10.004)), not on the fitted one. In other words, secondary extinctions occurred according to the degree distribution the mainland food web was supposed to resemble, not according to the degree distribution that best fitted the generated degree sequence.

3.2.4 Effects on Food Web Connectance and Average Degree on the Island

For all parameter sets, directed connectance on islands decreases with the local size of the network and rapidly reaches the same value as the mainland web at intermediate to large network sizes. The proportion of basal species in the mainland web affects this relation. Indeed, with a larger proportion of basal species in the mainland web, the local directed connectance is lower and reaches the mainland connectance faster. Overall, connectance is roughly constant in large enough island food webs.

Comparing the connectance of the island food webs obtained with or without taking into account the trophic interactions (i.e. comparing results following the TTIB vs the TIB), the observed connectance does not seem to be only driven by size. Indeed, island food webs built following the TTIB always show a higher connectance than those following the null model. This difference is slightly more important when the mainland food web displays a heavy-tailed degree distribution.

We find that the TTIB agrees roughly with Martinez' constant connectance hypothesis (Martinez, 1992; Fig. 10). The scaling of links per species, however, tends to look more like the Cohen Link-Species scaling law for mainland food webs of low connectance (Briand and Cohen, 1984; Cohen and Briand, 1984), especially in island networks with few primary producers (see Fig. S5 (doi:10.1016/bs.aecr.2016.10.004) for comparison).

3.2.5 Effects on the Number of Trophic Levels on the Island

When the mean trophic level on the islands is compared to the mean trophic level on the mainland, several patterns are observable (Fig. 11):

- When mainland connectance and/or species richness take low to intermediate values, and island diversity is low, the mean trophic level on the island is below that on mainland;
- When diversity on the island increases, mean trophic level increases too and quickly exceeds mean trophic level of the mainland food web. This difference with the mainland web increases as connectance and size of mainland food web increase;

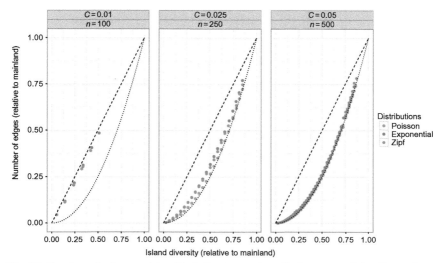

Fig. 10 Number of edges on island (relative to number of edges on mainland), as a function of the diversity on island (relative to mainland) and for different size (n) and connectance (C) along *columns*; points are slightly jittered for clarity. The *dashed line* represents the link-species scaling law and the dotted line, the constant connectance hypothesis. Here, proportion of primary producers is 5%. Point colour refers to mainland degree distribution.

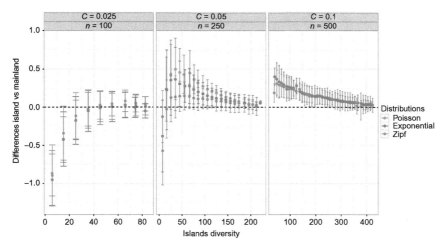

Fig. 11 Island mean trophic level minus mainland mean trophic level for different size (n) and connectance (C) of mainland food web along *columns*. Here, the proportion of primary producers is set at 10%. Points are slightly jittered for clarity and their colour refers to mainland degree distribution. There is no Poisson distribution in the rightmost panel, because we could not generate such a network with a single connected component.

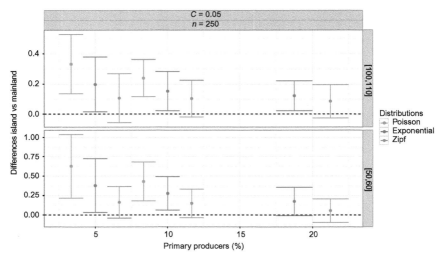

Fig. 12 Difference between island and mainland mean trophic levels as a function of the proportion of primary producers within the food web (*x*-axis), for different mainland degree distributions (colours), fixed characteristics of the mainland food web ($C = 0.05$ and 250 species), and different values of species richness on the island (*top panel*: ca. 105 species; *bottom panel*: ca. 55 species).

- When diversity on the island approaches that of the mainland, the mean trophic level also becomes similar to the one on mainland. Thus, mean trophic level changes with island species richness according to a hump-shaped, right-skewed curve.
- Absolute difference between island and mainland mean trophic levels is higher when the proportion of primary producers decreases (see also Fig. 12);
- In the same vein, the absolute difference with the mainland web is weaker for heavy-tailed distributions of degree in the mainland food web.

Thus, in a first phase of colonization, the mean trophic level is lower than in the mainland food web. Shorter trophic chains at low species diversity on the island could be explained by: (i) the sequential dependence of predators on their prey, because at low local diversity, the number of prey species present for a predator is only a subset on the predator's prey on mainland, and thus, it is more difficult for the former to colonize and stay sustainably on the island, thus eventually leading to lower mean trophic level; and, (ii) similarly, because only a subset of prey is present, predators are more prone to undergo cascading extinctions, and this is even more pronounced when the predator trophic level is high. These two effects are already known to limit the length of food chains in the metacommunity framework (Calcagno et al., 2011). Point (i) matches hypotheses about resource limitation, energetic constraints (Hutchinson, 1959) and level of perturbation. Indeed, low colonization-to-extinction ratios correspond to small and/or isolated islands or strongly disturbed habitats (i.e. high extinction rates). The size of island could also be related to notions of productive space and habitat heterogeneity which are known to limit food chain length (Post, 2002, but see Warfe et al., 2013). Isolated islands should present the same pattern (i.e. low species richness should lead to less productive ecosystem, and should reduce habitat heterogeneity—e.g. fewer engineer species, fewer microhabitats, etc.). Point (ii) relies more on the dynamical constraint hypothesis (May, 1972; Pimm and Lawton, 1977), under which longer food chains are less stable. However, here, like in Calcagno et al. (2011), these effects arise through spatial processes.

Mean trophic level on islands rapidly exceeds the one on mainland (especially in big and well-connected mainland webs) as species diversity on the islands increases. This could be explained by the presence of generalists, which are often also omnivorous in our framework. Such generalist/omnivores can readily colonize islands (Holt et al., 1999; Piechnik et al., 2008), even with a small subset of their usual prey, and then, could present higher realized trophic levels because prey at low trophic levels were absent (see also Pillai et al., 2011). The same effect exists in principle for all species, but decreases with the level of specialization. Moreover, mean degree on islands increases linearly with the number of species, initially (Fig. 10), so that species are ever more connected as diversity increases on the island. Thus, with more links, a species has more chance of being connected to a low trophic level species, which tends to reduce its trophic level. When species richness tends towards the level of the mainland, island webs increasingly resemble the mainland webs and therefore all properties also tend to be similar.

The fact that webs built from heavy-tailed mainland degree distributions deviate less than others from the mainland mean trophic level could be explained by the presence of intermediate trophic level super-generalists, which prey on a large proportion of species on mainland and are in turn consumed by a large proportion of species. When one or more such super-generalists are present, they can link many species together in a way very similar to the mainland web because super-generalists are always associated with primary producers and they assume trophic levels similar to their trophic level in the mainland web.

3.2.6 Effects on Food Web Modularity on the Island

In the first set of simulations, modularity was not controlled when generating the food webs. However, interesting results arise by looking at changes in modularity among network topologies. In mainland food webs, modularity scales with connectance and size of the mainland webs (Fig. S6 (doi:10.1016/bs.aecr.2016.10.004)). Networks characterized by heavy-tailed distributions are less modular than those characterized by lighter-tailed distributions, and the proportion of primary consumers has no effect on modularity (results not shown). This highlights the fact that numerous properties of networks are correlated (Fortuna et al., 2010; Orsini et al., 2015; Vermaat et al., 2009), and disentangling the effects of each property, or deducing that a property results from a particular process is a difficult task, which needs appropriate tools (Orsini et al., 2015).

Modularity on island food webs is generally superior or equal to the one on mainland, except when island food webs are obtained from a poorly connected and small mainland web. Island food web modularity displays a hump-shaped curve along the gradient of species richness (Fig. 13). The number of species (relative to the mainland) required to reach mainland modularity seems to decrease with the size and the connectance of the

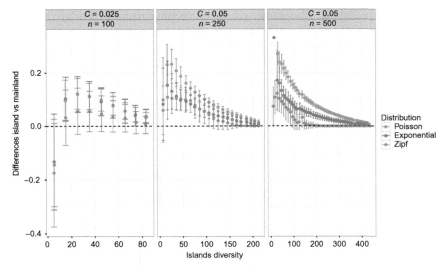

Fig. 13 Differences between island and mainland modularity as species richness increases on island (abscissas), for different size (*n*) and connectance (*C*) along *columns*. Here, the proportion of primary producers is 5%. Points are slightly jittered for clarity and their colour refers to mainland degree distribution.

mainland web, especially with Power law degree distributions. Food webs constructed from complex mainland webs $(C=0.05)$ with heavy-tailed degree distributions deviate less and reach the mainland web modularity faster. Again, this is only true when the mainland web is large and well connected. A large proportion of primary producers increases the difference in modularity between mainland and island, but only for small and poorly connected mainland webs (Fig. S7 (doi:10.1016/bs.aecr.2016.10.004)).

It should be noted that modularity is always lower under the TTIB than under the classic TIB (results not shown). Indeed, species samples obtained through the classic TIB are not necessarily simply connected; every connected component of the food web then automatically amounts to another module, hence increasing modularity. However, as empirical measures of modularity are always obtained on simply connected networks (i.e. a single connected component with no disconnected species), results obtained with the null model are not at all comparable with empirical measures on island food webs.

It is remarkable that modularity has more or less the same response to island diversity as mean trophic level. In the first stages, only a few numbers of primary producers are present, and then, trophic chains are constructed independently from each basal vertex by adding new species with higher trophic levels. This leads to several compartments based on different primary producers (different "channels") and could be seen as a niche-based construction of food webs, assimilating each channel as a different niche. When the diversity on the island increases, with higher colonization-to-extinction ratio, the different trophic "chains" tend to progressively merge together, thus finally leading to a unique component.

Food webs built from heavily tailed mainland degree distributions deviate less and reach the mainland modularity faster, as noted earlier. This could be a consequence of the presence of super-generalist species, which tend to merge the different compartments faster than expected under a more homogeneous degree distribution.

3.3 Results Obtained Using the SBM to Generate Mainland Food Web

3.3.1 Effects of Modularity on Species Richness

Modularity is often considered as a property that enhances the stability and robustness of trophic networks by limiting the diffusion of perturbations through the web (Thébault and Fontaine, 2010) and is often found in empirical data (e.g. Krause et al., 2003). Here, we investigated its effects on food

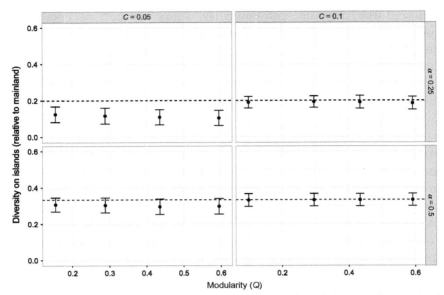

Fig. 14 Species richness on islands as a function of network modularity on the mainland (from Newman and Girvan's method), with mainland food webs built using a stochastic block model algorithm. Mainland connectance changes along *columns* and colonization-to-extinction ratio (α) along *rows*.

web assembly on islands. When looking at the influence of modularity on species richness under the TTIB framework, modularity, computed either with Newman and Girvan's or with Rosvall and Bergstrom's method, does not have any effect on species richness observed on the island (Fig. 14). Thus, more modular mainland food webs are not more easily reconstructed on islands (see also Fig. S7). This observation relates to early studies of Pimm (1979) which suggested that high levels of compartmentalization are not sta-bilizing. Thus, the origin of compartmentalization in food webs could be unrelated to a stability/robustness issue, but rather to the niche-based orga-nization of food webs (Guimera et al., 2010).

3.3.2 Effects of Modularity on Secondary Extinctions

As for species richness, the modularity of the mainland food web does not really affect the frequency or the size of cascading extinctions on islands (Figs 15 and S8 (doi:10.1016/bs.aecr.2016.10.004)). Contrary to what we expected, frequencies and sizes of secondary extinctions even show a slight increase with modularity. Thus, in our framework, food web compartmen-talization does not assure against cascading extinctions.

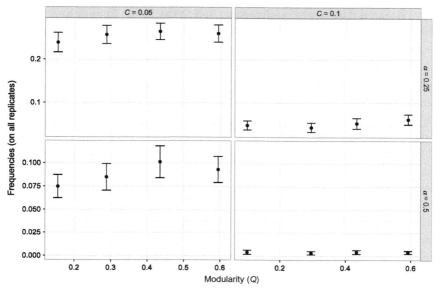

Fig. 15 Frequencies of secondary extinctions as a function of modularity (from Newman and Girvan's method) of mainland food webs. Mainland connectance changes along *columns* and colonization-to-extinction ratio (α) along *rows*.

4. DISCUSSION

4.1 Legacies of Island Biogeography Theory

MacArthur and Wilson's TIB (1963, 1967) has contributed to several conceptual revolutions that are still important in current theoretical ecology. These were originally proposed because of the particularities of insular habitats, but now it is well recognized that they also concern other ecological systems (Warren et al., 2015). First, because islands are discrete habitats harbouring small populations, they are subject to important demographic stochasticity and elevated extinction risk. Consequently, it is essential to consider the impacts of colonization and extinction dynamics and that biodiversity is never at equilibrium. Second, because of their isolation, community composition on islands is strongly contingent on the regional species pool and the ability of species to reach the islands. These two features of insular habitats lead to a different perspective of community assembly, in which macroscopic properties of communities are deterministic, despite considerable stochasticity in local community properties.

The neutral theory of biodiversity (Bell, 2000, 2001; Etienne and Alonso, 2005; Hubbell, 2001; Volkov et al., 2003) is perhaps the best example to illustrate the legacy of the TIB on today's theoretical ecology. As with the TIB, the neutral theory of biodiversity assumes that some meaningful predictions about community organization can be made without detailed knowledge of the natural history (Wennekes et al., 2012). The neutral theory integrates the dynamical balance between colonization and extinction dynamics and then adds competitive interactions. While it maintains the main predictions of the TIB (species abundance–rank curves and species turnover), it also expands the range of predictions owing to its individual-based formulation (e.g. species abundance distributions, range–abundance relationship). Neutral theory is, however, not the endpoint of the story, as we have argued in this chapter in which we developed theory for nonneutral interactions. While the TIB and neutral theory were originally concerned with competitive interactions, recent developments have also paved the way to a network theory of biogeography. Extending the TIB to several trophic levels also reveals different species–area relationships for prey and predators (Holt et al., 1999) and the reciprocal feedback between insular dynamics and network structure (Gravel et al., 2011b). We discuss here how the TIB can be used as a template to add complexity in species interactions and contribute to the integration of community ecology and biogeography.

4.2 The Future of Island Biogeography Theory

4.2.1 A General Theory for Sampling the Regional Species Pool

The TIB establishes the principle that a local community is a sample of the regional species pool. The size (i.e. species richness) and the composition of this sample depend on the probability of each species to be sampled from the regional pool. In the TIB, and incidentally in neutral theory as well, this probability is only influenced by two abiotic variables, that is island size and isolation. The model supposes that all species are functionally equivalent and have the same probability to compose the regional pool. All species are treated as equal and consequently they all have the same occupancy. At any time, a local community of a random sample of S species drawn from the regional pool can be described using the following general stochastic model:

$$P(R_i, X_i = 1) = P(R_i)P(X_i = 1 | R_i) \tag{67}$$

where $P(R_i, X_i = 1)$ is the probability of observing a species i with characteristics R_i on an island, $P(R_i)$ represents the probability of sampling a species with such characteristics and $P(X_i = 1|R_i)$ is the occupancy function for such a species, conditional on its characteristics. The classic TIB assumes that $P(X_i = 1|R_i)$ is independent of R_i and consequently the distribution of species characteristics on an island is only driven by their distribution in the regional species pool.

The trophic extension of the TIB introduced the idea that the sampling could be biased by trophic interactions, and indirectly by the traits driving these interactions (Gravel et al., 2016). Here, we derived a model in which diet breadth and trophic position with respect to primary resources determine species–specific occurrence probabilities. In this model, generalist species located at low trophic levels are more likely to find prey species at colonization and thereafter persist. Compared to the mainland species pool, local community composition is biased towards more generalist species, as described by:

$$P(X_i = 1|R_i) = \frac{c(R_i)}{c(R_i) + e(R_i)} \tag{68}$$

where c and e are colonization and extinction functions based on the characteristics of the species, R_i. In addition, we have shown that these characteristics are not stationary. In other words, the diet of a species on an island, and so its colonization and extinction probabilities, is a function not only of its position in the mainland food web, but also of the local community composition (see e.g. Fig. 11 on trophic levels). The inclusion of species–specific equilibrium occurrence probabilities, developed in the TTIB, therefore paves the way to the integration of functional ecology, community ecology and biogeography. This approach could indeed be extended to further sets of biogeographical models that incorporate biotic variables related to species dispersal capacities (e.g. symbionts, see Amsellem et al., 2017) and extinction probabilities.

Such an example is provided by the investigation of body mass distribution on islands. Body mass appears to be a fundamental trait that percolates through all aspects of community dynamics and ecosystem processes, from productivity to energy flows (Brown et al., 2004; Cohen et al., 2003; Etienne and Olff, 2004; Otto et al., 2007; Peters, 1986; Woodward et al., 2005). Body mass has also been successfully used to parameterize models of food web topology (Allesina et al., 2008; Eklöf et al., 2013; Gravel et al., 2013; Petchey et al., 2008; Rohr et al., 2010; Williams et al., 2010). At the biogeographical scale, body mass is likely to influence both colonization and extinction rates (Bradbury et al., 2008; Soininen, 2016). The scaling of colonization and extinction rates with body mass could be easily integrated into the TIB and the TTIB, leading to an allometric theory of island biogeography with species-specific and mass-dependent occurrence probabilities. Owing to the negative relationship between abundance and body mass (Blackburn and Gaston, 1999; Cohen et al., 2003; Damuth, 1981; Nee et al., 1991; White et al., 2007), we would expect extinction rate to correlate positively with body mass (Gotelli and Taylor, 1999; Petchey et al., 2004). However, empirical evidence for this assumption is weak (Gaston, 2000), while the supposed higher abundance of small-bodied species might be balanced by other factors that make them more prone to extinction, like their higher dependence on habitat complexity (Graham et al., 2011). In contrast, the scaling of colonization rate, dispersal and space use with body size is supported by many empirical studies, in particular in marine systems (Jetz et al., 2004; Peters and Wassenberg, 1983; Wieters et al., 2008). All life-history traits increasing the rate of colonization, such as fecundity, home range, mobility or diet generality are indeed positively correlated to fish body size (Kulbicki et al., 2015; Luiz et al., 2013; Mora et al., 2003; Nash et al., 2015). The scaling of colonization rate with body size is, however, less straightforward in taxonomic groups in which these traits scale differently. For example, fecundity decreases with body size for birds and mammals while mobility increases (Marquet et al., 2005). In terrestrial and semiterrestrial animals, a positive relationship between dispersal and body size is generally supported by existing data (Stevens et al., 2014), although exceptions do exist. Even if single large individuals might disperse further away than small ones, the larger number of offspring produced by populations of smaller organisms could also lead to inverse scaling relationships. Allee effects might also affect the establishment success of immigrants, and we do not have clear expectations of how it might scale with body size. While intuitive at first, the integration of allometric constraints into the TIB and the TTIB will require direct empirical investigation of colonization and extinction rates.

4.2.2 Abundance and Energy

In the TTIB, the consequences of ecological interactions are only perceptible when a predator cannot find any prey. However, the consequences of interactions may actually be perceptible for any change in the composition of the community when analysing species abundances. Thus, extinction rate should vary more smoothly with community composition (i.e. not sharply increase when the underlying food web is weakened by extinction of a keystone species) because of the role of trophic regulation on population size (Hairston et al., 1960; Oksanen et al., 1981; Persson et al., 1992). Moreover, fluctuations of individual species numbers due to interactions may further influence extinction rates when driving populations to very low densities. It is also the rationale on which the results pertaining to the stability of ecological networks are based (Allesina and Tang, 2012; May, 1972, 1973). Hence, introducing species abundance in the TTIB, and their link to dynamical stability, is a promising avenue towards a more mechanistic and integrated TIB.

Species abundances are also key to introduce energetic constraints in a mechanistic framework (Trebilco et al., 2013). In the TIB, the existence of such limits lies in considering the extinction rate as a decreasing function of the island area. In the TTIB, energetic constraints actually restrict the diversity of primary producers which in turn reduces species richness at higher trophic levels. Therefore, as in the TIB, rich communities are precluded, but in the TTIB, species are not equally affected: the higher their trophic level, the more affected they are. Furthermore, the energetic consumption of an entire island community at equilibrium can be computed based on the relationships between the body mass and the metabolic rates. This measurement can then be used to determine if the community may persist on the island or if imminent extinctions are expected due to the lack of available energy. Integrating species abundances and energetic constraints in the TTIB may provide new insights into the abundance–body mass relationship that remains poorly understood (Blackburn and Gaston, 1999; Blackburn et al., 1993; Damuth, 2007; Trebilco et al., 2013). It also promises the investigation of biodiversity distribution over large temperature gradient and might prove useful at explaining the widely documented relationship between energy and species richness (Currie, 1991; Currie et al., 2004; Hurlbert, 2004). Also, taken from an evolutionary perspective, these considerations might possibly come up with new explanations of gigantism and dwarfism on islands.

4.2.3 Environmental Heterogeneity

Kadmon and Allouche (2007) proposed a model unifying niche and metapopulation theories under the area–heterogeneity trade-off hypothesis (see also Allouche et al., 2012). The niche theory proposes that species coexistence occurs because a differentiated response to the environment reduces the interspecific competition below intraspecific competition (Chesson, 2000). Assuming species packing along niche axes, species richness should be proportional to the range of these axes and inversely proportional to the niche width of individual species. The TIB, on the other hand, proposes that species richness in a locality is a dynamical equilibrium between stochastic colonization and extinction events. The TIB predicts a positive relationship between area and species richness because larger islands will have lower extinction rates. Kadmon and Allouche (2007) proposed to unification of these concepts with a single assumption: for a given region, the average area of favourable habitats should decrease with increasing environmental heterogeneity because each extra portion of a novel habitat reduces the surface occupied by other ones. There are, therefore, two contrasting forces acting on species richness with increasing environmental heterogeneity: a positive effect of niche diversity and a negative effect of increased extinction rate (Allouche et al., 2012). The theory consequently predicts a hump–shape relationship between environmental heterogeneity and equilibrium species richness, which has been partly supported by empirical data (Hurlbert, 2004; Tews et al., 2004). The last missing piece of information is the scaling of environmental heterogeneity with area. A positive relationship, which is generally expected, would further strengthen the scaling relationship between area and species richness.

It could be feasible to extend this theory further to account for species-specific colonization and extinction probabilities, contingent on the environmental heterogeneity. With such an approach, results might differ as some species could benefit from increasing environmental heterogeneity because they are more likely to find their niche, or alternatively they could be increasingly maladapted with more environmental heterogeneity (Farkas et al., 2015). Overall, it is likely that a large portion of model deviations from the classic TIB would be linked to the variance of environmental conditions, and the average conditions relative to the average niches of the mainland species pool. With such extensions, we could find flat, positive, negative and hump-shaped relationships between average conditions and species richness.

The TIB has been established and tested for islands, but the vision provided in the early sixties goes beyond the scope of islands. The TIB can indeed be regarded as the simplest metacommunity model with two patches.

The first patch has an infinite area (the continent) and contains all the species. The second patch has a finite area (the island) and contains a sample of species present in the first patch. The TIB describes the occupancy dynamics of the latter given colonization of species from the former. Although this vision is applicable in many situations, regardless of the insular nature of patches considered, it has often been restricted to islands and fragmentation studies (Losos and Ricklefs, 2009). Among the limitations hampering the application of the TIB to other contexts, there is the paucity of predictions regarding abiotic conditions, such as temperature and precipitation, which are at the core of the theory of biogeography (Peterson et al., 2011). This apparent gap in the TIB is not surprising given that on many islands, abiotic conditions can reasonably be assumed similar between islands and the continent, i.e. the latitudinal range of colonization from the continent is restricted. However, this assumption prevents the TIB from being used to study species richness along environmental gradients.

Recently, Cazelles et al. (2015b) have developed a framework inspired from the TTIB that integrates abiotic constraints. According to their work, introducing an environmental gradient in the TTIB requires an explicit statement for how the local conditions prevailing on the island affect species differently, i.e. modulating the extinction rate, the colonization rate or both according to the values of the environmental gradient. The authors highlight the potential of such an integration for examining the interplay between biotic and abiotic constraints, which is often neglected in models used to predict species distributions (Thuiller et al., 2013). For instance, they hinted at the importance of positive covariation in the response to the environment between interacting species. Independent environmental requirements between interacting species might result in a dislocation of local networks. Broadly speaking, such an extension of the TIB corresponds to the introduction of the abiotic niche in the TTIB and the possibility of studying the relationship between the fundamental niche and the realized one (Cazelles et al., 2015a).

4.2.4 Network Macroevolution

The isolation of oceanic islands provides an ideal setting to investigate diversification dynamics in a context of limited immigration and gene flow (Gillespie, 2004; Losos et al., 1998; Warren et al., 2015). Island biogeography theory has been influential to the study of macroevolution and community phylogenetics (Warren et al., 2015). Fairly simple extensions to the modelling approach described in this study might allow investigation of the macroevolution of network structure. The model of colonization-extinction dynamics is analogous to speciation-extinction models of macroevolution, with the distinction that in macroevolutionary models the regional pools emerges from the dynamics and is not fixed in time.

There is currently an intense debate about the effect of ecological interactions on diversification rates (Harmon and Harrison, 2015; Rabosky and Hurlbert, 2015), and it is possible that an evolutionary extension of trophic island biogeography could perhaps address these issues. Except for a few mechanistic endeavours (e.g. Hubert et al., 2015), the bulk of current quantitative approaches considers a phenomenological model to represent the effect of species diversity on diversification, under which the net diversification rate (speciation and extinction) slowly decays with diversity, in a manner similar to a logistic growth model (Ricklefs, 2007). Following the dynamics of the TTIB, one could consider that a successful speciation event will occur provided that a mutant finds at least one prey species, and accordingly it goes extinct on losing its last prey (see also Quince et al., 2002). Such a model would predict an exponentially increasing diversification rate in absence of top-down control because the more species there are, the more successful mutants will be and the lower the extinction rate. On the other hand, diversification is expected to saturate at a stable equilibrium species richness in the presence of negative effects of predators on the extinction rate of the preys since higher species richness will result in stronger top-down control (Rabosky and Hurlbert, 2015). Such a model could be used to investigate a much larger range of predictions than typical birth-death models do (Ricklefs, 2007), with predictions on diversification rates as well as phylogenetic structure of trophic interactions (Peralta, in press).

4.2.5 Biological Invasions

A recurring question in the field of biological invasions is to understand the determinants of invasion success and impacts (David et al., 2017; Mollot et al., 2017; Pantel et al., 2017). Historically, approaches to solve this question have first relied on species-centred, static attributes (i.e. not variable among populations of the same species and/or considered absolute measures of species "invasiveness") such as fecundity, generation time, growth and development trajectories, etc. (Van Kleunen et al., 2010, 2015) and include "classic laws" e.g. Baker's law (Baker, 1955, 1967)—self-compatible plants should have an edge as invaders of areas disconnected from their native distribution. However, multiple traits are often required to obtain good predictions of species invasiveness (Kuster et al., 2008). A second approach has been to compare trait values to those of native species within the invaded ecosystems, i.e. to look at the similarity (or dissimilarity) of the potential invader with the rest of the local community, based on trait values, functional groups and/or phylogeny (Gross et al., 2013; Leffler et al., 2014; Minoarivelo and Hui, 2016; Pearse and Altermatt, 2013b). Other approaches have also used the opposite approach, i.e. looked at native traits or functions which could prevent invasion by a repetitive invader (Byun et al., 2013). However, in all these approaches, the structure of the underlying interaction networks—the invaded network and the network which natively hosts the invading species—is rarely taken into account.

The TTIB could be used as a simple null model of the invasibility of food webs, insofar as it could account for the trophic levels of both native and invasive species, and thus generates predictions for the probability of species occurrence "on the island" given its position within the interaction network (mostly its trophic level and degree as predator) and colonization/extinction parameters. As a dynamic model of species occupancy, the TTIB could also be used to predict temporal invasion sequences based on the position of invaders within the network. Whereas most models tend to predict invasion success independently of other invasions in the system, the TTIB could help design a more realistic model of network invasion through its built-in mechanism of sequential colonization dependence. In place of predicting the probability of a single invasion success, it would therefore give the probability that a given series of invasions occurs, or the probability that a given species invades before another potential invader.

4.3 Using the TTIB to Model Species Distribution

Since the pioneering study of Davis et al. (1998), there is growing evidence for the role of ecological interactions in community response to climate change (Harmon et al., 2009; Suttle et al., 2007). Treating species as isolated entities when predicting their future geographic distribution will likely result in erroneous predictions (Clark et al., 2014). To make such an assumption, we must understand how interactions influence species distributions (Holt, 2009) and also how environmental gradients affect the strength of interactions (Poisot et al., 2012). Adding abiotic constraints into the TTIB yields an adequate theoretical model to scrutinize the relative contribution of trophic interactions and individuals response in the context of climate change (Cazelles et al., 2015b). In this framework, a species colonizes a patch according to its abiotic requirements and survives only if it becomes embedded in a network that is assumed to be feasible. Hence, a predator whose prey cannot survive during the warming up will go extinct. This provides a simple model that reflects observed extinctions under climate changes mainly caused by changes in network structure (Cahill et al., 2013; Pearse and Altermatt, 2013a; Säterberg et al., 2013). This application of an extended TTIB could be the key to renewal of the theoretical foundations of species distribution models, but it will demand significant effort to turn this theoretical model into a statistical tool (see Gravel et al., 2011b).

4.4 From Theory to Data

Extensions of the TIB, such as the incorporation of trophic interactions, not only introduce more realism but also considerably enrich the range of predictions and our ability to test theory with empirical observations. The species–area relationship is one of the best-documented observations in ecology (Preston, 1962), but it also has several competing explanations for which there is currently no agreement (Lomolino, 2000). In such a case of conflicting theories with common predictions, distinguishing them does require testing them from different perspectives (Chave et al., 2002; McGill, 2003). It is of course a challenge to increase model complexity without facing a trade-off with the number of parameters to evaluate. But, the development and evaluation of the TTIB with empirical data has shown it is possible to keep a model relatively simple, without the addition of extra parameters, owing to some realistic assumptions (Gravel et al., 2011b). As presented earlier, the biggest advantage of the approach developed in this chapter is to derive species-level predictions, in comparison to the community aggregated predictions of the classic TIB. It is possible to fit the occupancy functions using maximum likelihood methods (Gravel et al., 2011b). The next step is perhaps to evaluate the capacity of the theory to predict finer aspects of network structure, such as the distribution of motifs (Stouffer et al., 2007). The frequency distribution of the motifs could be derived using approximate Bayesian computing (Beaumont, 2010) or exponential random graph models (Snijders et al., 2006), and compared to empirical data. Unfortunately, we still lack appropriate data to go beyond what has been done so far. Doing so will require specific data on realized interactions at different locations and, for now, there are only few candidate datasets with replicated network sampling that could allow it.

Surprisingly, despite the simplicity of the approach, we are not aware of any attempt to directly measure colonization and extinction processes, and then comparing quantitative predictions of expected species richness to observations. Doing so would require time–series data, such as the one collected by Simberloff and Wilson (1969) in their classic experiment, measuring the colonization and extinction probabilities for different islands and then relating them to island area and isolation. One would simply seek to evaluate the probability of a colonization (and extinction) event occurring between time step t and step $t + dt$. The model could be rendered more complex, with detailed investigation of the species-specific colonization and extinction rates, trying to relate them to their position in the food web and other functional traits. This is indeed what Cirtwill and Stouffer (2016) performed for colonization. They found that in Simberloff's experiment, the variability of colonization rates among arthropods was best understood from species diet. Time–series data for entire communities combined with knowledge of trophic interactions are hard to gather but would have immense value for the advancement of island biogeography theory specifically, and for community assembly theory and ecology more generally.

ACKNOWLEDGEMENTS

We thank D.A. Bohan for organizing this issue and for editing this manuscript, the EEP unit and particularly S. Gallina for access to computational resources, the French ANR for funding (AFFAIRS project in the BIOADAPT program—P.I.: Dr P. David—grant no. 12-ADAP-005; ARSENIC project—P.I.: Dr F. Massol—grant no. 14-CE02-0012), the CESAB working group COREIDS for opportunities to develop this work and TOTAL and the FRB for funding COREIDS. We thank N. Loeuille and an anonymous referee for making insightful suggestions on the first version of the manuscript.

APPENDIX. BREADTH-FIRST DIRECTING OF THE LINKS IN THE GENERATED NETWORK

For a undirected connected simple graph $\mathcal{G} = (\mathcal{V}, \mathcal{E})$, with $\mathcal{V}$ the set of vertices and $\mathcal{E}$ the edges set. $\mathcal{V}_B$ (with $\mathcal{V}_B \subset \mathcal{V}$) represent the basal species set, drawn from the colouring algorithm. Capital letters are vertices set, upper-case letters represent one vertex. Finally, double arrow ($\leftrightarrow$) represents an edge (directed or not), simple right arrow ($v \mapsto \omega$) is a directed edge from v to ω and simple left arrow ($\leftarrow$) is an affectation.

$$\mathcal{V}_f \leftarrow \mathcal{V}_B$$
$$\mathcal{V}' \leftarrow \mathcal{V}_B$$

while $\mathcal{V}' \neq \mathcal{V}$ do   // While all nodes have not be done

  $\mathcal{V}_f' \leftarrow \varnothing$
  for $v\ in\ \mathcal{V}_f$ do
    for $\omega\ in\ A(v)$ do   // $A(v)$ is the set of vertices adjacent to vertex v
      if $v \leftrightarrow \omega$ is undirected then   // If edge between v and ω is undirected
        $v \mapsto \omega$ // Edge is direct from v to ω
        $\mathcal{V}_f' \leftarrow \text{add}(\omega)$
      done
    done
  done

  for $v\ in\ \mathcal{V}_f'$ do
   for $\omega\ in\ A(v)$ do
    if $v \leftrightarrow \omega$ is undirected AND $\omega \in \mathcal{V}_f'$ then
      $v \mapsto \omega$ OR $\omega \mapsto v$ // Edge can be in one direction or the other, with the
    done   // same probability
   done
  done
  $\mathcal{V}_f \leftarrow \mathcal{V}_f'$
  $\mathcal{V}' \leftarrow \mathcal{V}' \cup \mathcal{V}_f'$
done

The above-described algorithm is illustrated in the following figure. Starting from an undirected graph (A–M) with basal species (A–C), the different steps (1–10) describe how (i) the graph is explored (steps 2, 5 and 8), (ii) the links are directed from one "trophic level" to the next (steps 3, 6 and 9) and (iii) links within a given trophic level are randomly directed (steps 4, 7 and 10).

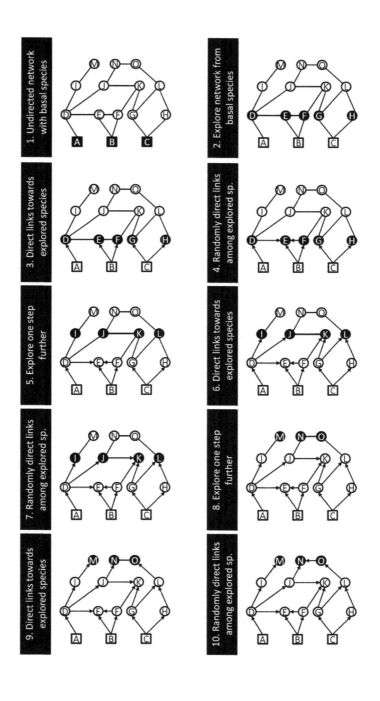

1. Undirected network with basal species

2. Explore network from basal species

3. Direct links towards explored species

4. Randomly direct links among explored sp.

5. Explore one step further

6. Direct links towards explored species

7. Randomly direct links among explored sp.

8. Explore one step further

9. Direct links towards explored species

10. Randomly direct links among explored sp.

REFERENCES

Albert, R., Jeong, H., Barabasi, A.-L., 2000. Error and attack tolerance of complex networks. Nature 406, 378–382.

Allesina, S., Tang, S., 2012. Stability criteria for complex ecosystems. Nature 483, 205–208.

Allesina, S., Alonso, D., Pascual, M., 2008. A general model for food web structure. Science 320, 658–661.

Allesina, S., Grilli, J., Barabas, G., Tang, S., Aljadeff, J., Maritan, A., 2015. Predicting the stability of large structured food webs. Nat. Commun. 6, http://dx.doi.org/10.1038/ncomms8842 (online journal).

Allhoff, K.T., Ritterskamp, D., Rall, B.C., Drossel, B., Guill, C., 2015. Evolutionary food web model based on body masses gives realistic networks with permanent species turnover. Sci. Rep. 5, http://dx.doi.org/10.1038/srep10955 (online journal).

Allouche, O., Kalyuzhny, M., Moreno-Rueda, G., Pizarro, M., Kadmon, R., 2012. Area–heterogeneity tradeoff and the diversity of ecological communities. Proc. Natl. Acad. Sci. U.S.A. 109, 17495–17500.

Amsellem, L., Brouat, C., Duron, O., Porter, S.S., Vilcinskas, A., Facon, B., 2017. Importance of microorganisms to macroorganisms invasions: is the essential invisible to the eye? (The little prince, A. de Saint-Exupery, 1943) Adv. Ecol. Res. 57, 99–146.

Arenas, A., Duch, J., Fernández, A., Gómez, S., 2007. Size reduction of complex networks preserving modularity. New J. Phys. 9, 176.

Astegiano, J., Guimarães Jr., P.R., Cheptou, P.-O., Vidal, M.M., Mandai, C.Y., Ashworth, L., Massol, F., 2015. Persistence of plants and pollinators in the face of habitat loss: Insights from trait-based metacommunity models. Adv. Ecol. Res. 53, 201–257. Academic Press.

Baker, H.G., 1955. Self compatibility and establishment after "long-distance" dispersal. Evolution 9, 347–349.

Baker, H.G., 1967. Support for Baker's law as a rule. Evolution 21, 853–856.

Barabási, A.L., Albert, R., 1999. Emergence of scaling in random networks. Science 286, 509–512.

Beaumont, M.A., 2010. Approximate Bayesian computation in evolution and ecology. Annu. Rev. Ecol. Evol. Syst. 41, 379–406.

Bell, G., 2000. The distribution of abundance in neutral communities. Am. Nat. 155, 606–617.

Bell, G., 2001. Neutral macroecology. Science 293, 2413–2418.

Berger, A., 2014. A note on the characterization of digraphic sequences. Discret. Math. 314, 38–41.

Blackburn, T.M., Gaston, K.J., 1999. The relationship between animal abundance and body size: a review of the mechanisms. Adv. Ecol. Res. 28, 181–210. Academic Press.

Blackburn, T.M., Lawton, J.H., Pimm, S.L., 1993. Non-metabolic explanations for the relationship between body size and animal abundance. J. Anim. Ecol. 62, 694–702.

Blöchliger, I., Zufferey, N., 2008. A graph coloring heuristic using partial solutions and a reactive tabu scheme. Comput. Oper. Res. 35, 960–975.

Bollobás, B., Riordan, O., Spencer, J., Tusnády, G., 2001. The degree sequence of a scale-free random graph process. Random Struct. Algorithm 18, 279–290.

Bonsall, M.B., Jansen, V.A.A., Hassell, M.P., 2004. Life history trade-offs assemble ecological guilds. Science 306, 111–114.

Boots, M., White, A., Best, A., Bowers, R., 2014. How specificity and epidemiology drive the coevolution of static trait diversity in hosts and parasites. Evolution 68, 1594–1606.

Bradbury, I.R., Laurel, B., Snelgrove, P.V.R., Bentzen, P., Campana, S.E., 2008. Global patterns in marine dispersal estimates: the influence of geography, taxonomic category and life history. Proc. R. Soc. Lond. B Biol. Sci. 275, 1803–1809.

Brännström, Å., Johansson, J., Loeuille, N., Kristensen, N., Troost, T.A., Lambers, R.H.R., Dieckmann, U., 2012. Modelling the ecology and evolution of communities: a review of past achievements, current efforts, and future promises. Evol. Ecol. Res. 14, 601–625.

Briand, F., Cohen, J.E., 1984. Community food webs have scale-invariant structure. Nature 307, 264–267.

Brown, J.H., Gillooly, J.F., Allen, A.P., Savage, V.M., West, G.B., 2004. Toward a metabolic theory of ecology. Ecology 85, 1771–1789.

Brualdi, R.A., Ryser, H.J., 1991. Combinatorial Matrix Theory. Cambridge University Press, Cambridge, UK.

Byun, C., de Blois, S., Brisson, J., 2013. Plant functional group identity and diversity determine biotic resistance to invasion by an exotic grass. J. Ecol. 101, 128–139.

Cahill, A.E., Aiello-Lammens, M.E., Fisher-Reid, M.C., Hua, X., Karanewsky, C.J., Yeong Ryu, H., Sbeglia, G.C., Spagnolo, F., Waldron, J.B., Warsi, O., Wiens, J.J., 2013. How does climate change cause extinction? Proc. R. Soc. B Biol. Sci. 280, 20121890.

Calcagno, V., Mouquet, N., Jarne, P., David, P., 2006. Coexistence in a metacommunity: the competition-colonization trade-off is not dead. Ecol. Lett. 9, 897–907.

Calcagno, V., Massol, F., Mouquet, N., Jarne, P., David, P., 2011. Constraints on food chain length arising from regional metacommunity dynamics. Proc. R. Soc. Lond. B Biol. Sci. 278, 3042–3049.

Camacho, J., Guimerà, R., Nunes Amaral, L.A., 2002. Robust patterns in food web structure. Phys. Rev. Lett. 88, 228102.

Cameron, R.A.D., Triantis, K.A., Parent, C.E., Guilhaumon, F., Alonso, M.R., Ibáñez, M., de Frias Martins, A.M., Ladle, R.J., Whittaker, R.J., 2013. Snails on oceanic islands: testing the general dynamic model of oceanic island biogeography using linear mixed effect models. J. Biogeogr. 40, 117–130.

Cazelles, K., Araújo, M.B., Mouquet, N., Gravel, D., 2015a. A theory for species co-occurrence in interaction networks. Theor. Ecol. 9, 39–48.

Cazelles, K., Mouquet, N., Mouillot, D., Gravel, D., 2015b. On the integration of biotic interaction and environmental constraints at the biogeographical scale. Ecography 39, 921–931.

Chase, J.M., Leibold, M.A., Downing, A.L., Shurin, J.B., 2000. The effects of productivity, herbivory, and plant species turnover in grassland food webs. Ecology 81, 2485–2497.

Chave, J., Muller-Landau, H.C., Levin, S.A., 2002. Comparing classical community models: theoretical consequences for patterns of diversity. Am. Nat. 159, 1–23.

Chesson, P., 2000. Mechanisms of maintenance of species diversity. Annu. Rev. Ecol. Syst. 31, 343–366.

Chung, F., Lu, L., Vu, V., 2003. Spectra of random graphs with given expected degrees. Proc. Natl. Acad. Sci. U.S.A. 100, 6313–6318.

Cirtwill, A.R., Stouffer, D.B., 2016. Knowledge of predator–prey interactions improves predictions of immigration and extinction in island biogeography. Glob. Ecol. Biogeogr. 25, 900–911.

Clark, J.S., Gelfand, A.E., Woodall, C.W., Zhu, K., 2014. More than the sum of the parts: forest climate response from joint species distribution models. Ecol. Appl. 24, 990–999.

Clauset, A., Shalizi, C.R., Newman, M.E.J., 2009. Power-law distributions in empirical data. SIAM Rev. 51, 661–703.

Cohen, J.E., Briand, F., 1984. Trophic links of community food webs. Proc. Natl. Acad. Sci. U.S.A. 81, 4105–4109.

Cohen, J., Łuczak, T., 1992. Trophic levels in community food webs. Evol. Ecol. 6, 73–89.

Cohen, J.E., Newman, C.M., 1985. A stochastic theory of community food webs: I. Models and aggregated data. Proc. R. Soc. B Biol. Sci. 224, 421–448.

Cohen, R., Erez, K., ben-Avraham, D., Havlin, S., 2000. Resilience of the internet to random breakdowns. Phys. Rev. Lett. 85, 4626–4628.

Cohen, R., Erez, K., ben-Avraham, D., Havlin, S., 2001. Breakdown of the internet under intentional attack. Phys. Rev. Lett. 86, 3682–3685.

Cohen, J.E., Jonsson, T., Carpenter, S.R., 2003. Ecological community description using the food web, species abundance, and body size. Proc. Natl. Acad. Sci. U.S.A. 100, 1781–1786.

Condit, R., Pitman, N., Leigh, E.G., Chave, J., Terborgh, J., Foster, R.B., Nunez, P., Aguilar, S., Valencia, R., Villa, G., Muller-Landau, H.C., Losos, E., Hubbell, S.P., 2002. Beta-diversity in tropical forest trees. Science 295, 666–669.

Csardi, G., Nepusz, T., 2006. The igraph software package for complex network research. InterJournal Complex Syst. 1695, 1–9.

Currie, D.J., 1991. Energy and large-scale patterns of animal- and plant-species richness. Am. Nat. 137, 27–49.

Currie, D.J., Mittelbach, G.G., Cornell, H.V., Field, R., Guégan, J.-F., Hawkins, B.A., Kaufman, D.M., Kerr, J.T., Oberdorff, T., O'Brien, E., Turner, J.R.G., 2004. Predictions and tests of climate-based hypotheses of broad-scale variation in taxonomic richness. Ecol. Lett. 7, 1121–1134.

Curtsdotter, A., Binzer, A., Brose, U., de Castro, F., Ebenman, B., Eklöf, A., Riede, J.O., Thierry, A., Rall, B.C., 2011. Robustness to secondary extinctions: comparing trait-based sequential deletions in static and dynamic food webs. Basic Appl. Ecol. 12, 571–580.

Damuth, J., 1981. Population density and body size in mammals. Nature 290, 699–700.

Damuth, J., Associate Editor: Claire de, M., Editor: Donald, L.D., 2007. A Macroevolutionary Explanation for Energy Equivalence in the Scaling of Body Size and Population Density. The American Naturalist, 169, 621–631.

Daufresne, T., Hedin, L.O., 2005. Plant coexistence depends on ecosystem nutrient cycles: extension of the resource-ratio theory. Proc. Natl. Acad. Sci. U.S.A. 102, 9212–9217.

David, P., Thébault, E., Anneville, O., Duyck, P.-F., Chapuis, E., Loeuille, N., 2017. Impacts of invasive species on food webs: a review of empirical data. Adv. Ecol. Res. 56, 1–60.

Davis, A.J., Jenkinson, L.S., Lawton, J.H., Shorrocks, B., Wood, S., 1998. Making mistakes when predicting shifts in species range in response to global warming. Nature 391, 783–786.

De Solla Price, D., 1976. A general theory of bibliometric and other cumulative advantage processes. J. Am. Soc. Inf. Sci. 27, 292–306.

Dieckmann, U., Marrow, P., Law, R., 1995. Evolutionary cycling in predator-prey interactions: population dynamics and the red queen. J. Theor. Biol. 176, 91–102.

Doebeli, M., Dieckmann, U., 2000. Evolutionary branching and sympatric speciation caused by different types of ecological interactions. Am. Nat. 156, S77–S101.

Downing, A.L., Leibold, M.A., 2002. Ecosystem consequences of species richness and composition in pond food webs. Nature 416, 837–841.

Drossel, B., McKane, A.J., Quince, C., 2004. The impact of nonlinear functional responses on the long-term evolution of food web structure. J. Theor. Biol. 229, 539–548.

Duggins, D.O., Simenstad, C.A., Estes, J.A., 1989. Magnification of secondary production by kelp detritus in coastal marine ecosystems. Science 245, 170–173.

Dunne, J.A., Williams, R.J., Martinez, N.D., 2002a. Food-web structure and network theory: the role of connectance and size. Proc. Natl. Acad. Sci. U.S.A. 99, 12917–12922.

Dunne, J.A., Williams, R.J., Martinez, N.D., 2002b. Network structure and biodiversity loss in food webs: robustness increases with connectance. Ecol. Lett. 5, 558–567.

Dunne, J.A., Williams, R.J., Martinez, N.D., 2004. Network structure and robustness of marine food webs. Mar. Ecol. Prog. Ser. 273, 291–302.

Eklöf, A., Jacob, U., Kopp, J., Bosch, J., Castro-Urgal, R., Chacoff, N.P., Dalsgaard, B., de Sassi, C., Galetti, M., Guimarães, P.R., Lomáscolo, S.B., Martín González, A.M.,

Pizo, M.A., Rader, R., Rodrigo, A., Tylianakis, J.M., Vázquez, D.P., Allesina, S., 2013. The dimensionality of ecological networks. Ecol. Lett. 16, 577–583.

Erdős, P., Rényi, A., 1959. On random graphs I. Publ. Math. Debr. 6, 290–297.

Estes, J.A., Duggins, D.O., 1995. Sea otters and Kelp forests in Alaska: generality and variation in a community ecological paradigm. Ecol. Monogr. 65, 75–100.

Estes, J.A., Tinker, M.T., Williams, T.M., Doak, D.F., 1998. Killer whale predation on sea otters linking oceanic and nearshore ecosystems. Science 282, 473–476.

Etienne, R.S., Alonso, D., 2005. A dispersal-limited sampling theory for species and alleles. Ecol. Lett. 8, 1147–1156.

Etienne, R.S., Olff, H., 2004. How dispersal limitation shapes species-body size distributions in local communities. Am. Nat. 163, 69–83.

Farkas, T.E., Hendry, A.P., Nosil, P., Beckerman, A.P., 2015. How maladaptation can structure biodiversity: eco-evolutionary island biogeography. Trends Ecol. Evol. 30, 154–160.

Fortuna, M.A., Bascompte, J., 2006. Habitat loss and the structure of plant-animal mutualistic networks. Ecol. Lett. 9, 278–283.

Fortuna, M.A., Stouffer, D.B., Olesen, J.M., Jordano, P., Mouillot, D., Krasnov, B.R., Poulin, R., Bascompte, J., 2010. Nestedness versus modularity in ecological networks: two sides of the same coin? J. Anim. Ecol. 79, 811–817.

Fortuna, M.A., Krishna, A., Bascompte, J., 2013. Habitat loss and the disassembly of mutalistic networks. Oikos 122, 938–942.

Gandon, S., Buckling, A., Decaestecker, E., Day, T., 2008. Host–parasite coevolution and patterns of adaptation across time and space. J. Evol. Biol. 21, 1861–1866.

Gaston, K.J., 2000. Global patterns in biodiversity. Nature 405, 220–227.

Gillespie, D.T., 1976. A general method for numerically simulating the stochastic time evolution of coupled chemical reactions. J. Comput. Phys. 22, 403–434.

Gillespie, D.T., 1977. Exact stochastic simulation of coupled chemical reactions. J. Phys. Chem. 81, 2340–2361.

Gillespie, R., 2004. Community assembly through adaptive radiation in Hawaiian spiders. Science 303, 356–359.

Gkantsidis, C., Mihail, M., Zegura, E., 2003. The markov chain simulation method for generating connected Power law random graphs. In: Proceedings of the Fifth Workshop on Algorithm Engineering and Experiments. SIAM, pp. 16–25.

Gotelli, N.J., Taylor, C.M., 1999. Testing macroecology models with stream-fish assemblages. Evol. Ecol. Res. 1, 847–858.

Govaert, G., Nadif, M., 2008. Block clustering with Bernoulli mixture models: comparison of different approaches. Comput. Stat. Data Anal. 52, 3233–3245.

Graham, N.A.J., Chabanet, P., Evans, R.D., Jennings, S., Letourneur, Y., Aaron MacNeil, M., McClanahan, T.R., Öhman, M.C., Polunin, N.V.C., Wilson, S.K., 2011. Extinction vulnerability of coral reef fishes. Ecol. Lett. 14, 341–348.

Gravel, D., Guichard, F., Loreau, M., Mouquet, N., 2010. Source and sink dynamics in meta-ecosystems. Ecology 91, 2172–2184.

Gravel, D., Canard, E., Guichard, F., Mouquet, N., 2011a. Persistence increases with diversity and connectance in trophic metacommunities. PLoS One 6. e19374. http://dx.doi.org/10.1371/journal.pone.0019374.

Gravel, D., Massol, F., Canard, E., Mouillot, D., Mouquet, N., 2011b. Trophic theory of island biogeography. Ecol. Lett. 14, 1010–1016.

Gravel, D., Poisot, T., Albouy, C., Velez, L., Mouillot, D., 2013. Inferring food web structure from predator–prey body size relationships. Methods Ecol. Evol. 4, 1083–1090.

Gravel, D., Albouy, C., Thuiller, W., 2016. The meaning of functional trait composition of food webs for ecosystem functioning. Philos. Trans. R. Soc. B 371, 20150268.

Gross, N., Börger, L., Duncan, R.P., Hulme, P.E., 2013. Functional differences between alien and native species: do biotic interactions determine the functional structure of highly invaded grasslands? Funct. Ecol. 27, 1262–1272.

Guilhaumon, F., Gimenez, O., Gaston, K.J., Mouillot, D., 2008. Taxonomic and regional uncertainty in species-area relationships and the identification of richness hotspots. Proc. Natl. Acad. Sci. U.S.A. 105, 15458–15463.

Guimera, R., Stouffer, D.B., Sales-Pardo, M., Leicht, E.A., Newman, M.E.J., Amaral, L.A.N., 2010. Origin of compartmentalization in food webs. Ecology 91, 2941–2951.

Haegeman, B., Loreau, M., 2014. General relationships between consumer dispersal, resource dispersal and metacommunity diversity. Ecol. Lett. 17, 175–184.

Hairston, N.G., Smith, F.E., Slobodkin, L.B., 1960. Community structure, population control, and competition. Am. Nat. 44, 421–425.

Hannan, L.B., Roth, J.D., Ehrhart, L.M., Weishampel, J.F., 2007. Dune vegetation fertilization by nesting sea turtles. Ecology 88, 1053–1058.

Hanski, I., 1999. Metapopulation Ecology. Oxford University Press, Oxford.

Harmon, L.J., Harrison, S., 2015. Species diversity is dynamic and unbounded at local and continental scales. Am. Nat. 185, 584–593.

Harmon, J.P., Moran, N.A., Ives, A.R., 2009. Species response to environmental change: impacts of food web interactions and evolution. Science 323, 1347–1350.

Hastings, A., 1980. Disturbance, coexistence, history, and competition for space. Theor. Popul. Biol. 18, 363–373.

Hastings, A., 1987. Can competition be detected using species co-occurrence data? Ecology 68, 117–123.

Havens, K., 1992. Scale and structure in natural food webs. Science 257, 1107–1109.

Helfield, J.M., Naiman, R.J., 2002. Salmon and alder as nitrogen sources to riparian forests in a boreal Alaskan watershed. Oecologia 133, 573–582.

HilleRisLambers, J., Adler, P.B., Harpole, W.S., Levine, J.M., Mayfield, M.M., 2012. Rethinking community assembly through the lens of coexistence theory. Annu. Rev. Ecol. Evol. Syst. 43, 227.

Holt, R.D., 1977. Predation, apparent competition, and structure of prey communities. Theor. Popul. Biol. 12, 197–229.

Holt, R.D., 2002. Food webs in space: on the interplay of dynamic instability and spatial processes. Ecol. Res. 17, 261–273.

Holt, R.D., 2009. Bringing the Hutchinsonian niche into the 21st century: ecological and evolutionary perspectives. Proc. Natl. Acad. Sci. 106, 19659–19665.

Holt, R.D., Lawton, J.H., Polis, G.A., Martinez, N.D., 1999. Trophic rank and the species-area relationship. Ecology 80, 1495–1504.

Holt, R.D., Gomulkiewicz, R., Barfield, M., 2003. The phenomology of niche evolution via quantitive traits in a "black-hole" sink. Proc. R. Soc. Lond. Ser. B 270, 215–224.

Hubbell, S.P., 2001. The Unified Neutral Theory of Biodiversity and Biogeography. Princeton University Press, Princeton.

Hubert, N., Calcagno, V., Etienne, R.S., Mouquet, N., 2015. Metacommunity speciation models and their implications for diversification theory. Ecol. Lett. 18, 864–881.

Huffaker, C.B., 1958. Experimental studies on predation: dispersion factors and predator-prey oscillations. Hilgardia 27, 343–383.

Huffaker, C.B., Shea, K.P., Herman, S.G., 1963. Experimental studies on predation: complex dispersion and levels of food in an acarine predator-prey interaction. Hilgardia 34, 305–329.

Hurlbert, A.H., 2004. Species–energy relationships and habitat complexity in bird communities. Ecol. Lett. 7, 714–720.

Hutchinson, G.E., 1959. Homage to Santa Rosalia, or why are there so many kinds of animals? Am. Nat. 93, 145–159.

Jefferies, R.L., Rockwell, R.F., Abraham, K.F., 2004. Agricultural food subsidies, migratory connectivity and large-scale disturbance in Arctic coastal systems: a case study. Integr. Comp. Biol. 44, 130.

Jetz, W., Carbone, C., Fulford, J., Brown, J.H., 2004. The scaling of animal space use. Science 306, 266–268.

Kadmon, R., Allouche, O., 2007. Integrating the effects of area, isolation, and habitat heterogeneity on species diversity: a unification of island biogeography and niche theory. Am. Nat. 170, 443–454.

Kawecki, T.J., 2004. Ecological and evolutionary consequences of source-sink population dynamics. In: Hanski, I., Gaggiotti, O.E. (Eds.), Ecology, Genetics, and Evolution of Metapopulations. Elsevier Academic Press, Burlington, pp. 387–414.

Kéfi, S., Berlow, E.L., Wieters, E.A., Navarrete, S.A., Petchey, O.L., Wood, S.A., Boit, A., Joppa, L.N., Lafferty, K.D., Williams, R.J., Martinez, N.D., Menge, B.A., Blanchette, C.A., Iles, A.C., Brose, U., 2012. More than a meal … integrating non-feeding interactions into food webs. Ecol. Lett. 15, 291–300.

Kim, H., Toroczkai, Z., Erdős, P.L., Miklós, I., Székely, L.A., 2009. Degree-based graph construction. J. Phys. A Math. Theor. 42, 392001.

Krause, A.E., Frank, K.A., Mason, D.M., Ulanowicz, R.E., Taylor, W.W., 2003. Compartments revealed in food-web structure. Nature 426, 282–285.

Kulbicki, M., Parravicini, V., Mouillot, D., 2015. Patterns and processes in reef fish body size. In: Mora, C. (Ed.), Ecology of Fishes on Coral Reefs. Cambridge University Press, Cambridge, UK, pp. 104–115.

Kuster, E.C., Kuhn, I., Bruelheide, H., Klotz, S., 2008. Trait interactions help explain plant invasion success in the German flora. J. Ecol. 96, 860–868.

Leffler, A.J., James, J.J., Monaco, T.A., Sheley, R.L., 2014. A new perspective on trait differences between native and invasive exotic plants. Ecology 95, 298–305.

Leibold, M.A., 1996. A graphical model of keystone predators in food webs: trophic regulation of abundance, incidence, and diversity patterns in communities. Am. Nat. 147, 784–812.

Leibold, M.A., Holyoak, M., Mouquet, N., Amarasekare, P., Chase, J.M., Hoopes, M.F., Holt, R.D., Shurin, J.B., Law, R., Tilman, D., Loreau, M., Gonzalez, A., 2004. The metacommunity concept: a framework for multi-scale community ecology. Ecol. Lett. 7, 601–613.

Leicht, E.A., Newman, M.E.J., 2008. Community structure in directed networks. Phys. Rev. Lett. 100, 118703.

Levins, R., 1969. Some demographic and genetic consequences of environmental heterogeneity for biological control. Bull. Entomol. Soc. Am. 15, 237–240.

Levins, R., Heatwole, H., 1963. On the distribution of organisms on islands. Caribb. J. Sci. 3, 173–177.

Loeuille, N., Leibold, M.A., 2008. Ecological consequences of evolution in plant defenses in a metacommunity. Theor. Popul. Biol. 74, 34–45.

Loeuille, N., Leibold, M.A., 2014. Effects of local negative feedbacks on the evolution of species within metacommunities. Ecol. Lett. 17, 563–573.

Loeuille, N., Loreau, M., 2005. Evolutionary emergence of size-structured food webs. Proc. Natl. Acad. Sci. U.S.A. 102, 5761–5766.

Lomolino, M.V., 1985. Body size of mammals on islands: the island rule reexamined. Am. Nat. 125, 310–316.

Lomolino, M.V., 2000. Ecology's most general, yet protean pattern: the species-area relationship. J. Biogeogr. 27, 17–26.

Loreau, M., Holt, R.D., 2004. Spatial flows and the regulation of ecosystems. Am. Nat. 163, 606–615.

Loreau, M., Daufresne, T., Gonzalez, A., Gravel, D., Guichard, F., Leroux, S.J., Loeuille, N., Massol, F., Mouquet, N., 2013. Unifying sources and sinks in ecology and Earth sciences. Biol. Rev. 88, 365–379.

Losos, J.B., Ricklefs, R.E., 2009. The Theory of Island Biogeography Revisited. Princeton University Press, Princeton, NJ.

Losos, J.B., Jackman, T.R., Larson, A., de Queiroz, K., Rodriguez-Schettino, L., 1998. Contingency and determinism in replicated adaptive radiations of island lizards. Science 279, 2115–2118.

Luiz, O.J., Allen, A.P., Robertson, D.R., Floeter, S.R., Kulbicki, M., Vigliola, L., Becheler, R., Madin, J.S., 2013. Adult and larval traits as determinants of geographic range size among tropical reef fishes. Proc. Natl. Acad. Sci. U.S.A. 110, 16498–16502.

MacArthur, R.H., Wilson, E.O., 1963. An equilibrium theory of insular zoogeography. Evolution 17, 373–387.

MacArthur, R.H., Wilson, E.O., 1967. The Theory of Island Biogeography. Princeton University Press, Princeton.

Marquet, P.A., Quiñones, R.A., Abades, S., Labra, F., Tognelli, M., Arim, M., Rivadeneira, M., 2005. Scaling and Power-laws in ecological systems. J. Exp. Biol. 208, 1749–1769.

Martinez, N.D., 1992. Constant connectance in community food webs. Am. Nat. 139, 1208–1218.

Massol, F., Cheptou, P.O., 2011. When should we expect the evolutionary association of self-fertilization and dispersal? Evolution 65, 1217–1220.

Massol, F., Petit, S., 2013. Interaction networks in agricultural landscape mosaics. Adv. Ecol. Res. 49, 291–338.

Massol, F., Gravel, D., Mouquet, N., Cadotte, M.W., Fukami, T., Leibold, M.A., 2011. Linking ecosystem and community dynamics through spatial ecology. Ecol. Lett. 14, 313–323.

May, R.M., 1972. Will a large complex system be stable? Nature 238, 413–414.

May, R.M., 1973. Qualitative stability in model ecosystems. Ecology 54, 638–641.

McCann, K.S., Rasmussen, J.B., Umbanhowar, J., 2005. The dynamics of spatially coupled food webs. Ecol. Lett. 8, 513–523.

McGill, B.J., 2003. A test of the unified neutral theory of biodiversity. Nature 422, 881–885.

Miklós, I., Erdos, P.L., Soukup, L., 2013. Towards random uniform sampling of bipartite graphs with given degree sequence. Electron. J. Comb. 20. P16.

Minoarivelo, H.O., Hui, C., 2016. Invading a mutualistic network: to be or not to be similar. Ecol. Evol. 6, 4981–4996.

Mollot, G., Pantel, J.H., Romanuk, T.N., 2017. The effects of invasive species on the decline in species richness: a global meta-analysis. Adv. Ecol. Res. 56, 61–83.

Mora, C., Chittaro, P.M., Sale, P.F., Kritzer, J.P., Ludsin, S.A., 2003. Patterns and processes in reef fish diversity. Nature 421, 933–936.

Morton, R.D., Law, R., 1997. Regional species pools and the assembly of local ecological communities. J. Theor. Biol. 187, 321–331.

Nakano, S., Murakami, M., 2001. Reciprocal subsidies: dynamic interdependence between terrestrial and aquatic food webs. Proc. Natl. Acad. Sci. U.S.A. 98, 166–170.

Nash, K.L., Welsh, J.Q., Graham, N.A.J., Bellwood, D.R., 2015. Home-range allometry in coral reef fishes: comparison to other vertebrates, methodological issues and management implications. Oecologia 177, 73–83.

Nee, S., May, R.M., 1992. Dynamics of metapopulations: habitat destruction and competitive coexistence. J. Anim. Ecol. 61, 37–40.

Nee, S., Read, A.F., Greenwood, J.J.D., Harvey, P.H., 1991. The relationship between abundance and body size in British birds. Nature 351, 312–313.

Newman, M.E.J., 2004. Detecting community structure in networks. Eur. Phys. J. B. 38, 321–330.

Newman, M.E.J., 2005. Power laws, Pareto distributions and Zipf's law. Contemp. Phys. 46, 323–351.

Newman, M.E.J., 2006a. Finding community structure in networks using the eigenvectors of matrices. Phys. Rev. E 74, 036104.

Newman, M.E.J., 2006b. Modularity and community structure in networks. Proc. Natl. Acad. Sci. U.S.A. 103, 8577–8582.

Newman, M.E.J., Girvan, M., 2004. Finding and evaluating community structure in networks. Phys. Rev. E 69, 026113.

Newman, M.E.J., Strogatz, S.H., Watts, D.J., 2001. Random graphs with arbitrary degree distributions and their applications. Phys. Rev. E 64, 026118.

Nordbotten, J.M., Stenseth, N.C., 2016. Asymmetric ecological conditions favor Red-Queen type of continued evolution over stasis. Proc. Natl. Acad. Sci. U.S.A. 113, 1847–1852.

Nuismer, S.L., Jordano, P., Bascompte, J., 2013. Coevolution and the architecture of mutualistic networks. Evolution 67, 338–354.

Ojanen, S.P., Nieminen, M., Meyke, E., Pöyry, J., Hanski, I., 2013. Long-term metapopulation study of the Glanville fritillary butterfly (Melitaea cinxia): survey methods, data management, and long-term population trends. Ecol. Evol. 3, 3713–3737.

Oksanen, L., Fretwell, S.D., Arruda, J., Niemelä, P., 1981. Exploitation ecosystems in gradients of primary productivity. Am. Nat. 118, 240–261.

Orsini, C., Dankulov, M.M., Colomer-de-Simon, P., Jamakovic, A., Mahadevan, P., Vahdat, A., Bassler, K.E., Toroczkai, Z., Boguna, M., Caldarelli, G., Fortunato, S., Krioukov, D., 2015. Quantifying randomness in real networks. Nat. Commun. 6, http://dx.doi.org/10.1038/ncomms9627.

Otto, S.B., Rall, B.C., Brose, U., 2007. Allometric degree distributions facilitate food-web stability. Nature 450, 1226–1229.

Pantel, J.H., Bohan, D.A., Calcagno, V., David, P., Duyck, P.-F., Kamenova, S., Loeuille, N., Mollot, G., Romanuk, T.N., Thébault, E., Tixier, P., Massol, F., 2017. 14 questions for invasion in ecological networks. Adv. Ecol. Res. 56, 293–340.

Pearse, I.S., Altermatt, F., 2013a. Extinction cascades partially estimate herbivore losses in a complete Lepidoptera-plant food web. Ecology 94, 1785–1794.

Pearse, I.S., Altermatt, F., 2013b. Predicting novel trophic interactions in a non-native world. Ecol. Lett. 16, 1088–1094.

Peralta, G., in press. Merging evolutionary history into species interaction networks. Funct. Ecol. http://dx.doi.org/10.1111/1365-2435.12669.

Persson, L., Diehl, S., Johansson, L., Andersson, G., Hamrin, S.F., 1992. Trophic interactions in temperate lake ecosystems: a test of food chain theory. Am. Nat. 140, 59–84.

Petchey, O.L., Downing, A.L., Mittelbach, G.G., Persson, L., Steiner, C.F., Warren, P.H., Woodward, G., 2004. Species loss and the structure and functioning of multitrophic aquatic systems. Oikos 104, 467–478.

Petchey, O.L., Beckerman, A.P., Riede, J.O., Warren, P.H., 2008. Size, foraging, and food web structure. Proc. Natl. Acad. Sci. U.S.A. 105, 4191–4196.

Peters, R.H., 1986. The Ecological Implications of Body Size. Cambridge University Press, Cambridge, UK.

Peters, R.H., Wassenberg, K., 1983. The effect of body size on animal abundance. Oecologia 60, 89–96.

Peterson, A.T., Soberón, J., Pearson, R.G., Anderson, R.P., Martínez-Meyer, E., Nakamura, M., Araújo, M.B., 2011. Ecological Niches and Geographic Distributions. Princeton University Press, Princeton, NJ.

Piechnik, D.A., Lawler, S.P., Martinez, N.D., 2008. Food-web assembly during a classic biogeographic study: species' "trophic breadth" corresponds to colonization order. Oikos 117, 665–674.

Pillai, P., Guichard, F., 2012. Competition-colonization trade-offs, competitive uncertainty, and the evolutionary assembly of species. PLoS One 7. e33566.

Pillai, P., Loreau, M., Gonzalez, A., 2010. A patch-dynamic framework for food web metacommunities. Theor. Ecol. 3, 223–237.

Pillai, P., Gonzalez, A., Loreau, M., 2011. Metacommunity theory explains the emergence of food web complexity. Proc. Natl. Acad. Sci. U.S.A. 108, 19293–19298.

Pimm, S.L., 1979. The structure of food webs. Theor. Popul. Biol. 16, 144–158.

Pimm, S.L., Lawton, J.H., 1977. Number of trophic levels in ecological communities. Nature 268, 329–331.

Pimm, S.L., Lawton, J.H., Cohen, J.E., 1991. Food web patterns and their consequences. Nature 350, 669–674.

Poisot, T., Gravel, D., 2014. When is an ecological network complex? Connectance drives degree distribution and emerging network properties. PeerJ 2. e251.

Poisot, T., Canard, E., Mouillot, D., Mouquet, N., Gravel, D., 2012. The dissimilarity of species interaction networks. Ecol. Lett. 15, 1353–1361.

Polis, G.A., Hurd, S.D., 1995. Extraordinarily high spider densities on islands: flow of energy from the marine to terrestrial food webs and the absence of predation. Proc. Natl. Acad. Sci. U.S.A. 92, 4382–4386.

Polis, G.A., Power, M.E., Huxel, G.R., 2004. Food Webs at the Landscape Level. University of Chicago Press, Chicago and London.

Pons, P., Latapy, M., 2006. Computing communities in large networks using random walks. J. Graph Algorithms Appl. 10, 191–218.

Post, D.M., 2002. The long and short of food-chain length. Trends Ecol. Evol. 17, 269–277.

Post, D.M., Conners, M.E., Goldberg, D.S., 2000. Prey preference by a top predator and the stability of linked food chains. Ecology 81, 8–14.

Preston, F.W., 1962. The canonical distribution of commonness and rarity: part I. Ecology 43, 185–215.

Quince, C., Higgs, P.G., McKane, A.J., 2002. Food web structure and the evolution of ecological communities. In: Lässig, M., Valleriani, A. (Eds.), Biological Evolution and Statistical Physics. Springer-Verlag, Berlin, pp. 281–298.

Rabosky, D.L., Hurlbert, A.H., 2015. Species richness at continental scales is dominated by ecological limits. Am. Nat. 185, 572–583.

Ricklefs, R.E., 2007. Estimating diversification rates from phylogenetic information. Trends Ecol. Evol. 22, 601–610.

Ricklefs, R.E., Renner, S.S., 2012. Global correlations in tropical tree species richness and abundance reject neutrality. Science 335, 464–467.

Rohr, R.P., Scherer, H., Kehrli, P., Mazza, C., Bersier, L.F., 2010. Modeling food webs: exploring unexplained structure using latent traits. Am. Nat. 176, 170–177.

Romanuk, T.N., Zhou, Y., Valdovinos, F.S., Martinez, N.D., 2017. Robustness trade-offs in model food webs: invasion probability decreases while invasion consequences increase with connectance. Adv. Ecol. Res. 56, 263–291.

Rooney, N., McCann, K., Gellner, G., Moore, J.C., 2006. Structural asymmetry and the stability of diverse food webs. Nature 442, 265–269.

Rossberg, A.G., Matsuda, H., Amemiya, T., Itoh, K., 2006. Food webs: experts consuming families of experts. J. Theor. Biol. 241, 552–563.

Rosvall, M., Bergstrom, C.T., 2008. Maps of random walks on complex networks reveal community structure. Proc. Natl. Acad. Sci. U.S.A. 105, 1118.

Rousset, F., 1999. Reproductive value vs sources and sinks. Oikos 86, 591–596.

Rozdilsky, I.D., Stone, L., 2001. Complexity can enhance stability in competitive systems. Ecol. Lett. 4, 397–400.

Rummel, J.D., Roughgarden, J., 1985. A theory of faunal buildup for competition communities. Evolution 39, 1009–1033.

Ryser, H.J., 1957. Combinatorial properties of matrices of zeros and ones. Can. J. Math. 9, 371–377.

Salathé, M., Kouyos, R.D., Bonhoeffer, S., 2008. The state of affairs in the kingdom of the Red Queen. Trends Ecol. Evol. 23, 439–445.

Säterberg, T., Sellman, S., Ebenman, B., 2013. High frequency of functional extinctions in ecological networks. Nature 499, 468–470.

Shurin, J.B., Amarasekare, P., Chase, J.M., Holt, R.D., Hoopes, M.F., Leibold, M.A., 2004. Alternative stable states and regional community structure. J. Theor. Biol. 227, 359–368.

Simberloff, D., 1976. Experimental zoogeography of islands: effects of island size. Ecology 57, 629–648.

Simberloff, D.S., Abele, L.G., 1976. Island biogeography theory and conservation practice. Science 191, 285–286.

Simberloff, D.S., Wilson, E.O., 1969. Experimental zoogeography of islands: the colonization of empty islands. Ecology 50, 278–296.

Slatkin, M., 1974. Competition and regional coexistence. Ecology 55, 128–134.

Snijders, T.A.B., Pattison, P.E., Robins, G.L., Handcock, M.S., 2006. New specifications for exponential random graph models. Sociol. Methodol. 36, 99–153.

Soininen, J., 2016. Spatial structure in ecological communities—a quantitative analysis. Oikos 125, 160–166.

Stanley, R.P., 1973. Acyclic orientations of graphs. Discret. Math. 5, 171–178.

Stevens, V.M., Whitmee, S., Le Galliard, J.-F., Clobert, J., Böhning-Gaese, K., Bonte, D., Brändle, M., Matthias Dehling, D., Hof, C., Trochet, A., Baguette, M., 2014. A comparative analysis of dispersal syndromes in terrestrial and semi-terrestrial animals. Ecol. Lett. 17, 1039–1052.

Stouffer, D.B., Bascompte, J., 2011. Compartmentalization increases food-web persistence. Proc. Natl. Acad. Sci. U.S.A. 108, 3648–3652.

Stouffer, D.B., Camacho, J., Jiang, W., Amaral, L.A.N., 2007. Evidence for the existence of a robust pattern of prey selection in food webs. Proc. R. Soc. B Biol. Sci. 274, 1931–1940.

Stumpf, M.P.H., Porter, M.A., 2012. Critical truths about Power laws. Science 335, 665–666.

Suttle, K.B., Thomsen, M.A., Power, M.E., 2007. Species interactions reverse grassland responses to changing climate. Science 315, 640–642.

Takimoto, G., Post, D.M., Spiller, D.A., Holt, R.D., 2012. Effects of productivity, disturbance, and ecosystem size on food-chain length: insights from a metacommunity model of intraguild predation. Ecol. Res. 27, 481–493.

Tews, J., Brose, U., Grimm, V., Tielbörger, K., Wichmann, M.C., Schwager, M., Jeltsch, F., 2004. Animal species diversity driven by habitat heterogeneity/diversity: the importance of keystone structures. J. Biogeogr. 31, 79–92.

Thébault, E., Fontaine, C., 2010. Stability of ecological communities and the architecture of mutualistic and trophic networks. Science 329, 853–856.

Thuiller, W., Münkemüller, T., Lavergne, S., Mouillot, D., Mouquet, N., Schiffers, K., Gravel, D., 2013. A road map for integrating eco-evolutionary processes into biodiversity models. Ecol. Lett. 16, 94–105.

Tilman, D., 1994. Competition and biodiversity in spatially structured habitats. Ecology 75, 2–16.

Tokita, K., Yasutomi, A., 2003. Emergence of a complex and stable network in a model ecosystem with extinction and mutation. Theor. Popul. Biol. 63, 131–146.

Trebilco, R., Baum, J.K., Salomon, A.K., Dulvy, N.K., 2013. Ecosystem ecology: size-based constraints on the pyramids of life. Trends Ecol. Evol. 28, 423–431.

Triantis, K.A., Guilhaumon, F., Whittaker, R.J., 2012. The island species–area relationship: biology and statistics. J. Biogeogr. 39, 215–231.

Ulanowicz, R.E., 1997. Ecology, the Ascendent Perspective. Columbia University Press, New York, NY.

Van Kleunen, M., Weber, E., Fischer, M., 2010. A meta-analysis of trait differences between invasive and non-invasive plant species. Ecol. Lett. 13, 235–245.

Van Kleunen, M., Dawson, W., Maurel, N., 2015. Characteristics of successful alien plants. Mol. Ecol. 24, 1954–1968.

Vanoverbeke, J., Urban, M.C., De Meester, L., 2016. Community assembly is a race between immigration and adaptation: eco-evolutionary interactions across spatial scales. Ecography 39, 858–870.

Vermaat, J.E., Dunne, J.A., Gilbert, A.J., 2009. Major dimensions in food-web structure properties. Ecology 90, 278–282.

Viger, F., Latapy, M., 2005. Random generation of large connected simple graphs with prescribed degree distribution. In: 11th International Conference on Computing and Combinatorics, Kunming, Yunnan, Chine.

Volkov, I., Banavar, J.R., Hubbell, S.P., Maritan, A., 2003. Neutral theory and relative species abundance in ecology. Nature 424, 1035–1037.

Warfe, D.M., Jardine, T.D., Pettit, N.E., Hamilton, S.K., Pusey, B.J., Bunn, S.E., Davies, P.M., Douglas, M.M., 2013. Productivity, disturbance and ecosystem size have no influence on food chain length in seasonally connected rivers. PLoS One 8, e66240.

Warren, B.H., Simberloff, D., Ricklefs, R.E., Aguilée, R., Condamine, F.L., Gravel, D., Morlon, H., Mouquet, N., Rosindell, J., Casquet, J., Conti, E., Cornuault, J., Fernández-Palacios, J.M., Hengl, T., Norder, S.J., Rijsdijk, K.F., Sanmartín, I., Strasberg, D., Triantis, K.A., Valente, L.M., Whittaker, R.J., Gillespie, R.G., Emerson, B.C., Thébaud, C., 2015. Islands as model systems in ecology and evolution: prospects fifty years after MacArthur-Wilson. Ecol. Lett. 18, 200–217.

Wennekes, P.L., Rosindell, J., Etienne, R.S., 2012. The neutral-niche debate: a philosophical perspective. Acta Biotheor. 60, 257–271.

White, E.P., Ernest, S.K.M., Kerkhoff, A.J., Enquist, B.J., 2007. Relationships between body size and abundance in ecology. Trends Ecol. Evol. 22, 323–330.

Whittaker, R.J., Fernández-Palacios, J.M., 2007. Island Biogeography: Ecology, Evolution, and Conservation. Oxford University Press, Oxford, UK.

Wieters, E.A., Gaines, S.D., Navarrete, S.A., Blanchette, C.A., Menge, B.A., 2008. Scales of dispersal and the biogeography of marine predator-prey interactions. Am. Nat. 171, 405–417.

Williams, R.J., Martinez, N.D., 2000. Simple rules yield complex food webs. Nature 404, 180–183.

Williams, R.J., Berlow, E.L., Dunne, J.A., Barabási, A.L., Martinez, N.D., 2002. Two degrees of separation in complex food webs. Proc. Natl. Acad. Sci. U.S.A. 99, 12913–12916.

Williams, R.J., Anandanadesan, A., Purves, D., 2010. The probabilistic niche model reveals the niche structure and role of body size in a complex food web. PLoS One 5. e12092.

Wilson, E.O., Simberloff, D.S., 1969. Experimental zoogeography of islands: defaunation and monitoring techniques. Ecology 50, 267–278.

Wood, S.A., Russell, R., Hanson, D., Williams, R.J., Dunne, J.A., 2015. Effects of spatial scale of sampling on food web structure. Ecol. Evol. 5, 3769–3782.

Woodward, G., Ebenman, B., Ernmerson, M., Montoya, J.M., Olesen, J.M., Valido, A., Warren, P.H., 2005. Body size in ecological networks. Trends Ecol. Evol. 20, 402–409.

Wright, J.P., Gurney, W.S.C., Jones, C.G., 2004. Patch dynamics in a landscape modified by ecosystem engineers. Oikos 105, 336–348.

CHAPTER FIVE

Robustness Trade-Offs in Model Food Webs: Invasion Probability Decreases While Invasion Consequences Increase With Connectance

T.N. Romanuk*,†,1, Y. Zhou†,‡, F.S. Valdovinos§, N.D. Martinez†,§

*Dalhousie University, Halifax, NS, Canada
†Pacific Ecoinformatics and Computational Ecology Lab, Berkeley, CA, United States
‡School of Public Policy, University of Maryland, College Park, MD, United States
§University of Arizona, Tucson, AZ, United States
1Corresponding author: e-mail address: tromanuk@gmail.com

Contents

Abstract

The invasion of ecosystems by nonnative species is widely considered the greatest threat to biodiversity after habitat loss. Given limited theoretical and empirical understanding of ecological robustness to such perturbations, we simulated invasions of complex ecological networks by integrating the 'niche model' of food web structure and a nonlinear bioenergetic model of population dynamics. Overall, 7958 successful invasions by 100 different invaders in 150 food webs with 15–29 original species (mean 20) and 5–38% connectance (mean 16%) showed that most (61%) communities were robust to invasion in that they experienced no species loss. The distribution of robustness in terms of the fraction of native species that persisted (mean 94%) was skewed with a long tail reaching to values as low as 20%. Loss of a single species occurred less frequently (14% of cases) than 'extinction cascades' involving the loss of two or more

Advances in Ecological Research, Volume 56
ISSN 0065-2504
http://dx.doi.org/10.1016/bs.aecr.2016.11.001

species (25% of cases). These cascades were often caused by invaders with many prey species and few predator species. While low-connectance webs and webs invaded by omnivores were most likely to lose at least one additional species, high-connectance webs experiencing extinction cascades lost the most species, especially when invaded by secondary consumers. These and earlier simulation results suggest how the structure of invaded communities and the properties of invaders involve trade-offs among robustness and resistance to invasion. For example, high-connectance communities are highly resistant and robust to invasion overall but lose the most species in the relatively few cases when extinctions occur. Low-connectance webs are the least resistant and more often lack robustness but lose the fewest species in the relatively many cases when extinctions occur. Broadly speaking, these findings suggest that high connectance makes food webs rigidly resistant to invasion but more brittle once such rigidity is breached. Low-connectance webs are less rigid while more flexibly suffering fewer extinctions when extinctions occur.

1. INTRODUCTION

Species invasions are widely considered to be one of the greatest threats to global biodiversity after habitat loss (Vitousek et al., 1997). Evidence linking invasions to extinctions independent of other anthropogenic perturbations remains the subject of considerable debate (Clavero and Garcia-Berthou, 2005; Gurevitch and Padilla, 2004). Still, a few well-documented cases such as the invasion of the predatory brown tree snake, *Boiga irregularis*, which triggered an extinction cascade affecting over 20 species on Guam (Fritts and Rodda, 1998) and the invasion of the yellow crazy ant, *Anoplolepis gracilipes*, which caused a spate of extinctions affecting at least three trophic levels (O'Dowd et al., 2003) have shown that species invasions can lead to multiple extinctions and reorganization of native food webs (Fritts and Rodda, 1998; O'Dowd et al., 2003). The long time scales of such events, the relatively short scales of observations, and the many uncontrolled anthropogenic and other factors jointly affecting ecosystems make understanding community resistance, robustness, and resilience to invaders in terms of native species' extirpation problematic.

To better explore these critical aspects of invasion ecology and circumvent the above-mentioned problems, we simulated species invasions in models of complex ecological networks. Our models integrate a stochastic model of food web structure with a nonlinear bioenergetic dynamic model (Romanuk et al., 2009; Williams and Martinez, 2004b) to determine how connectance ($C = L/S^2$, the realized fraction of feeding links, where L is the number of links and S is the number of species in the web) and the trophic position of invaders affect the robustness and resilience of communities to invasion by a single species. This investigation builds on our earlier study (Romanuk et al., 2009) using the same simulations which explored how *resistance* to invasions varied with food web and invader properties. That work showed that food webs were least resistant to invaders that were herbivorous, omnivorous, possessed broad diets, ate species low on a trophic niche axis, and had few predators. Least-resistant webs were those with low connectance, high mean biomass, and greater species richness. Such work is part of a larger class of modelling studies that explore how complex communities emerge from ecological (e.g. invasion) and evolutionary (e.g. speciation) mechanisms (Brännström et al., 2012).

While our earlier study focused on invasion resistance, this study focuses on the robustness of food webs *after* invasion has occurred. We follow Levin and Lubchenco (2008) in considering robustness in terms of 'the maintenance of functioning in the face of disturbance,' where functioning of the food web is the maintenance of biodiversity and disturbance is a single successful invasion. We follow Dunne et al. (2002a) by measuring robustness in terms of the fraction of species within a community lost due to a perturbation. Food webs are robust if invasion results in no loss of species. When species loss does occur, less-than-full robustness is quantitatively by how many species are lost due to an invasion.

2. METHODS

Our invasion simulations were constructed in four steps (see Romanuk et al., 2009 for additional details). The first step specifies the structure of food webs (Dunne, 2006) and the fundamental niche of the invaders (Morlon et al., 2014) using the niche model (Williams and Martinez, 2000). The second step computes the population dynamics of species within the network for 2000 time steps ($t = 2000$). Species that persist to $t = 2000$ are then deemed to constitute a dynamically persistent web. The third step replicates that persistent web and adds an invader to it at $t = 2000$. Then population dynamics of both versions of the persistent web, one with the invader and the other 'control' web without the invader, are then simulated for another 2000 time steps until $t = 4000$.

2.1 The Niche Model

The niche model (see Williams and Martinez, 2000, 2008 for additional details) generates food webs based on two input parameters following an algorithm that arranges each species along a trophic hierarchy and a contiguous range of species within that hierarchy on which each species feed. One input parameter is species richness, S, and the other is the faction of all possible feeding links (L) that are realized which is known as directed connectance, C ($C = L/S^2$; Fig. 1A). Species are arranged along the hierarchy by assigning each of the S species a uniformly random 'niche value' (n_i for species i) within the interval $(0,1)$. Each species is then constrained to consume all prey species within a stochastically chosen range (r_i) whose mean

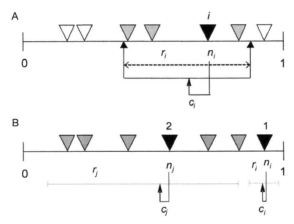

Fig. 1 Diagram of the niche model and the invasion sequence. (A) S (trophic species richness) and C (connectance) are set at the desired values. Each of S species (here $S = 7$, shown by *inverted triangles*) is assigned a 'niche value' (n_i) drawn uniformly from the interval $(0,1)$. Species i consumes all species falling in a range (r_i) that is placed by uniformly by drawing the center of the range (c_i) from the interval ($r_i/2$, n_i). Thus, in this diagram, species i consumes four species (*shaded* and *black triangles*) including itself. The size of r_i is assigned by using a beta distribution to randomly draw values from the interval $(0,1)$ whose expected value is $2C$ and then multiplying that value by n_i to obtain a web with C that matches the desired C. These rules stochastically assign each invader three fundamental niche values (n_i, r_i, c_i). These values determine the invader's fundamental niche and, in concert with the fundamental niches of species in the invaded web, determine the realized niche of the invader. Thus, for example, an invader i with specific r_i and c_i has higher generality in an invaded web when relatively many species' n_j fit within i's feeding range than when invading a web with relatively few species' n_j fitting within r_i. (B) Example of attempted invasions by two different invaders into the same web. Invader 1 cannot invade because no species falls within its feeding range. Invader 2 can invade as it has prey (*five grey triangles* and itself) and therefore is allowed.

size is C. This range is placed on the niche axis by randomly choosing its center (c_i) to be less than the ith species' niche value (n_i). The placement of the feeding range allows up to half a consumer's range to include species with higher niche values than the consumer, thus allowing looping and cannibalism. While several modifications of the niche model have been described (Allesina et al., 2008; Beckerman et al., 2006; Cattin et al., 2004; Stouffer et al., 2005; Williams and Martinez, 2008), we continue to use the original niche model because it is the only one to have no free parameters and appears to provide the most accurate overall fit to the empirical structure of complex food webs both in terms of the central tendencies and the variability of food web structure (Dunne et al., 2004; Martinez and Cushing, 2006; Stouffer et al., 2006; Williams and Martinez, 2008). We generated invaders by parameterizing the niche model with $C=0.15$ and then generating 100 species with this model to define each invader's corresponding n_i, r_i, and c_i without ever constructing the web that would result from these parameters. Essentially, these three parameters determine each invader's fundamental niche which determines its realized niche only upon introduction into a web (Morlon et al., 2014).

2.2 Nonlinear Dynamic Model

The dynamic model closely follows previous work (McCann and Yodzis, 1995; McCann and Hastings, 1997; McCann et al., 1998; Yodzis and Innes, 1992) that has been generalized to n species and arbitrary functional responses (Brose et al., 2006; Martinez et al., 2006; Williams and Martinez, 2004b). Extending the earlier notation (Yodzis and Innes, 1992) to n-species systems, the variation of B_i, the biomass of species i, over time t, is given by

$$\frac{dB_i}{dt}(t) = G_i(\mathbf{B}) - x_i B_i(t) + \sum_{j}^{n} \left(x_i \gamma_{ij} F_{ij}(B) B_i(t) - x_j \gamma_{ji} F_{ji}(B) B_i(t) / e_{ji} \right) \quad (1)$$

The first term on the right $G_i(B) = r_i B_i(t)(1 - B_i(t)/K_i)$ is the gross primary production rate of a basal species i. r_i is the intrinsic growth rate that is nonzero only for basal species, and K_i is species i's carrying capacity. The second term is metabolic loss where x_i is the mass-specific metabolic rate. The third and fourth terms are gains from resources and losses to consumers, respectively, where y_{ij} is the maximum rate at which species i assimilates species j per unit metabolic rate of species i. The nondimensional functional response that may depend on resource and consumer species' biomasses, $F_{ij}(B)$, gives the realized fraction of the maximum ingestion rate of predator species i consuming prey species j; e_{ij} is the conversion efficiency with which the biomass of species j lost due to consumption by species i is converted into the biomass of species i. Dividing the last term by e_{ij} converts the biomass assimilated by consumer j into biomass lost by resource i. Realistic ranges for the parameters in these equations have been estimated from empirical measurements (Brose et al., 2005, 2006; Yodzis and Innes, 1992). We simplified the dynamical model through our choice of parameter values. First, we chose a single value for each of the parameters $K_i = 1$, $r_i = 1$, $x_i = 0.5$, $y_{ij} = 6$, $e_{ij} = 1$, and $B_0 = 0.5$ for each set of a model's iterations. Previous analyses using this integrated structural-dynamic approach (Martinez et al., 2006) have shown that simulations that draw these parameters from normal distributions with specified means and standard deviations ($e_{ij} > 1$ not being allowed) gave similar results to fixed parameter simulations.

While a wide variety of functional responses have been proposed in the literature, our model uses a modified 'type II.2' functional response that is very close to, and highly consistent with empirical support for, the type II response while being intermediate between type II and III functional responses (Holling, 1959; Williams and Martinez, 2004b). This response models consumption of resource j by consumer i as:

$$F_{ij}(B) = \frac{B_j^{1+q}}{\sum_k B_k^{1+q} + B_{oij}^{1+q}} \tag{2}$$

where B_{0ji} is the half-saturation density and $q = 0.2$. A consumer consumes all available resources at a rate equal to its maximum consumption rate times the functional response. The amount of biomass that each resource loses to that consumer is equal to the resource's density divided by the sum of all the densities of the consumer's resources times the consumer's rate of consumption. This assumes a generalist can consume at its maximal rate even if only one of its prey species is extremely abundant (Williams, 2008).

2.3 Generating Persistent Webs

We constructed three sets of 250 niche models webs with $S = 30$ and $C = 0.05$, 0.15, or 0.30 (750 webs in total) as input parameters. The webs were then analysed for biologically implausible energy flow patterns, such as loops with no external energy source (e.g. cannibals with no other food source). Only webs without this structural looping were considered further. In our case, this yielded 171, 162, and 147 webs, respectively, for each of the three connectance categories (low, $C=0.05$; medium, $C=0.10$; and high, $C=0.30$). We then assigned each species a uniformly random initial biomass between 0.5 and 1 and simulated the dynamics of these webs for 2000 time steps. Species were eliminated from the web if their abundance decreased below an extinction threshold of 10^{-10}. Mean species richness (standard deviations or 'SD') of the persistent webs at $t=2000$ is 22 (3.66), 15 (4.44), and 10 (4.70) for each group of webs among low, medium, and high C webs, respectively. Also within these categories, C increased by 0.02 from their initial values at $t=0$ to those at $t=2000$ from 0.05 to 0.07 (0.01), 0.15 to 0.17 (0.02), and 0.30 to 0.32 (0.04), respectively (Romanuk et al., 2009). Fifty webs with $S \geq 15$ were randomly selected from each connectance category resulting in a total of 150 persistent webs which were used for further simulations. Out of 30 initial species in these 150 webs, mean S (SD) is 19 (2.4) at $t=2000$. For the 50 webs within each connectance category, mean S at $t=2000$ is 22 (3.6), 19 (2.2), and 17 (2.0) for low, medium, and high C webs, respectively.

2.4 Invasion Simulations

We introduced a potential invader into persistent food webs at $t=2000$ and determined its feeding interactions according to its niche model parameters (n_i, r_i, and c_i) in concert with the parameters of those variables for the other species in the food web. The invader was introduced with a uniformly random initial biomass between 0.5 and 1, after which we simulated the dynamics for 2000 additional time steps ($t=4000$; Fig. 1B) while maintaining the extinction threshold at 10^{-10}. We repeated this procedure for each of the 150 persistent webs with each of the 100 model invaders yielding a total of 15,000 introductions. We analysed each web after introduction and before running the dynamics and automatically discarded the 961 scenarios in which the invader was trophically disconnected from the invaded web or caused biologically implausible energy flows (e.g. a heterotrophic species that eats only itself). Of the remaining 14,039 introductions, in 57% of the cases, the invaders did not persist to $t=4000$ (Romanuk et al., 2009). We only consider here those 7958 cases of successful invasions. These include 2257 invasions of primary producers, all of which were successful as noted previously (Romanuk et al., 2009).

We focused our analyses on the robustness of food webs (Dunne et al., 2002a) to invasions in terms of the number of species that persist despite successful introductions that we call 'invasions.' We call extinctions of more than one species 'extinction cascades,' which are of great applied and basic interest. Robustness was measured as a proportion equal to the number of species in the web persisting at $t = 2000$ minus the invader divided by the number of species in the web at $t = 4000$ minus the invader. Communities with a robustness of one are maximally robust. We grouped invaders into five categories based on their prey-averaged trophic level (TL; Williams and Martinez, 2004a) and shortest chain length (SCL). TL equals one plus the mean trophic level of all the invader's resource species. SCL equals the shortest chain length from the invader to a basal taxon. The five categories of invaders were: basal species ($SCL = 0$, $TL = 1$), herbivores ($SCL = 1$, $TL = 2$), herbivorous omnivores ($SCL = 1$, $TL > 2$), carnivores ($SCL = 2$), and tertiary and higher consumers ($SCL > 2$). The percentage (number) of invaders in each trophic category was 18% (2257) basal species, 15% (1016) herbivores, 41% (3721) herbivorous omnivores, 24% (651) secondary consumers, and 0.6% (13) tertiary and higher consumers. We did not consider the last category further because none successfully invaded high C webs and very few invaded the other webs. In addition to describing the effects of various categories of invaders on the robustness of webs distinguished primarily in terms of connectance classes, we also used generalized linear models (GLMs) to explore the topological properties that best predicted robustness across all categories of species loss (no loss, one species lost, cascades) as well as within cascades to determine the invader and web properties that best explained the magnitude of extinction cascades.

We use 17 properties to describe food web structure (Briand and Cohen, 1984; Cohen and Briand, 1984; Martinez, 1991, Williams and Martinez, 2000). One property is simply the number of species within the food web (S). Two other properties are standard measures of the richness of trophic interactions within food webs: links per species (L/S) also referred to as link density; and directed connectance ($C = L/S^2$). Five more properties describe the fractions of the following types of species in a food web: top (%T, species that have resource species but lack any consumer species), intermediate (%I, species that have both resource and consumer species), basal species (%B, species that have consumer species but lack resources species e.g. plants), cannibals (%C, species that eat themselves), and omnivores (%Omn, species that eat species at different trophic levels). Omnivores include herbivorous and all other omnivores. Four more properties indicate the fraction of links between the following types of species: top–basal (%T–B), top–intermediate (%T–I), intermediate–intermediate (%I–I), and intermediate–basal (%I–B). The mean trophic level of a food web was calculated in two ways (Williams and Martinez, 2004a). One (*MSCL*) is the mean *SCL* of all species within the food web. The other (*MTL*) is the mean of all species' *TL*. Two additional properties are the standard deviation of mean generality (*GenSD*) and vulnerability (*VulSD*) among species which quantify the variabilities of species' normalized predator and prey counts, respectively (Schoener, 1989; Williams and Martinez, 2000). Trophic similarity of a pair of species is the number of predators and prey shared in common divided by the pair's total number of predators and prey (Martinez, 1991, 1993; Solow and Beet, 1998; Yodzis and Winemiller, 1999). The average of all species' similarity indices is the property called mean similarity (*MSim*). The final web property is a 'small-world' (Watts and Strogatz, 1998) property called the clustering coefficient (*CC*), which is the mean fraction of species pairs connected to the same species and each other (Camacho et al., 2002; Dunne et al., 2002b; Williams et al., 2002).

Six properties of the invader were evaluated: generality (*InvGen*), vulnerability (*InvVul*), and omnivory (*InvOmn*), are defined as, respectively, the invaders' normalized prey (generality) and predator (vulnerability) counts (Williams and Martinez, 2000), and the standard deviation of the prey-averaged trophic level of the invader's prey (Williams and Martinez, 2004a). Generality and vulnerability are normalized, respectively, by dividing the number of each species' prey and predator species by the number of the species in the invaded food web (*S*). Prey-averaged trophic level (*Inv-TL*) equals 1 plus the mean *TL* of all of the invader's resource species (Williams and Martinez, 2004a). Shortest chain trophic level (*InvSCL*) equals the invader's *SCL* (Williams and Martinez, 2004a). The final invader property was the initial biomass of the invader (*InvBio*) which was randomly assigned from a uniform distribution between 0.5 and 1. Initial invader biomass was, on average, 15 times higher than the mean biomass of species within the webs at $t = 2000$ but was equivalent to the average initial biomass of each species in the webs (0.75) at $t = 0$.

To determine whether the best predictors of robustness differed according to trophic category of the invader, we conducted Gaussian GLM analyses of all invaders lumped together as well as invaders separated into their four different trophic categories. GLMs were performed in JMP (2007, JMP, Version 6. SAS Institute Inc., Cary, NC, 1989–2007) using stepwise regression and minimum AIC to select the best model. We evaluate relative importance of each selected independent variable using its *F*-value. Final invader and web properties ($t = 4000$) were independent variables, and robustness was the dependent variable in these analyses (Table 1).

Table 1 GLM Results for Robustness to Invasion: (A) Across all simulations where one or more species was lost, and (B) Across extinction cascades where two or more species were lost.

	ALL		Basal		Herbivores		Herbivorous Omnivore		Carnivore	
	F	*p*	*F*	*p*	*F*	*p*	*F*	*p*	*F*	*p*
(A) Across all invasions										
Food web properties										
C	337.074	<0.001	70.173	<0.001	556.215	<0.001	236.585	<0.001	122.955	<0.001
L/S	4454.793	<0.001	1710.494	<0.001			1860.021	<0.001	406.667	<0.001
%B										
%O	11.904	0.001	14.362	<0.001			35.880	<0.001	14.698	<0.001
%I	20.673	<0.001	12.014	0.001			13.491	<0.001		
%C	65.229	<0.001	87.153	<0.001						
%T	0.126	0.722							10.209	0.001
%T-B			14.123	<0.001			6.679	0.010		
%T-I					13.338	<0.001				
%I-I	37.955	<0.001					13.409	<0.001		
%I-B	92.786	<0.001	43.497	<0.001	40.251	<0.001	36.200	<0.001		
GenSD	311.738	<0.001	83.766	<0.001			137.062	<0.001		
VulSD	52.356	<0.001	17.602	<0.001	57.026	<0.001	39.190	<0.001	9.096	0.003
MSCL	73.106	<0.001	15.078	<0.001	11.923	0.001	55.317	<0.001		

	Adj. R²	p	Adj. R²	p	Adj. R²	p	Adj. R²	p	Adj. R²	p
MTL					21.456	<0.001	9.924	0.002		
Msim	381.629	<0.001	143.199	<0.001	22.154	<0.001	130.083	<0.001	46.845	<0.001
CC	195.063	<0.001	16.910	<0.001			104.224	<0.001	73.944	<0.001
Invader properties										
InvOmn	165.601	<0.001	41.872	<0.001	161.433	<0.001			7.634	0.006
InvGen	123.048	<0.001	57.465	<0.001	132.979	<0.001	82.892	<0.001		
InvVul	16.344	<0.001			17.483	<0.001			20.160	<0.001
InvTL	116.257	<0.001					59.107	<0.001		
InvBio	196.678	<0.001					108.367	<0.001	8.581	0.004
Adj. R²	0.564		0.520		0.520		0.655		0.474	
Full model		>0.001		<0.001		<0.001		<0.001		<0.001
(B) *Across invasion cascades*										
Food web properties C	285.045	<0.001	32.402	<0.001	490.828	<0.001	374.221	<0.001	89.969	<0.001
L/S	1753.650	>0.001	564.871	<0.001	356.918	<0.001	1249.517	<0.001	95.261	<0.001
%B										
%O	21.229	<0.001			17.648	<0.001				
%I	13.411	<0.001								
%C	19.824	<0.001	4.628	0.032					9.687	0.002

Continued

Table 1 GLM Results for Robustness to Invasion: (A) Across all simulations where one or more species was lost, and (B) Across extinction cascades where two or more species were lost.—cont'd

	ALL		Basal		Herbivores		Herbivorous Omnivore		Carnivore	
	F	p	F	P	F	p	F	p	F	p
$\%T$							4.672	0.031		
$\%T–B$	24.962	<0.001					49.646	<0.001		
$\%T–I$	29.768	<0.001								
$\%I–I$							37.318	<0.001		
$\%I–B$	75.769	<0.001	50.895	<0.001					115.434	<0.001
$GenSD$	117.909	<0.001	40.229	<0.001			36.110	<0.001		
$VulSD$	29.471	<0.001	15.241	<0.001					57.659	<0.001
$MSCL$	31.522	<0.001	39.874	<0.001			8.817	0.003		
MTL									43.044	<0.001
$Msim$	118.887	<0.001	48.580	<0.001			27.350	<0.001	6.799	0.011
CC	14.590	<0.001					13.488	<0.001	5.159	0.025
Invader properties										
$InvOmn$	30.721	<0.001					13.282	<0.001		
$InvGen$	27.064	<0.001					29.980	<0.001	7.333	0.008
$InvVul$					13.674	<0.001			19.746	<0.001

	Adj. R^2	p	Adj. R^2	p	Adj. R^2	p	Adj. R^2	p	Adj. R^2	p
InvTL	33.998	<0.001					12.305	<0.001		
InvBio	6.670	0.010					10.898	0.001		
Full model	0.712	<0.001	0.777	<0.001	0.644	<0.001	0.731	<0.001	0.851	<0.001

Shown is the F-value and p-value for the contribution of food web properties and invader properties (using values at the $t = 4000$) to the model. Separate analyses were done for all invaders regardless of trophic level (ALL), as well as for basal, herbivore, herbivorous omnivore, and carnivore invaders. Adjusted R^2 and p-value for full model shown at the bottom of the table. Definitions of abbreviations are described in the text. Blanks indicate results for which $p > 0.05$.

3. RESULTS

Several strong trends emerged involving robustness to invasion both overall and also with respect to the trophic category of the invader and the complexity of the invaded webs. We present robustness both in terms of the fractions of food webs that lost no species due to invasion, which we call 'qualitative robustness,' and in terms of the fractions of species within a food web lost due to invasion, which we call 'quantitative robustness.'

Overall, communities were remarkably robust to invasions as evidenced by a qualitative robustness of 0.61 (Fig. 2) and a quantitative robustness of 0.94 due to webs losing only an average of 0.06 (SD = 0.11) of the species in the community due to invasion. The least robust community maintained only 20% of its diversity due to losing 15 species. The number of webs within bins of quantitative robustness increased roughly exponentially from beginning at 0.2 and progressing towards one with the exception of high C

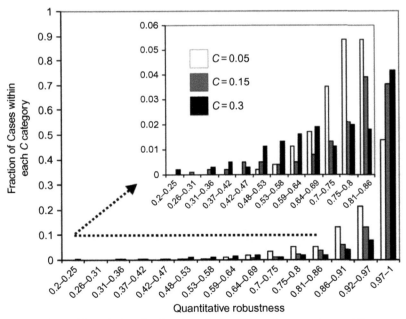

Fig. 2 Fraction of invaded webs (y-axis) within different connectance (C) categories (three different *grey* levels) according to their quantitative robustness (ranges indicated on the x-axis). The inset is a close-up (with different y-scale) on webs with quantitative robustness between 0.2 and 0.86.

communities, which were more bimodally distributed with a secondary mode appearing between 0.6 and 0.8 (Fig. 2, inset). Qualitative robustness increased with connectance (Fig. 3A) from that in low C webs which maintained all species least frequently (0.48 of cases) to medium (0.70) and high (0.76) C webs. However, the ratio of extinction cascades to single extinctions shows a somewhat mixed trend with C. Cascades were about twice as likely as single extinctions in low (0.35 cascades, 0.17 single extinctions) and high (0.16 cascades, 0.08 single extinctions) C webs.

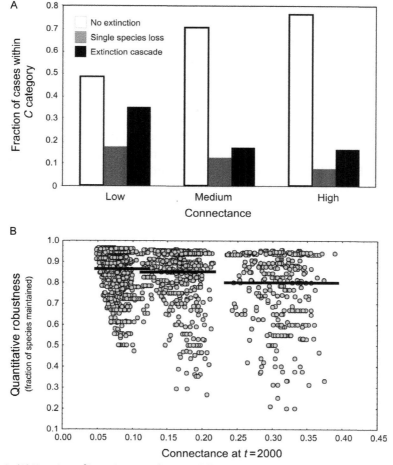

Fig. 3 (A) Fraction of invasions resulting in different levels of species loss among different connectance categories. (B) Fraction of species maintained among communities experiencing species loss as a function of connectance.

Such extinctions were more equitable (0.17 cascades, 0.13 single extinctions) in medium C webs (Fig. 3A). Perhaps most intriguing is the marked decrease in quantitative robustness with C of webs experiencing extinction cascades which decreased from low (0.83) to medium (0.79) to high (0.71) C webs. This decrease mirrors the decrease in quantitative robustness of all webs experiencing any extinctions and the minimal robustness which went from 0.5 to 0.3 to 0.2 in low, medium, and high C webs, respectively (Fig. 3B). While overall qualitative and quantitative robustness increased with C, quantitative robustness decreased with increasing C within various subsets of webs experiencing extinctions. These results broadly describe a trade-off between extinction probabilities decreasing with increasing C while extinction consequences increase with C among webs experiencing extinctions.

Relative to the number of invaders in each trophic category (fraction of cases in each category), webs were least qualitatively robust to invasions by herbivorous omnivores, which extirpated species most frequently, allowing all species to be maintained in just over half the cases (0.54) followed by invasions by herbivores (0.58), secondary consumers (0.67), and basal species (0.73). In line with these results, herbivores and herbivorous omnivores triggered extinction cascades more frequently than basal or secondary consumers (Fig. 4A). Herbivore and herbivorous omnivore invaders were approximately twice as likely (0.27 and 0.32) to trigger extinction cascades than single extinctions (0.15 and 0.14). This skew towards low robustness was less pronounced for basal and carnivore invaders which caused extinction cascades (0.15 for basal, 0.17 for carnivores) and single extinctions (0.11 for basal, 0.16 for carnivores) more equitably (Fig. 4A). While qualitative robustness to herbivorous omnivores was low, overall quantitative robustness was very similar among different trophic categories within each C category (Fib 4b). The most prominent exception is that, among webs experiencing extinctions (Fig. 4B, solid symbols), quantitative robustness to omnivores invading medium C webs (mean 0.77) was significantly lower than that in medium C webs invaded by secondary consumers (mean 0.86) and basal species (mean 0.84; Fig. 4B).

As described earlier, increasing trophic complexity, as measured by C, tended to increase qualitative robustness. This complexity–robustness effect was remarkably consistent for herbivorous omnivores and carnivores for which the probability of no extinctions went up and the probability of single extinctions and extinction cascades went down as C increased (Fig. 4A). This effect, while still present, was less consistent for basal and herbivore invaders. While the qualitative robustness was highest in high C webs,

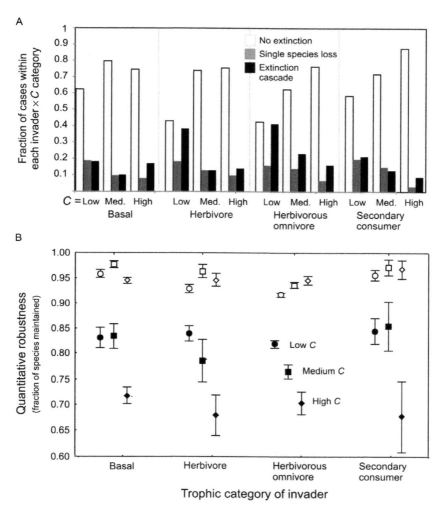

Fig. 4 (A) Fraction of simulations experiencing different levels of species loss according to the trophic category of the invader and connectance level of the invaded community. (B) Quantitative robustness in terms of fraction of species maintained among communities overall (open symbols) and among only those experiencing species losses greater than 1 (closed symbols) according to the trophic category of the invader and connectance level (symbol shape) of the invaded community.

quantitative robustness was lowest among those relatively few high C webs experiencing extinction cascades irrespective of the invader's trophic category (Fig. 4B). This remarkably consistent trade-off between the probability and consequence of invasions, both among aggregated (Fig. 3) and disaggregated (Fig. 4) trophic categories, means that the overall averages of quantitative robustness, illustrated as open symbols in Fig. 4B, exhibit less consistent complexity–robustness effects.

Relatively few of the 100 invaders caused relatively many extinctions among all simulated invasions (Fig. 5). Invaders caused on average 119 extinctions among all species in all 150 webs with each invader's average ranging from 40 to 632. The most destructive invader, which accounts for 5% of all species that went extinct, was an omnivore that, among the 150 webs it invaded, had a mean TL (SD) of 2.5 (0.38), a high mean generality of 2.69 (1.61), and a low mean vulnerability of 0.27 (0.38). Its highest generality equaled 8.61 times the average generality (1.00) in the invaded web. The other top four destructive invaders had a similar pattern of high mean generality 3.80 (1.95) and low mean vulnerability 0.27 (0.40) but a higher mean TL of 3.23 (0.61).

GLM analyses reinforce and extend our above analyses from a complementary perspective focused on the ability of model input variables to explain quantitative robustness (percent variability of robustness explained; Table 1). Because quantitative robustness is a fraction of initial S, GLM results showing that higher S significantly increases quantitative robustness could be considered spurious due to the presence of S in both the dependent and independent variables. Another view of this is that diversity enhances stability by making species' extinction probability go down with increasing S.

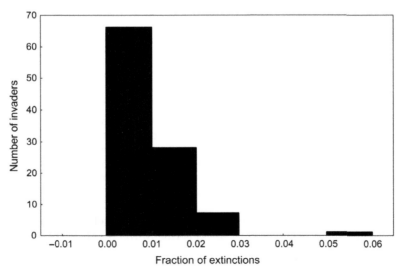

Fig. 5 Number of invaders causing different fractions of all extinctions within all invaded communities.

However, in absolute terms, there was no relationship between S and the number of species that went extinct (0.03%, $p=0.109$). We sidestep this debate by excluding S from all but the last of the GLM results in this section. Across all invasions, high L/S best explained (38%) quantitative robustness. While including 13 additional web properties (51%) and 5 invader properties (5%) enabled the GLM to explain over half (56%) of the variance in robustness, L/S by far best explained quantitative robustness greatly exceeding the next best variable, mean maximum similarity (3%). A similar trend was observed within trophic categories, where high L/S again best explained robustness against basal (38%; full model adjusted $R^2=0.52$), herbivore (34%; full model adjusted $R^2=0.52$), omnivore (39%; full model adjusted $R^2=0.66$), and carnivore (27%; full model adjusted $R^2=0.47$) invaders.

Among only cases experiencing extinction cascades, quantitative robustness was again best explained by a high L/S (47%) across all trophic categories of invaders (full model adjusted $R^2=0.71$). Similarly, quantitative robustness to basal invaders was also best described by high L/S (54%, full model adjusted R^2 0.78). For herbivores, low C (37%) followed by high L/S (27%) explained most of the variance in quantitative robustness (full model adjusted $R^2=0.64$), whereas for herbivorous omnivores, high L/S (49%) and low C (15%) explained most of the variance in quantitative robustness (full model adjusted $R^2=0.73$). Quantitative robustness against carnivore invaders was best explained by higher fractions of links between intermediate and basal species (22%), high L/S (18%) and low C (17%) and low $VulSD$ (11%; full model adjusted $R^2=0.85$). Taken together, these results suggest the webs with large S, which can have the somewhat contradictory properties of high L/S and low C, are the most resilient to extinction cascades. Such properties are not actually contradictory and instead have many species and low C. While these webs still experience multiple extinctions, the fraction of species extirpated are attenuated compared to webs without high L/S and low C. Even when S is included in GLM analyses, L/S overwhelmingly best explains overall robustness.

4. DISCUSSION

Few issues in ecology have been more compelling than understanding the stability of ecosystems. Early studies of stability of model ecosystems considered effects of very generic network properties on population dynamics (May, 1972). Later studies focused on the effects of changes in biodiversity on ecosystem function (Naeem et al., 1994). Throughout these studies, resistance and robustness have become perhaps the most enduring aspects of the many dimensions and scales of ecological stability in the face of change (Donohue et al., 2016). Similarly, species invasions are some of the earliest and most enduring ecological changes that scientists have studied (Elton, 1958; Levine and D'Antonio, 1999; Levine et al., 2004). Here, we combine these two traditions using computational methods applied to the structure and dynamics of complex ecological networks. These methods enable us to systematically study a wide range of communities and invaders with large enough sample sizes to distinguish strong from weak effects in how invaders alter the systems they invade. We find that complexity strongly affects how ecosystems respond to change, yet another central theme in illustrious explorations of effects of change on ecological function (Cardinale et al., 2012; Holling, 2001; May, 1972). More specifically, our results highlight the role of network complexity in maintaining diversity in the face of a successfully invading species.

Our earlier work based on the same simulations (Romanuk et al., 2009) found that high connectance (C) enabled communities to better resist invasion. This finding is consistent with another theoretical study (Baiser et al., 2010; but see Lurgi et al., 2014) and corroborated by a synthesis of empirical studies showing that food web connectance increases the resistance of communities to species invasion (Smith-Ramesh et al., 2016). Here, we shift our focus from the properties of invasive species and habitats that promote successful invasion to exploring the consequences of successful invasions. We extend our earlier results by showing that, in addition to being highly resistant to invasions, high C communities are also highly robust in that they are more likely to maintain all their component species despite being successfully invaded.

However, there does appear to be a cost to this high resistance and robustness in high-connectance webs. The relatively few high C communities in which extinctions occur tend to experience the largest extinction cascades. Low C communities, which poorly resist invasions (Romanuk et al., 2009), lose species more often due to successful invasions. The benefit of having low C appears to be a higher minimum and mean robustness of communities experiencing extinctions. Broadly speaking of extinction, increasing C trades off higher probability and lower consequence in low C webs for lower probability and higher consequence in high C webs.

This contribution of complexity to robustness extends to each group of invaders from plants through carnivores, but GLM analyses point towards a more prominent role of L/S than connectance (Martinez, 1991).

While these two measures of connectivity (Martinez, 1991) are mathematically related, L/S is much more node-based, while C is much more system-based. In other words, it appears that what other species are doing is more important for invasion effects than the system level density of the links. This issue may be as important as it is subtle and deserves more attention.

Perhaps, the most prominent theory relating ecological complexity to robustness or 'the opposite of vulnerability' is panarchy (Gunderson and Holling, 2002; Holling, 2001). Panarchy hypothesizes robustness to be a function of connectedness which increases as adaptive systems mature. This in turn increases system control in terms of its ability to maintain itself and become more resistant and resilient in the face of external perturbations. However, this increased connectedness and coincident control also creates 'accidents waiting to happen' by enabling disturbances to strongly propagate in the highly connected system and causing a collapse. Our results corroborate this view of highly connected systems being more resistant and broadly robust while also being more susceptible to a specific collapse of ecosystem function that maintains only small fractions of the previous levels of biodiversity. Species diversity and L/S within our invasion based food web theory are, respectively, analogous, if not homologous, to system variation and connectedness within panarchy theory and system uncertainty and average mutual information within another theory of resilience, ascendency theory (Ulanowicz, 1997). Over time since establishment after a significant disturbance, all these properties increase as systems mature which increases robustness according to their respective theories.

More specific to invasion ecology, our results highlight five hypotheses that warrant particular attention. One is that ecosystems are relatively robust to invasions in terms of resisting extinctions caused by invasions. This is suggested by the low fractions of cases in which invasions extirpate any species (0.39) and more than one species (0.25). However, when extinction cascades occur, they tend to be large as indicated by a mean cascade size of 22% of the species in the community. Third, herbivorous omnivores and herbivores lead to extinctions in more cases than basal species or carnivores, but carnivore invaders in high-connectance webs result in the highest magnitude extinction cascades. Fourth, invasions lead to extinctions more often in low-connectance webs, while the magnitude of extinction cascades is higher in high-connectance webs. Finally, a few highly destructive invaders cause a disproportionate amount of species loss, with the top five invaders leading to the extinction of ~18% of the total number of extinctions. Later, we discuss each of these hypotheses in more detail.

While invasive species can be highly destructive, only a small fraction of introduced species successfully establish. There has been considerable debate on whether species invasions are a leading cause of global extinctions (Clavero and Garcia-Berthou, 2005; Gurevitch and Padilla, 2004; see also Mollot et al. 2017). Clavero and Garcia-Berthou (2005) have suggested that extinctions of 65 of 129 birds, 11 of 23 world fish, 27 of 40 North American fish, and 12 of 25 mammals can be linked to species invasions. However, work by Gurevitch and Padilla (2004) suggests that invasions can conclusively be linked to extinctions in less than 2% of cases of species extinctions in the IUCN red list database. More recently, Murphy and Romanuk (2014) conducted a meta-analysis of 245 studies analysing the impact of human disturbances on species richness. The authors found that species invasions together with land-use change had the largest impact on species richness resulting in a 23.7% and 24.8% decline, respectively, followed by habitat loss (14%), nutrient addition (8.2%), and increases in temperature (3.6%). In our simulations of niche model, food webs are consistent with invasions causing significant numbers of extinctions. While assigning proximate causes of extinctions is fraught with the usual observational difficulties due to interactive effects of, for example, disturbance, climate change, and land-use change (Sax and Gaines, 2008), a number of high-profile cases have conclusively shown that species invasions do lead to extinctions of native species. Arguably the best-known case is that of the invasion of the brown tree snake, *B. irregularis*, on Guam which led to the direct extermination of between 12 and 15 species of birds and reptiles and may have played a role in the local extinction of at least 21 species (Fritts and Rodda, 1995). One of the few studies to document diffuse food web consequences due to the invasion of a single species has been the invasion of the yellow crazy ant, *A. gracilipes*, which caused a rapid, catastrophic shift in the rain forest ecosystem of a tropical oceanic island, affecting at least three trophic levels. The primary trigger for meltdown was the extirpation of the dominant consumer, the red land crab, *Gecarcoidea natalis*, by the yellow crazy ant (O'Dowd et al., 2003).

Our result suggesting that on average invasions lead to more extinctions in low-connectance webs than in high-connectance webs would be corroborated by Murphy and Romanuk's (2014) study if one accepts that connectance is higher in terrestrial food webs than in aquatic food webs (Chase, 2000; Smith-Ramesh et al., 2016) because invaders are thought to produce more extinctions in terrestrial than aquatic ecosystems (Murphy and Romanuk, 2014). Evidence that trophic cascades are more common in aquatic ecosystems than in terrestrial ecosystems (Chase, 2000; Strong, 1992) further supports an increased potential for invasion-caused extinctions in aquatic than in terrestrial systems.

On average overall, the fraction of species extirpated by invasion was relatively low in our simulations, with only 6% or 1.3 species in a web going extinct. While it is tempting to see the invasion-extinction dynamics as one of simple replacement, extinction cascades were generally more likely than single extinctions, with 25% of invasions leading to the cascades and only 14% of invasions leading to single extinctions. When an extinction cascade occurred, 22% of the species on average went extinct. In a few cases, invasions resulted in almost the complete collapse of the web with up to 80% of the species became extinct following the invasion. In these cases, the invader was typically a highly general omnivore that was relatively invulnerable to predation. While catastrophic extinctions following invasions have been observed in a number of ecosystems (O'Dowd et al., 2003; Fritts and Rodda, 1995), most invasion-related extinctions have been observed in island ecosystems (Sax et al., 2007).

The resistance of the webs to extinction, the magnitude of secondary extinctions, and the likelihood of extinction cascades vs single extinctions depended on both the trophic category of the invader and the complexity of the web. Invasions lead to extinctions more often in the low-connectance webs; however, the magnitude of extinction cascades was higher in high-connectance webs. Invasions resulting in extinction cascades of more than six species occurred in only 1% of cases in low-connectance webs but in 27.5% of cases in the high-connectance webs, suggesting that as food web complexity increases, food webs become more resistant to extinction per se but when extinctions do occur following invasions they often lead to very high-magnitude extinction cascades. Intriguingly, in the low- and high-connectance webs, the average magnitude of the cascades was relatively unaffected by the trophic position of the invader. In contrast, in medium-connectance webs, the trophic position of the invader led to a wide range of magnitudes in extinction cascades ranging from 14% for basal species to 24% for omnivores.

A few highly destructive invaders caused a disproportionate amount of the species loss, with the top five of 100 invaders being responsible for ~18% of the total number of extinctions. Recent reviews have called for an increased attention to highly cosmopolitan destructive invaders. Thus, it is of interest to determine whether some invaders cause consistently high extinction magnitudes or whether the consequences of invasions depend on the fine nuances of each invasion scenario. Support for the former option is suggested by the consistency in the effects of the highly destructive invaders in our simulations. For example, the invader that caused the highest number of extinctions also caused extinctions in the most webs (79%), suggesting that invader type, rather than network structure, was responsible for this effect. This compares to the average invaders, which caused extinctions in only 33% of the 150 webs they invaded. The consequence of this pattern is that, when an invader causes secondary extinctions, it does so across many webs and through extirpating many species within webs. This supports one of the few accepted generalities in invasion ecology: invasion success in other locations is one of the few good predictors of whether a species will be invasive or not in a given locality (Williamson, 1996).

The most destructive invaders in terms of species loss were generalist omnivores that were relatively invulnerable to predation. For animals, omnivory appears to be a common property associated with high invasibility (Romanuk et al., 2009; Ruesink, 2005), and our simulations suggest that generalist omnivores lead to high magnitude extinction cascades. An example of this type of invader is the yellow crazy ant, *A. gracilipes*, which feeds on grains, seeds, small isopods, myriapods, molluscs, arachnids, land crabs, and insects, decaying matter and vegetation (O'Dowd et al., 2003). Highly destructive invaders were also identified in other trophic categories. For example, the fourth most destructive invader was a generalist secondary consumer with low vulnerability. In general, the highest magnitude extinctions cascades resulted from invasions by secondary consumers in highly connected webs. Species extinctions have been strongly linked to invasions particularly for invasions by predators (Sax et al., 2007). Successfully introduced generalist fish predators such as the Nile perch, *Lates niloticus*, which was introduced to Lake Victoria in 1954 to supplement fisheries, have been implicated in the disappearance or threat of extinction of nearly two-thirds of the lakes 300+ haplochromine cichlid species.

The situation for plants is somewhat different. Due to the constant growth term in the dynamic model, all introductions of basal species lead to successful invasion (Romanuk et al., 2009), whereas only 37% of introductions of consumers lead to successful invasion (Romanuk et al., 2009). Overall, basal invaders were the least likely to cause extinctions, particularly in the medium- and high-connectance webs. This overall result contrasts with Murphy and Romanuk's (2014) result showing that, among the empirical studies they analysed, basal invaders produce the highest richness decline among all taxa. This discrepancy between our model results and the empirical data may be explained by the unrealistic way we are modelling carrying capacity for plants. Instead of a community-level carrying capacity (Boit et al., 2012), our model assumes individual–species carrying capacity, which means that plants compete for resources only within their own species. This prevents competitive exclusion among plant species. Another point about basal invaders is that they were also more equitable in terms of their probability of causing single extinctions vs extinctions cascades, particularly in the low- and medium-connectance webs.

Several issues limit the generality of our results and interpretations. A key limitation is our assumption that body sizes do not vary. Variation of body size is a critical issue in food web dynamics in particular (Berlow et al., 2009; Brose et al., 2006; Petchey et al., 2008) and ecosystem function more generally (Hildrew et al., 2007). While more realistic bioenergetic models that account for observed variation of body size and other factors could have been used for this study (e.g. Boit et al., 2012; Brose et al., 2006; Gilarranz et al., 2016; Kuparinen et al., 2016) along with a nutrient model (Brose et al., 2005) or community-level carrying capacity (Boit et al., 2012) to more realistically model plant invasions, we used identical models to our earlier invasion simulations to more consistently address questions about both invasibility and the consequences of invasion. Future invasion studies could address these and another important limitation, i.e., the lack of nutrient competition between plants, by using Berlow et al.'s (2009) extension of our methods called the allometric trophic network model (e.g. Lurgi et al., 2014). A more empirical limitation is the anecdotal nature of data on invasion effects. With so few cases and consequent lack of strong patterns in the data, the similarity between our results and the cited case histories could be mere coincidence. Still, our analyses overall show how effects of invasions by even a single species can diffuse throughout food webs, in some cases resulting in the near complete collapse of the native community. Our results suggest that viewing invasions and extinctions from a food web perspective can help better conceptualize and predict the consequences of invasions. Our methods show how such complex ecological networks can be computationally explored further to advance both invasion biology and the understanding of robustness, resilience, and resistance in complex systems.

ACKNOWLEDGEMENTS

Comments from E.L. Berlow and B. Bauer improved the manuscript. Intel Corporation, Hewlett-Packard Corporation, IBM Corporation, and the National Science Foundation Grant EIA-0303575 made hardware and software available for the CITRIS Cluster which was used in producing these research results. H. Zhang provided access to the CITRIS Cluster and helped manage and store data. R.J. Williams wrote the initial java code used in the simulations. This research was supported by US National Science Foundation Grants DBI-0234980 from the Biological Databases and Informatics program, ITR-0326460 from the Information Technology Research Program, DEB-1241253 from the Dimensions of Biodiversity Program and DEB-1212243 from the Dynamics of Coupled Natural-Human Systems program to N.D.M and by a National Science and Engineering Research Council of Canada (NSERC) Discovery Grant to T.N.R.

REFERENCES

Allesina, S., Alonso, D., Pascual, M., 2008. A general model for food web structure. Science 320, 658–661.

Baiser, B., Russell, G.J., Lockwood, J.L., 2010. Connectance determines invasion success via trophic interactions in model food webs. Oikos 119, 1970–1976.

Beckerman, A., Petchey, O., Warren, P., 2006. Foraging biology predicts food web complexity. Proc. Natl. Acad. Sci. U.S.A. 103, 13745–13749.

Berlow, E.L., Dunne, J.A., Martinez, N.D., Stark, P.B., Williams, R.J., Brose, U., 2009. Simple prediction of interaction strengths in complex food webs. Proc. Natl. Acad. Sci. U.S.A. 106, 187–191.

Boit, A., Martinez, N.D., Williams, R.J., Gaedke, U., 2012. Mechanistic theory and modelling of complex food-web dynamics in Lake Constance. Ecol. Lett. 15 (6), 594–602.

Brännström, Å., Johansson, J., Loeuille, N., Kristensen, N., Troost, T.A., Lambers, R.H.R., Dieckmann, U., 2012. Modelling the ecology and evolution of communities: a review of past achievements, current efforts, and future promises. Evol. Ecol. Res. 14, 601–625.

Briand, F., Cohen, J.E., 1984. Community food webs have scale-invariant structure. Nature 307, 264–267.

Brose, U., Berlow, E.L., Martinez, N.D., 2005. Scaling up keystone effects from simple to complex ecological networks. Ecol. Lett. 8, 1317–1325.

Brose, U., Williams, R.J., Martinez, N.D., 2006. Allometric scaling enhances stability in complex food webs. Ecol. Lett. 9, 1228–1236.

Camacho, J., Guimerà, R., Amaral, L.A.N., 2002. Robust patterns in food web structure. Phys. Rev. Lett. 88, 228102.

Cardinale, B.J., Duffy, J.E., Gonzalez, A., Hooper, D.U., Perrings, C., Venail, P., Narwani, A., Mace, G.M., Tilman, D., Wardle, D.A., Kinzig, A.P., 2012. Biodiversity loss and its impact on humanity. Nature 486, 59–67.

Cattin, M.F., Bersier, L.F., Banasek-Richter, C., Baltensperger, R., Gabriel, J.P., 2004. Phylogenetic constraints and adaptation explain food-web structure. Nature 427, 835–839.

Chase, J.M., 2000. Are there real differences among aquatic and terrestrial food webs? Trends Ecol. Evol. 15, 408–412.

Clavero, M., Garcia-Berthou, E., 2005. Invasive species are a leading cause of animal extinctions. Trends Ecol. Evol. 20, 110.

Cohen, J.E., Briand, F., 1984. Trophic links of community food webs. Proc. Natl. Acad. Sci. U.S.A. 81, 4105–4109.

Donohue, I., Hillebrand, H., Montoya, J.M., Petchey, O.L., Pimm, S.L., Fowler, M.S., Healy, K., Jackson, A.L., Lurgi, M., McClean, D., O'Connor, N.E., O'Gorman, E.J., Yang, Q., 2016. Navigating the complexity of ecological stability. Ecol. Lett. 19, 1172–1185.

Dunne, J.A., 2006. The network structure of food webs. In: Pascual, M., Dunne, J.A. (Eds.), Ecological Networks: Linking Structure to Dynamics in Food Webs. Oxford University Press, Oxford, pp. 27–86.

Dunne, J.A., Williams, R.J., Martinez, N.D., 2002a. Network structure and biodiversity loss in food webs: robustness increases with connectance. Ecol. Lett. 5, 558–567.

Dunne, J.A., Williams, R.J., Martinez, N.D., 2002b. Food-web structure and network theory: the role of connectance and size. Proc. Natl. Acad. Sci. 99 (20), 12917–12922.

Dunne, J.A., Williams, R.J., Martinez, N.D., 2004. Network structure and robustness of marine food webs. Mar. Ecol. Prog. Ser. 273, 291–302.

Elton, C.S., 1958. Ecology of Invasions by Animals and Plants. Chapman & Hall, London.

Fritts, T.H., Rodda, G.H., 1995. Invasions of the brown tree snake. In: LaRoe, E.T., Farris, C.S., Puckett, C.E., Doran, P.D., Mac, M.J. (Eds.), Our Living Resources: A Report to the Nation on the Distribution, Abundance, and Health of US Plants, Animals and Ecosystems. US Department of the Interior, National Biological Service. Washington, DC, pp. 454–456.

Fritts, T.H., Rodda, G.H., 1998. The role of introduced species in the degradation of island ecosystems: a case history of Guam. Annu. Rev. Ecol. Syst. 29, 113–140.

Gilarranz, L.J., Mora, C., Bascompte, J., 2016. Anthropogenic effects are associated with a lower persistence of marine food webs. Nat. Commun. 7, 10737.

Gunderson, L.H., Holling, C.S., 2002. Panarchy: Understanding Transformations in Human and Natural Systems. Island Press, Washington, DC.

Gurevitch, J., Padilla, D., 2004. Are invasive species a major cause of extinctions? Trends Ecol. Evol. 19, 470–474.

Hildrew, A., Raffaelli, D., Edmonds-Brown, R., 2007. Body Size: The Structure and Function of Aquatic Ecosystems. Cambridge University Press, New York.

Holling, C.S., 1959. Some characteristics of simple types of predation and parasitism. Can. Entomol. 7, 385–398.

Holling, C.S., 2001. Understanding the complexity of economic, social and ecological systems. Ecosystems 4, 390–405.

Kuparinen, A., Boit, A., Valdovinos, F.S., Lassaux, H., Martinez, N.D., 2016. Fishing-induced life-history changes degrade and destabilize harvested ecosystems. Sci. Rep. 6, 22245. http://dx.doi.org/10.1038/srep22245.

Levin, S.A., Lubchenco, J., 2008. Resilience, robustness and marine ecosystem-based management. Bioscience 58, 27–32.

Levine, J.M., D'Antonio, C.M., 1999. Elton revisited: a review of evidence linking diversity and invasibility. Oikos 87, 15–26.

Levine, J.M., Adler, P.B., Yelenik, S.G., 2004. A meta-analysis of biotic resistance to exotic plant invasions. Ecol. Lett. 7, 975–989.

Lurgi, M., Galiana, N., Lopez, B.C., Joppa, L.N., Montoya, J.M., 2014. Network complexity and species traits mediate the effects of biological invasions on dynamic food webs. Front. Ecol. Evol. 2, 36.

Martinez, N.D., 1991. Artifacts or attributes? Effects of resolution on the Little Rock Lake food web. Ecol. Monogr 61 (4), 367–392.

Martinez, N.D., 1993. Effects of resolution on food web structure. Oikos 66, 403–412.

Martinez, N.D., Cushing, L.J., 2006. Additional model complexity reduces fit to complex food-web structure. In: Pascual, M., Dunne, J.A. (Eds.), Ecological Networks: Linking Structure to Dynamics in Food Webs. Oxford University Press, Oxford, pp. 87–89.

Martinez, N.D., Williams, R.J., Dunne, J.A., 2006. Diversity, complexity, and persistence in large model ecosystems. In: Pascual, M., Dunne, J.A. (Eds.), Ecological Networks: Linking Structure to Dynamics in Food Webs. Oxford University Press, Oxford, pp. 167–185.

May, R.M., 1972. Will a large complex system be stable? Nature 238, 413–414.

McCann, K., Hastings, A., 1997. Re-evaluating the omnivory-stability relationship in food webs. Proc. R. Soc. Lond. B. Biol. Sci 264, 1249–1254.

McCann, K., Yodzis, P., 1995. Bifurcation structure of a three-species food chain model. Theor. Popul. Biol. 48, 93–125.

McCann, K., Hastings, A., Huxel, G.R., 1998. Weak trophic interactions and the balance of nature. Nature 395, 794–798.

Mollot, G., Pantel, J.H., Romanuk, T.N., 2017. The effects of invasive species on the decline in species richness: a global meta-analysis. Adv. Ecol. Res. 56, 61–83.

Morlon, H., Kefi, S., Martinez, N.D., 2014. Effects of trophic similarity on community composition. Ecol. Lett. 17, 1495–1506.

Murphy, G.E., Romanuk, T.N., 2014. A meta-analysis of declines in local species richness from human disturbances. Ecol. Evol. 4, 91–103.

Naeem, S., Thompson, L.J., Lawler, S.P., Lawton, J.H., Woodfin, R.M., 1994. Declining biodiversity can affect the functioning of ecosystems. Nature 368, 734–737.

O'Dowd, D.J., Green, P.T., Lake, P.S., 2003. Invasional 'meltdown' on an oceanic island. Ecol. Lett. 6 (9), 812–817.

Petchey, O.L., Beckerman, A.P., Riede, J.O., Warren, P.H., 2008. Size, foraging, and food web structure. Proc. Natl. Acad. Sci. U.S.A. 105, 4191–4196.

Romanuk, T.N., Zhou, Y., Brose, U., Berlow, E.L., Williams, R.J., Martinez, N.D., 2009. Predicting invasion success in complex ecological networks. Philos. Trans. R. Soc. Lond. B 364, 1743–1754.

Ruesink, J.L., 2005. Global analysis of factors affecting the outcome of freshwater fish introductions. Conserv. Biol. 19 (6), 1883–1893.

Sax, D.F., Gaines, S.D., 2008. Species invasions and extinction: the future of native biodiversity on islands. Proc. Natl. Acad. Sci. U.S.A. 105, 11490–11497.

Sax, D.F., Stachowicz, J.J., Brown, J.H., Bruno, J.F., Dawson, M.N., Gaines, S.D., Grosberg, R.K., Hastings, A., Holt, R.D., Mayfield, M.M., O'Connor, M.I., 2007. Ecological and evolutionary insights from species invasions. Trends Ecol. Evol. 22 (9), 465–471.

Schoener, T.W., 1989. Food webs from the small to the large. Ecology 70, 1559–1589.

Smith-Ramesh, L.M., Moore, A.C., Schmitz, O.J., 2016. Global synthesis suggests that food web connectance correlates to invasion resistance. Glob. Chang. Biol. http://dx.doi.org/10.1111/gcb.13460.

Solow, A.R., Beet, A.R., 1998. On lumping species in food webs. Ecology 79, 2013–2018.

Stouffer, D.B., Camacho, J., Guimera, R., Ng, C.A., Amaral, L.A.N., 2005. Quantitative patterns in the structure of model and empirical food webs. Ecology 86, 1301–1311.

Stouffer, D.B., Camacho, J., Amaral, L.A.N., 2006. A robust measure of food web intervality. Proc. Natl. Acad. Sci. U.S.A. 103, 19015–19020.

Strong, D.R., 1992. Are trophic cascades all wet? Differentiation and donor-control in speciose ecosystems. Ecology 73, 747–754.

Ulanowicz, R.E., 1997. Ecology: The Ascendent Perspective. Columbia University Press, New York. 201 p.

Vitousek, P.M., Mooney, H.A., Jubchenco, J., Melillo, J.M., 1997. Human domination of Earth's ecosystems. Science 277, 494–499.

Yodzis, P., Innes, S., 1992. Body size and consumer-resource dynamics. Am. Nat. 139, 1151–1175.

Yodzis, P., Winemiller, K.O., 1999. In search of operational trophospecies in a tropical aquatic food web. Oikos 87, 327–340.

Williams, R.J., Martinez, N.D., 2000. Simple rules yield complex food webs. Nature 404, 180–183.

Watts, D.J., Strogatz, S.H., 1998. Collective dynamics of 'small-world' networks. Nature 393 (6684), 440–442.

Williams, R.J., 2008. Effects of network and dynamical model structure on species persistence in large model food webs. Theor. Ecol. 1 (3), 141–151.

Williams, R.J., Martinez, N.D., 2004a. Limits to trophic levels and omnivory in complex food webs: theory and data. Am. Nat. 163, 458–468.

Williams, R.J., Martinez, N.D., 2004b. Stabilization of chaotic and non-permanent food web dynamics. Eur. Phys. J. B 38, 297–303.

Williams, R.J., Martinez, N.D., 2008. Success and its limits among structural models of complex food webs. J. Anim. Ecol. 77, 512–519.

Williams, R.J., Berlow, E.L., Dunne, J.A., Barabási, A.-L., Martinez, N.D., 2002. Two degrees of separation in complex food webs. Proc. Natl. Acad. Sci. U.S.A. 99, 12913–12916.

Williamson, M., 1996. Biological Invasions, vol. 15. Springer Science & Business Media.

CHAPTER SIX

14 Questions for Invasion in Ecological Networks

J.H. Pantel[*,†,1,2], **D.A. Bohan**[‡,2], **V. Calcagno**[§,2], **P. David**[*,2],
P.-F. Duyck[¶,2], **S. Kamenova**[∥,2], **N. Loeuille**[#,2], **G. Mollot**[**,2],
T.N. Romanuk[††,2], **E. Thébault**[#,2], **P. Tixier**[‡‡,§§,2], **F. Massol**[¶¶,2]

*Centre d'Ecologie Fonctionnelle et Evolutive, UMR 5175, CNRS-Université de Montpellier-UMIII-EPHE, Montpellier, France
†Centre for Ecological Analysis and Synthesis, Foundation for Research on Biodiversity, Bâtiment Henri Poincaré, Rue Louis-Philibert, 13100 Aix-en-Provence, France
‡UMR1347 Agroécologie, AgroSup/UB/INRA, Pôle Gestion des Adventices, Dijon Cedex, France
§Université Côte d'Azur, CNRS, INRA, ISA, France
¶CIRAD, UMR PVBMT, 97410 Saint Pierre, La Réunion, France
∥University of Guelph, Guelph, ON, Canada
#Institute of Ecology and Environmental Sciences UMR 7618, Sorbonne Universités-UPMC-CNRS-IRD-INRA-Université Paris Diderot-UPEC, Paris, France
**SupAgro, UMR CBGP (INRA/IRD/CIRAD/Montpellier SupAgro), Montferrier-sur-Lez, France
††Dalhousie University, Halifax, NS, Canada
‡‡CIRAD, UR GECO, Montpellier Cedex, France
§§CATIE , Cartago, Costa Rica
¶¶CNRS, Université de Lille, UMR 8198 Evo-Eco-Paleo, SPICI Group, Lille, France
¹Corresponding author: e-mail address: jelena.pantel@cefe.cnrs.fr

Contents

---

² All authors except the first and last are in alphabetical order.

Advances in Ecological Research, Volume 56
ISSN 0065-2504
http://dx.doi.org/10.1016/bs.aecr.2016.10.008

293

Abstract

Why do some species successfully invade new environments? Which of these invasive species will alter or even reshape their new environment? The answers to these questions are simultaneously critical and complex. They are critical because invasive species can spectacularly alter their new environment, leading to native species extinctions or loss of important ecosystem functions that fundamentally reduce environmental and societal services. They are complex because invasion success in a novel environment is influenced by various attributes embedded in natural landscapes— biogeographical landscape properties, abiotic environmental characteristics, and the relationship between the invasive species and the resident species present in the new environment. We explore whether a condensed record of the relationships among species, in the form of a network, contains the information needed to understand and predict invasive species success and subsequent impacts. Applying network theory to study invasive species is a relatively novel approach. For this reason, much research will be needed to incorporate existing ecological properties into a network framework and to identify which network features hold the information needed to understand and predict whether or not an invasive species is likely to establish or come to dominate a novel environment. This paper asks and begins to answer the 14 most important questions that biologists must address to integrate network analysis into the study of invasive species. Answering these questions can help ecologists produce a practical monitoring scheme to identify invasive species before they substantially alter native environments or to provide solutions to mitigate their harmful impacts.

1. INTRODUCTION

Question 1 What is an Invader from the Perspective of Ecological Networks?

There is no single widely accepted definition of invasive species because the attributes of the invader and the consequences for the invaded habitat often vary. Colautti and MacIsaac (2004) offer a helpful process-based scheme to understand invasions. Their definition of an invasive species depends on which stage—from stage 0 (a resident in a potential donor region), stage I (after uptake into a transport vector, such as human contact), stage II (survived transport and released from vector), stage III (successful reproduction and establishment), stage IV (dispersed from new local habitat or became locally numerically dominant), and stage V (widespread and dominant)—a species is experiencing. In their scheme, any given species can be classified according to the stage they occupy. However, this classification scheme does not clearly define what is unique to the process of invasion. Valéry et al. (2008) provide another definition: 'A biological invasion consists of a species' acquiring a competitive advantage following the disappearance of natural obstacles to its proliferation, which allows it to spread rapidly and to conquer novel areas within recipient ecosystems in which it becomes a dominant population'. Presently, the Global Invasive Species Database (http://www.iucngisd.org/gisd/) lists alien species, which are defined as 'nonnative, nonindigenous, foreign, exotic ... species, subspecies, or lower taxon occurring outside of its natural range (past or present) and dispersal potential (i.e. outside the range it occupies naturally or could not occupy without direct or indirect introduction or care by humans) and includes any part, gametes or propagule of such species that might survive and subsequently reproduce', and alien invasive species, which are defined as 'an alien species which becomes established in natural or seminatural ecosystems or habitat, is an agent of change, and threatens native biological diversity' (Shine et al., 2000).

Ecological networks are defined as a set of nodes, representing species or trophic groups, connected by edges defining interactions between the nodes, and can be used to describe any type of ecological interaction, including metabolic interactions, trophic interactions, or mutualistic relationships such as pollinator–plant interactions. Studying properties of species interaction networks in general, including food webs, has revealed many important

insights that could not have been gained from considering all pairs of interactions in isolation. Network properties influence the robustness of food
webs to extinction. For example, increased interaction nestedness, whereby
specialist species are connected only to subsets of the connections of generalist species (e.g. Bascompte et al., 2003), can buffer against secondary
extinctions in mutualistic networks because extinctions of specialists still
leave interaction partners for the remaining species (Memmott et al.,
2004). Increased connectance, the number of observed interactions between
species divided by the maximum possible number of interactions in the network, can buffer against secondary extinctions after removal of most
connected species (Dunne et al., 2002a). Compartmentalization (or modularity), whereby subsets of species interact frequently with one another but
infrequently with species outside the compartment, can minimize the
impacts of disturbance in a food web because the disturbance can be contained within a single compartment (Stouffer and Bascompte, 2011). In contrast, modularity reduces community stability in mutualistic networks
(Thébault and Fontaine, 2010).

The focus of this paper, and of issues 56 and 57 of *Advances in Ecological
Research*, is to consider how viewing species in the framework of interaction networks can provide novel insight into the causes and consequences
of invasive species. Once an assemblage of species is defined as a network,
many different attributes can be studied to determine their impacts on
the causes or consequences of invasibility. Here, we draw attention to the
most important questions that have emerged from research that views
invasive species through the lens of ecological networks, as identified by
a diverse group of theoretical and empirical researchers studying species
interactions in systems ranging from human microbiomes to agricultural
systems. The questions can be divided into four main categories relating
to what aspects of network attributes are under consideration—structural,
functional, evolutionary, or dynamical properties of networks. One interesting question about the structure of ecological networks and the role
these structural attributes may play in the success or impact of invasive
species is (Question 2): How does empty niche space influence network
invasibility? The architecture of connections between species can play an
important role in network stability, so it is also interesting to ask
(Question 3): Are invasive species connected to invaded food webs in
particular ways relative to native species? Ecological networks may be
defined by different types of interactions, so it is important to compare
the impacts of network structure across different network types. We ask

(Question 4): Do distinct types of species interactions influence invasion success in different ways? Additionally, networks of interacting species also perform various ecosystem functions and we seek to better understand (Question 5): How do invaders affect the distribution of biomass in invaded networks? and (Question 6): How do invaders change patterns of ecosystem functioning, particularly nutrient cycling, in invaded networks?

Few ecological systems are evolutionarily static, and this is likely the case for ecological networks as well. Species are dynamically evolving even as they interact with one another and additional evolutionary information associated with nodes and edges, such as phenotypic traits or phylogenetic relatedness, may simultaneously influence and be influenced by network architecture and by the process of invasion. We ask (Question 7): Does the relationship between evolved phenotypic traits of invasive species and invaded networks influence the likelihood of invasions?, (Question 8): How can we use phylogenetic similarity to better understand invasions in networks?, and (Question 9): How do invasive species affect subsequent network evolutionary dynamics? Network dynamics, or how properties of the network change over time, may also be influenced by invasive species, but the fundamental question (Question 10): How do the impacts of invaders on invaded networks change over time and how does network structure influence this? remains to be researched thoroughly. 'Invasion meltdown', whereby nonindigenous species facilitate one another's invasion (Simberloff and Von Holle, 1999), may be influenced by network properties and so it is important to ask (Question 11): Does previous invasion influence the probability of subsequent invasion in ecological networks? Likewise, another attribute of ecological networks, stability, may inform the likelihood of invasion success so we ask (Question 12): How does the [in]stability of the network facilitate or prevent invasions?

Understanding the causes and consequences of species invading ecological networks requires incorporating conceptual and theoretical advances in ecology and evolutionary biology. For example, spatial connectivity among habitats can structure species assemblages, and it will be important to consider how spatial connectivity influences network structure and relate this to the success or impacts of invasive species. We consider this by asking (Question 13): How does spatial connectivity influence ecological networks and what are the implications for the study of invasive species? We also look towards advances in high-throughput sequencing (HTS) to improve data quality and monitoring of ecological networks by asking (Question 14): How can we integrate HTS tools to monitor networks before and after invasions?

The questions we introduce here do not have clear answers at the moment. This paper is meant to introduce these questions as novel and promising research directions, discuss the findings of relevant literature, and to highlight which aspects of the questions are strong candidates for future study. Considering species in the context of their relationships with one another increases the chance of understanding and predicting the consequences of alteration in networks due to invasion.

2. STRUCTURAL CONSIDERATIONS FOR INVASION IN ECOLOGICAL NETWORKS

Question 2 How does Empty Niche Space Influence Network Invasibility?

Predicting the occurrence and severity of invasions has been at the heart of invasion biology research since its inception (Elton, 1958). Predictive approaches have typically taken one of two directions: either a species-centric focus on traits or attributes that increase or decrease a species' likelihood to become invasive (Kolar and Lodge, 2001; Van Kleunen et al., 2010) or an ecosystem-centric focus on properties of ecosystems or communities that make them more or less resistant to invasions (Lonsdale, 1999; Richardson and Pyšek, 2006). These approaches have been useful starting points but they do not reflect the expectation that invasion is likely to depend more on the match between the potential invasive species and the recipient ecosystem than on the intrinsic properties of either (Facon et al., 2006). A given species is not an ideal invader everywhere. For example, *r*-selected traits such as high growth rate or large numbers of offspring are often thought to promote invasibility (Rejmánek and Richardson, 1996; Sakai et al., 2001), yet many invasions are limited by the presence of competitors in the recipient ecosystem and therefore by the presence of *K*-selected traits such as producing fewer but larger offspring that enhance competitive strength (Duyck et al., 2006). Likewise, a given ecosystem is not identically sensitive to all possible alien species. Perturbed systems may be considered generally susceptible to invasions (Chytrý et al., 2008; Davis et al., 2000), but these environments may only favour invaders with particular characteristics such as the ability to rapidly exploit available resources under low-competition conditions. Low-diversity systems are thought to be susceptible to invasions because fewer species in the ecosystem means more unexploited niche axes (Levine and D'Antonio, 1999), but more species might also promote invasion because it corresponds to more potential prey or mutualists for invading newcomers. A network-based view of the ecological niche space represented by potential invading species and recipient communities may bridge the gap left by looking only at attributes of both entities alone (Hui et al., 2016).

Using the perspective of ecological networks, the definition of invasion success becomes the ability of the species to establish links within the recipient community and maintain a positive growth rate. The likelihood of invasion success then depends on the number and nature of the potential established links and the degree of energy available for the introduced species to persist. One of the various aspects of food web structure that might influence the likelihood of invasions includes the degree of saturation along niche axes in the recipient community and the invader's position along these niche axes (reviewed by David et al., this issue). In a network, this might be reflected in the connectance level of the resident community and the links between the resident community and the invasive species (see Question 3: Are invasive species connected to invaded food webs in particular ways relative to native species?), in the number of species in the same trophic level as the invader, or in the absence of apex predators observed in recently disturbed or historically isolated systems such as islands. Isolating the mechanism of the influence of niche saturation on invasion likelihood in food web networks can prove difficult because these factors can be confounded. Simulating networks with different connectance levels, for example, leads to characteristic values for other structural properties such as nestedness and modularity (Fortunato, 2010; Hui et al., 2016; Staniczenko et al., 2013). Nevertheless, there is evidence that invasion of an ecosystem by a particular species can be predicted using matrices of trait-mediated interactions between species, either quantitative or even simplified to qualitative ($+$, $-$, or 0) directions of interactions between species (Hui et al., 2016).

Some limitations remain that must be overcome for a network perspective to appropriately capture the information associated with niche availability. Networks often focus on a single type of interaction—either trophic interactions in food webs or pollination in plant–insect mutualistic networks (see Question 4: Do distinct types of species interactions influence invasion success in different ways?). Defining network nodes and edges in these ways may mean that some aspects of niche availability are not captured, such as the ability of some species to modify or create new habitat structure (e.g. ecosystem engineering; Jones et al., 1994). The existence of a vacant niche for an ecosystem engineer is often only realized a posteriori, after they have profoundly modified their invaded habitat. Perhaps, the best-known example of this is the invasive freshwater zebra mussel, *Dreissena polymorpha*, which is capable of filtering large volumes of water, dramatically reducing phytoplankton biomass, and ultimately rechannelling energy pathways from pelagic–profundal to benthic–littoral zones in lakes they have invaded (Higgins and Vander Zanden, 2010; Karatayev et al., 2002). Zebra mussel beds provide substrate for attachment of sessile organisms, biodeposition, and shelter, and these aspects of habitat alteration have facilitated invasion by additional species in the lakes they inhabit (Ricciardi, 2001; Ricciardi and MacIsaac, 2000). Ecosystem engineering and invasion facilitation are difficult to map onto an interaction network framework (but see Question 11: Does invasion influence the probability of subsequent invasion in ecological networks?), but their incorporation is critically important for understanding the biology of invasions. Invasion facilitation (e.g. 'invasion meltdown'; Simberloff and Von Holle, 1999) presents an additional complication to the theory that empty niche space promotes invasion because in this scenario, instead of successive invaders gradually contributing to saturate the niche space and therefore decrease the probability of further invasion, the opposite occurs. Invasive species extend, rather than fill, available niche space.

Ecological networks defined in terms of species interactions also lack a potentially helpful predictor of empty niche space, species traits. Syndromes of biological traits are often synthesized into functional groups, and absent functional groups might point towards invasion risk. In this manner, the lack of sessile filter-feeding bivalves attached to hard substrates in various lakes could have indicated a priori their susceptibility to *D. polymorpha* invasion. Traits can indicate additional subtleties of niche complementarity (Hui et al., 2016). For example, Azzurro et al.'s (2014) study about Lessepsian invasions of fish (i.e. invasions of the Mediterranean Sea by taxa from the Red Sea via the Suez Canal) suggests that invasive species that differ morphologically from native species may have a higher likelihood of establishment (see Question 7: Does the relationship between evolved phenotypic traits of invasive species and invaded networks influence the likelihood of invasions?). Life history traits indicative of competitive hierarchies can inform predictions about the likelihood of invasive species displacing natives (e.g. in Tephritid fruit flies; Duyck et al., 2007). While informative, these trait-based comparative approaches are limited to sets of species that are similar enough to have homologous traits, typically those occupying the same trophic level.

The data needed to integrate trophic interactions, facilitation interactions such as habitat construction and ecosystem engineering, and species functional traits, as well as other axes of interactions such as competition are rarely observed for a group of species but these multilayered datasets are increasing (i.e. Melián et al., 2009) and methods to analyse them are being developed (i.e. Mucha et al., 2010). Kéfi et al. (2016) applied a probabilistic clustering method to a unique dataset of rocky intertidal communities on a stretch of the central Chilean coast (Kéfi et al., 2015) that encompassed trophic feeding interactions, interference competition for space, and habitat creation by sessile species. They observed that the three-dimensional multiplex network, with 106 species and more than 4600 interaction links, could be described by a small subset of functional groups that could not be identified using any single data layer in isolation. Interaction networks that encompass multidimensional interaction types will provide new information about the functional compartments observed in various diverse ecosystems and may suggest potential areas of empty niche space that are vulnerable to invasion in the future.

Some existing theoretical approaches may be ideal to predict invasibility as a function of niche availability. Invasions first require introduction of the alien species into a new environment, so the process is by definition a spatial one. As a null model, a framework based on the theory of island biogeography (MacArthur and Wilson, 1967) could constitute a baseline to identify factors that operate within this spatial context irrespective of interspecific interactions. These factors include invasive organismal properties such as propagule pressure (Colautti et al., 2006; Lockwood et al., 2005) and recipient ecosystem properties such as spatial isolation. Deviations from predictions of neutral spatial models (e.g. those based on island area or distance from mainland in the MacArthur and Wilson, 1967 model) can be interpreted as effects of abiotic and biotic filtering processes. The trophic theory of island biogeography (Gravel et al., 2011; also see Massol et al., this issue) is a promising candidate for an alternative model that considers niche-based processes. In this model, colonization and extinction probabilities depend on the presence of predator and prey species, instead of colonization and extinction events occurring as stochastically independent processes. This dependency thus embodies the concept of available niche space and provides a first attempt to explicitly integrate both local network properties and landscape characteristics (e.g. distance to mainland and island area) in the prediction of successful invasions.

Question 3 Are Invasive Species Connected to Invaded Food Webs in Particular Ways Relative to Native Species?

Ecological networks have some characteristic signatures for structural properties such as connectance, which is the number of observed interactions between species divided by the total number of potential links between species in the network, and degree distribution, which is the distribution of the number of links for all species in the network. Observed connectance values vary but might be influenced by properties such as temperature (Petchey et al., 2010) and constraints on diet breadth imposed by optimal foraging (Beckerman et al., 2006). Food webs display a range of degree distributions that deviate from the distribution expected if connections assembled at random (i.e. Poisson; Dunne et al., 2002b), suggesting self-organizing properties, and these distributions are related to properties such as network connectance and size (Dunne et al., 2002b). Mutualistic networks often display heavy-tailed distributions of degrees, characterized by a large number of species with low connectivity and a small number of 'super-generalist' species that interact with many other taxa (Jordano et al., 2003).

Connectivity patterns can influence invasion success. From an invader's perspective, an increase in generalism (the number of prey species fed on) and in omnivory (calculated as the standard deviation around the weighted average of the trophic levels of an invader's prey species; Williams and Martinez, 2004a) was associated with an increased likelihood of invasion success in theoretical food webs (Romanuk et al., 2009). From the invaded network's perspective, the success of invasive species tended to decrease as the connectance of theoretical food webs increased (Baiser et al., 2010; Romanuk et al., 2009). However, additional factors can interact with connectance to influence the success or failure of invasive species. Vulnerability to predation, i.e., an increase in the number of connections to predator taxa, decreases establishment success of invasive prey (Romanuk et al., 2009), while carnivorous species are actually more successful at invading highly connected food webs (Baiser et al., 2010). These results indicate that the relationship between invasion success and connectance depends on trophic position. Modularity, whereby subsets of species interact frequently with one another but infrequently with species outside the module, may buffer food webs against the consequences of invasion because the disturbance effects may be limited within a single module (Krause et al., 2003).

The type of network considered can also influence the role of connectance for invasion success and subsequent impacts. For example, a comparison of lowly and highly invaded forest and island plant–pollinator networks revealed overall similar connectance levels but a transfer of connections from native to alien generalists in the highly invaded networks (Aizen et al., 2008; but see Bartomeus et al., 2008 for an example where invasive plants facilitate an increase in connectance between native plants and pollinators). Mutualist networks with a few highly connected 'super-generalist' species may also facilitate invasions by incorporating novel aliens into the highly connected components of the interaction network (Olesen et al., 2002).

Some important questions on the role of food web connectivity properties for invasion success and susceptibility remain to be addressed. The degree of invaders relative to the degree distribution in the invaded network has not been evaluated systematically in theoretical or empirical studies. It would also be interesting to compare the connectance levels and the degree distributions between an invasive species' native vs invaded network to determine whether it is possible to predict invasion success based on network (rather than species) properties.

Question 4 Do Distinct Types of Species Interactions Influence Invasion Success in Different Ways?

Interspecific interactions can be antagonistic (beneficial for one partner but detrimental for the other one, as in predator–prey interactions), mutualistic (beneficial for both partners, as in plant–pollinator interactions), or competitive (detrimental to both species). Interactions can also be transitory (e.g. predation events), long-term and sustained (e.g. lifetime mutualistic symbioses), or can lie anywhere along the continuum between these two durations (Fig. 1). Invasive species spanning the entirety of this continuum have successfully integrated into ecological networks in the recipient community (i.e. Eastwood et al., 2007; Traveset and Richardson, 2014). To date, most studies of invasive species focus on a single type of interaction at a time, and the main hypotheses for how species interactions influence the likelihood of invasion success are related to either competition (e.g. the biotic resistance hypothesis), predation and parasitism (e.g. antagonistic interactions, the enemy release hypothesis), or mutualism (e.g. some examples of invasion facilitation, i.e. Green et al., 2011). However, invasion success and associated consequences for communities and ecosystems are likely to result from the joint effect of different types of interspecific interactions (Inderjit and van der Putten, 2010; Mitchell et al., 2006).

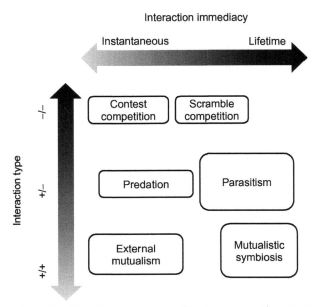

Fig. 1 Type and length of duration of interspecific interactions. The ways in which species interact with one another can vary in the direction and magnitude of their impact for participant species and in the length of time over which the species interact. Both of these axes can influence network properties and the probability of successful species invasions.

Several studies have shown that ecological networks have different structures depending on the type and strength of interaction (Bellay et al., 2015; Chagnon et al., 2016; Fontaine et al., 2011; Thébault and Fontaine, 2010). Interaction type also influences the relationship between network structure and community dynamics—greater connectance and nestedness increase species persistence in mutualistic webs while they decrease persistence in antagonistic networks (Thébault and Fontaine, 2010). It is thus likely that distinct types of interaction networks might respond differently to invasions due to their varying structures and dynamics. The nature of this response is easier to predict for some interaction types than others. Mutualisms have strong impacts on the success of potential invaders and subsequent population dynamics of species in the invaded network (see Amsellem et al., 2017 and Médoc et al., 2017). Invaders that successfully disrupt existing mutualisms may increase their likelihood of establishment in novel habitats and likely create important shifts in ecosystem functioning and effects on native species (Brouwer et al., 2015). Facultative mutualisms may promote invasions of novel species more easily than obligate mutualisms, which require strong dependencies between partners (Rodríguez-Echeverría and Traveset, 2015; Traveset and Richardson, 2014). On the other hand, the effects of antagonistic relationships for invasion success often vary. Exotic species may spread their parasites in novel ecosystems (Carpentier et al., 2007; Roy et al., 2008, 2011) or acquire novel parasites (Sheath et al., 2015). The enemy release hypothesis posits that alien species will experience increased invasion success in novel habitats that are devoid of the 'natural enemies' found in their original habitats. However, a review of studies of this hypothesis found mixed evidence for this—invasive species did encounter a reduced diversity of enemies in their introduced compared to native range, but the impact of enemies in the invaded community was similar for native and introduced species (Colautti et al., 2004).

Studies of ecological networks will therefore benefit from considering a diversity of trophic and nontrophic interactions (Kéfi et al., 2012), because interaction types are observed to combine in nonrandom ways and to influence community response to perturbation (Fontaine et al., 2011; Kéfi et al., 2012, 2015; Pocock et al., 2012; Sauve et al., 2014, 2016). For example, the effects of species invasions are more likely to propagate in highly connected interaction networks, and this connectedness can arise from either trophic interactions such as low-intimacy mutualisms (Fontaine et al., 2011) or nontrophic interactions such as competition for space or refuge provisioning (Kéfi et al., 2015). In another example, the dominance of superior competitors and exclusion of inferior competitors predicted by traditional resource ratio coexistence theory (León and Tumpson, 1975; Tilman, 1980, 1982) may not be observed if interference competition (and priority effects) is taken into consideration (Gerla et al., 2009).

Using networks to understand interactions between species requires defining edges, and this becomes complicated when multiple interaction types are combined. Kéfi et al. (2012) review modelling approaches to incorporate nontrophic interactions through their modification of trophic functional responses, and even to consider interactions that do not influence feeding. Kéfi et al. (2016) then successfully reduced a complex web of trophic, competitive, and facilitative interactions into a smaller subset of multilayer ecological functional groups. Another solution may be to use phylogenetic distances between species. Mitchell et al. (2006) developed a coherent framework that integrates enemy release, mutualist facilitation, competitive release, and abiotic environmental suitability, and how their relationship to invader success can be viewed via phylogenetic distance

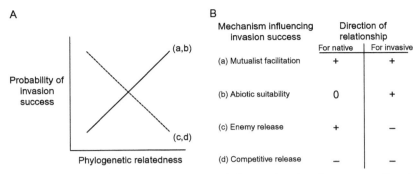

Fig. 2 Dependency of invasion success on phylogenetic relatedness as mediated by species interaction type. (A) Phylogenetic relatedness between an invasive species and species already inhabiting the novel environment can determine whether or not the invasive is likely to succeed in the new environment, but the direction of this relationship depends on (B) whether or not the invasive species receives positive benefits from ecological interactions in the novel environment. If the invasive species (a) has a mutualistic relationship with a species in the novel habitat or (b) benefits from abiotic resources, the invasive species is more likely to succeed in the new habitat if they are closely related to the native species because the invasive species is more likely to share the features that lead to benefits. If the invasive species has (c) enemies or (d) competitors in the novel habitat, the invasive species is less likely to succeed in the new habitat if they are closely related to the native species because the invasive species is more likely to share the features that lead to costs (after Mitchell et al., 2006, Fig. 3).

between invasive and native species (Fig. 2). In this framework, species introduced to communities including close relatives should experience high rates of enemy, mutualist, and competitor accumulation and encounter favourable abiotic conditions. Invasion success would therefore be limited by (the lack of) enemy or competitor release and enhanced by favourable habitat filtering or mutualist facilitation. Species introduced to communities without close relatives could experience successful invasions via enemy or competitor release or unsuccessful invasions via unfavourable habitat filtering or loss of beneficial mutualisms. The mechanistic basis of the direction of some of these relationships between interaction type and phylogenetic distance is better understood than others. Related prey species do tend to share more consumers than related consumers share prey species (Cagnolo et al., 2011; Elias et al., 2013; Naisbit et al., 2012), and related plant species tend to share more herbivores than they share pollinators (Fontaine and Thébault, 2015), supporting the hypothesized negative relationship between enemy release and phylogenetic relatedness. The mechanism for the negative relationship between phylogenetic relatedness and competitive release, that more closely related competitors are more likely to be pheno-typically similar and thus competitively exclude one another, has mixed support (Jones et al., 2013; see also Question 7: Does the relationship between evolved phenotypic traits of invasive species and invaded networks influence the likelihood of invasions?). However, phylogenetic relatedness remains a promising explanatory variable of the interactions among different members of a network that may help us understand how distinct types of species interactions influence invasion success (see Question 8: How can we use phylogenetic similarity to better understand invasions in networks?).

3. FUNCTIONAL CONSIDERATIONS FOR INVASION IN ECOLOGICAL NETWORKS

Question 5 How do Invaders Affect the Distribution of Biomass in Invaded Networks?

The consequences of biological invasions for recipient communities are often considered in terms of events such as extinctions and decreases in species richness (i.e. Gurevitch and Padilla, 2004; Ricciardi, 2004; Wilsey et al., 2009). However, shifts in the relative abundances and biomass distribution of the community may be substantial and are important to consider as well (Ehrenfeld, 2010). Abundance shifts typically precede extinction events and may therefore provide early signals of biodiversity loss in an invaded system, and altered biomass distributions should produce a consequent redistribution of energy fluxes in the ecosystem. In order to truly understand the novel functioning and services of the invaded ecosystem, it is therefore crucial to understand the changes in abundance and biomass that result after alien species have invaded (Walsh et al., 2016). Here, we consider the shifts in species abundances and biomass distributions expected after invasion of novel species from a trophic network perspective, including shifts in the same functional group as the invaded species, in lower trophic levels (e.g. top-down effects), and in higher trophic levels (e.g. bottom-up effects).

Species abundance distributions in communities generally indicate the coexistence of a few dominant and many rare species (e.g. 'hollow curve' species abundance distributions; McGill et al., 2007). This pattern, observed across different types of ecological systems regardless of the proportion of native and invasive species it holds, appears to be one of the few general laws in ecology (McGill et al., 2007). Because the general shapes of abundance distributions are unchanged by invasion and because invasive species tend to become one of the dominant species in the system (e.g. in order to be noticed and labelled as an invasive species), it follows that invasive species vastly decrease the abundance of and replace at least one dominant native species. This is consistent with empirical observations. A multihabitat study in the Bruce Peninsula National Park in Canada found that invasive plant species represented the majority of the standing biomass, particularly when total abundance in the sampled location was low (de Gruchy et al., 2005). A review of 64 grasslands in 13 countries also indicated invasive replacement of dominant species: exotic plant species were six times more likely than native species to be dominant in the observed system (Seabloom et al., 2015). If invasive species act as super-competitors that assume one of the positions of numerical dominance in the invaded system, they may in turn lead to either a null (if they simply acquire resources at the expense of competitors) or positive (if they extract additional resources not available to native species) effect on total biomass and a negative effect on the evenness of biomass distributions. The above-ground biomass in experimental grassland communities containing all exotic species was significantly higher than those containing all native species in one study (Wilsey et al., 2009). Another study compared vegetation of environmentally similar sites with either invasive-dominant or uninvaded plant communities and observed reduced evenness in the percent cover of species in the invaded plots (Hejda et al., 2009).

Invasive species that become abundant are potential energy sources for other species and may create bottom-up effects within trophic networks, but they may also consume more resources and thus create important top-down effects as well. The direction and magnitude of effects for the abundances of existing species in the network greatly depend on the trophic level of the invasive species. Lower trophic levels are strongly constrained by bottom-up effects (Loeuille and Loreau, 2004; Strong, 1992; White, 2005), so exotic plant or herbivore species may substantially increase abundances in higher trophic levels by relaxing key energy constraints. An exception might occur when, for example, an invading plant is so highly defended that its biomass is mostly unavailable for the rest of the food web (Loeuille and Leibold, 2008; Loeuille and Loreau, 2004; Strong, 1992). Top-down effects are more likely when invasive species occupy higher trophic levels, and numerous studies have observed decreasing prey abundances and increasing extinctions following invasions by exotic predators (David et al., this issue; Dorcas et al., 2012; Rodda et al., 1997). This idea is so widely accepted that trophic manipulation, via adding exotic predators to control biomass at lower trophic levels, is a commonly applied biocontrol technique (Demelo et al., 1992). We therefore hypothesize that in most instances, the addition of invasive species at lower trophic levels will increase abundances at higher trophic levels by increasing resource availability (Correia, 2001; Pintor and Byers, 2015), while invasive species introduced at higher trophic levels will create trophic cascades by decreasing biomass of the trophic level below and thus increasing biomass two trophic levels below (see David et al., this issue).

Additional complexities to determine how invasive species influence biomass distribution in ecological networks remain to be addressed. Invasive species can have indirect effects on recipient communities as well, such as niche construction or facilitation, that influence the distribution of biomass and functioning of the network (Ehrenfeld, 2010). These indirect effects may interact with the previously described bottom–up and top–down effects in complex ways. In one particularly striking example, the presence or absence of introduced arctic foxes on the Aleutian Islands determined whether or not the landscape remained a grassland or was transformed into a tundra system. In this instance, arctic foxes preying on seabird populations (a top–down effect) led to reduced nutrient transport from ocean to land and reduced soil fertility, promoting transformation from a grassland into a system dominated by shrubs and forbs (Croll et al., 2005). Until these indirect effects, especially those with large magnitude, are integrated into ecological network perspectives (see Question 11: Does invasion influence the probability of subsequent invasion in ecological networks?), it may be difficult to predict the range of effects that invasive species have on abundance and biomass distributions. One promising direction includes reconceptualizing existing metaecosystem models (Gounand et al., 2014; Gravel et al., 2010; Loreau et al., 2003) in the context of invasive species to consider their effects on nutrient cycling processes and consequent biomass distributions. In addition to these conceptual advances, improved empirical and experimental data to understand the importance of niche construction relative to direct top–down and bottom–up effects are greatly needed.

Question 6 How do Invaders Change Patterns of Ecosystem Functioning, Particularly Nutrient Cycling, in Invaded Networks?

Invaders affect the functioning of invaded ecosystems in a variety of ways. A primary mechanism of influence is via effects on primary production and nutrient cycling (Strayer, 2012), but other notable effects include shifts in fire regimes after invasion by fire-adapted plants (Brooks et al., 2004), changes in soil or marine substrate structure (Simberloff, 2011), and ecosystem engineering (Ehrenfeld, 2010). Invasions can trigger further invasions (see Question 11: Does previous invasion influence the probability of subsequent invasion in ecological networks?), and the mechanism for this in some instances is through shifts in ecosystem functioning, for example, after invasion by nitrogen-fixing plant species (Simberloff, 2011). Even after their removal, the impact of these ecosystem shifts can continue to linger and promote the persistence of other exotic species. This was observed in the instance of Australian bridal creeper, where decomposition of its phosphorus-enriched leaves increased local soil fertility and led to effects that persisted and promoted enhanced cover of some exotic species after the creeper was experimentally removed (Turner and Virtue, 2006; Turner et al., 2008; see Grove et al., 2015 for another example). Microbial ecosystems within a host's body can also be affected by foreign species. Consequences include perturbations or disruptions of the host's immune functioning (Khosravi and Mazmanian, 2013), release of toxins that influence other microbiota (Hecht et al., 2016; Wexler et al., 2016), and even manipulation of the host's metabolism (Brown et al., 2008; Martin et al., 2008), which can subsequently alter or 'ecosystem engineer' environmental conditions such as the host's diet (Ezenwa et al., 2012). The interesting case of invasions in microbiomes is discussed further in Murall et al. (this issue).

A comprehensive meta-analysis by Vilà et al. (2011) provides substantial insight into the consequences of invasive plants for nutrient cycling. Invasive species tend to increase carbon, nitrogen, and phosphorus pools in most invaded ecosystems, but on average, they decrease litter decomposition (though variation among studies is considerable). Examples of the mechanisms for alteration of nutrient cycling include significant increases in nitrogen pools after invasion of nitrogen-fixing plants (Grove et al., 2015), altered stream nutrient storage and remineralization rates after invasion of phosphorus-rich freshwater fish (Capps and Flecker, 2013), and disruption of native plant–fungi mutualistic interactions by invasive plants leading to decreased seedling growth of native plants (Stinson et al., 2006). These alterations of nutrient regimes may have substantial impacts across large spatial scales, spanning multiple trophic levels. For example, the invasion of a nitrogen-fixing tree, *Myrica faya*, in Hawaii led to vast changes in the stoichiometry not only of the forest but also of associated freshwater ecosystems (Asner and Vitousek, 2005). The sensitivity of ecosystem nutrient cycling to invasive species could be evaluated by applying network stability analysis to food webs with edges determined by nutrient transfer and weighted by transfer efficiency (Moe et al., 2005). Incorporation of ecological stoichiometric information into trophic networks (Lee et al., 2011) may provide the link between the consequences of invasive species for nutrient cycling (González et al., 2010).

4. EVOLUTIONARY CONSIDERATIONS FOR INVASION IN ECOLOGICAL NETWORKS

Question 7 Does the Relationship Between Evolved Phenotypic Traits of Invasive Species and Invaded Networks Influence the Likelihood of Invasions?

The notion that phenotypic traits of organisms may influence the success, failure, or subsequent impact after establishment of invasive species has been thoroughly explored. Researchers have made efforts to retrospectively analyze properties associated with harmful invasive species such as likelihood of human transport or ecological range tolerance (i.e. Marchetti et al., 2004; Nyberg and Wallentinus, 2005; Van Kleunen et al., 2010). The biological invasion process itself plays a role in favouring some phenotypic attributes. For example, species that interact commensally with humans or are associated with human recreational activities are more likely to have invasion opportunities (Jeschke and Strayer, 2006). Attributes such as small body size or propagule dormancy are also associated with movement and thus may play a role in invasion propensity (King and Buckney, 2001; Kolar and Lodge, 2001; Wainwright and Cleland, 2013; but see Blackburn et al., 2009). Phenotypic traits can also play an important role in structuring communities and food webs (Brose et al., 2006; Loeuille and Loreau, 2010; Rezende et al., 2009), and these phenotypic attributes of the recipient communities also influence their likelihood of being invaded (Thuiller et al., 2006).

However, the relationship between a species' invasion success or a recipient community's invasion susceptibility may not be fixed but instead may depend on the difference between the traits possessed by the invader and the recipient community. The mechanisms of this matching between the invader and the invaded community may be either that: (i) limiting similarity prevents niche overlap between the invader and the invaded community, and therefore, invasive species are more successful if they bring some novel phenotype or function to the recipient community (e.g. discrete trait invaders that add a new function to the invaded ecosystem or continuous trait invaders that differ substantially from natives in traits that are continuously distributed among species in the recipient community; Dukes and Mooney, 2004), or (ii) invaders are better able to succeed in novel environments if they share traits with natives that serve as preadaptations in the new environmental conditions (Duncan and Williams, 2002; mechanisms described in more detail in Strauss et al., 2006a). Some studies have found that the same phenotype in one species can lead to differential impacts on the rest of the species in the community depending on the phenotypic properties of that community. For example, plant species in the North American range of the invasive knapweed, *Centaurea diffusa*, were more susceptible to *C. diffusa* allelotoxin than plant species in the native Eurasian habitat (Vivanco et al., 2004).

Few studies have explicitly tested the influence of phenotypic matching between invasive species and recipient communities. Fargione et al. (2003) experimentally added plant species from different functional guilds to prairie-grassland plots with varying composition and found that introduced species were less successful when functionally similar species were present and abundant. Competitive inhibition of similar species, via resource consumption or soil nitrate reduction, was the primary mechanism of this exclusion. More studies used phylogenetic relatedness or presence of species in the same taxonomic group as a proxy for similarity in ecological niche use (discussed in more detail in Question 8: How can we use phylogenetic similarity to better understand invasions in networks?). From a network perspective, many traits are implicitly encoded by setting species as nodes in a food web, and many of these traits are correlated with body size (reviewed in Woodward et al., 2005). For example, dietary overlap is an important structural component of trophic networks (Williams and Martinez, 2000), and this is often determined by body size (Woodward and Hildrew, 2002). It may thus be possible to use body size to visualize empty niche space that suggests potential vulnerability to invasive species (see Question 2: How does empty niche space influence network invasibility?). Increased phenotypic matching that results from coevolution of interaction partners can influence the structure of mutualistic networks and increase their connectance (Nuismer et al., 2013), and connectance can influence the susceptibility of networks to invasion (see Question 3: Are invasive species connected to invaded food webs in particular ways relative to native species?).

It is important to note that phenotypes of invading species and recipient communities can evolve and change following invasion (Strauss et al., 2006b; see also Question 9: How do invasive species impact subsequent network evolutionary dynamics?). This can alter the information needed to compare invasive and native phenotypic distributions, and it may be important to take evolutionary potential into account when constructing Invasive Species Predictive Schemes (ISPS; Whitney and Gabler, 2008). Organisms of course also possess numerous traits, many of which may either contribute to fitness in a novel environment or may be correlated with other traits that are selected upon. For example, Strauss et al. (2006b) discuss instances where sexually selected traits such as mate signalling may face distinct selection pressures in novel environments when new predators or competitors are encountered. As with any study using trait-based approximations for niche use, caution must be exercised when choosing the phenotypes to be used as predictors.

Question 8 How can we Use Phylogenetic Similarity to Better Understand Invasions in Networks?

The evolutionary relationships between species in a food web can have an important, often under-appreciated role in determining the structure of that food web. For example, herbivores often consume a phylogenetically constrained set of hosts (Ødegaard et al., 2005; Weiblen et al., 2006). This constraint is observed for invasive species as well. Invasive European plants, which have not coevolved with native European butterfly and moth herbivores, were constrained to interact with particular phylogenetic groups of the Lepidopteran herbivores, and these invasive plant species were more likely to be consumed by native herbivores if they had a confamilial native plant already present in the system (Pearse and Altermatt, 2013). For this reason, phylogenetic similarity may be used to forecast likely interaction linkages when new species invade landscapes (Ives and Godfray, 2006; see also Kamenova et al., this issue).

While predicting interactions between novel invasive and native species is a relatively new direction of study, using evolutionary similarity to determine the likelihood of invasion success has a longer history. Darwin (1859) suggested that the relatedness of an exotic species to the native community could both: (i) confer an advantage to the exotic species by increasing the likelihood that the invasive species' traits match the novel environment, and (ii) lead to a disadvantage because the invasive species would be more likely to encounter direct competitors or shared enemies with the native species. Empirical evidence has yielded evidence for both directions of effect as well as no effect (reviewed in Diez et al., 2008). Mitchell et al. (2006) provide a useful framework that takes into account the different types of interactions between an exotic species and the new ecosystem that are important for determining invasion success (Fig. 2). Enemies, mutualists, competitors, and abiotic niche use all constrain invasion success and can be evaluated using measures of phylogenetic relatedness (see Question 4: Do distinct types of species interactions influence invasion success in different ways?). T.N. Romanuk et al. (personal communication) uncovered an intriguing new mechanism that may explain increased success when invaders are more similar to native species. In their series of simulated invasions into trophic food webs, invasive species that were more closely related to species in the novel food web experienced increased invasion success and displayed more links with both predator and prey species in the novel food web compared with invasive species that were less closely related. They hypothesize that more closely related species increased trophic overlap and may thus serve to stabilize novel food webs and increase their complexity and persistence. Their results support observations that invasive species with no close relatives in the novel habitat and with extremely novel niche use are associated with substantial alterations and simplifications of local trophic networks (i.e. invasion of the brown tree snake in Guam, Wiles et al., 2003; invasion of Burmese pythons in the Everglades, Dorcas et al., 2012). Their results also support the dichotomy between 'discrete trait invaders' (that add a novel niche or function to the invaded system) and 'continuous trait invaders' (that vary along a niche or function already present and continuously distributed among species in the invaded system; Dukes and Mooney, 2004) and, subject to further testing, these results would seem to indicate that discrete invaders may be rarer but have larger impacts than continuous invaders.

Some studies used phylogenetic relatedness or presence of species in the same taxonomic group as a proxy for similarity in ecological niche use. Ricciardi and Atkinson (2004) found that high-impact invaders in a diverse array of aquatic ecosystems are more likely to belong to genera that were not already present in the invaded location, while Strauss et al. (2006a) found that highly invasive California grass species were significantly less related to native grasses than introduced but noninvasive (i.e. nonpest) grass species were. However, an extensive survey of New Zealand plant species (Duncan and Williams, 2002) revealed the opposite pattern, i.e., that introduced and naturalized plant species were more likely to belong to genera that also contained native species. However, it is important to note that phylogenetic and taxonomic relatedness is only a proxy for trait similarity, and the assumption that the traits most relevant for niche use are phylogenetically conserved is not always borne out by observation (Gerhold et al., 2015; Gianuca et al., 2016).

Question 9 How do Invasive Species Affect Subsequent Network Evolutionary Dynamics?

Establishment of invasive species in a novel habitat requires, at a minimum, available resources for that species, which means that invasive species must insert themselves into interaction networks in the invaded habitat. This insertion can create an altered selective landscape for the invasive species and for the organisms the invasive species interacts with. In some instances, invasive species experience genetic alteration subsequent to their invasion (Bossdorf et al., 2005; Dlugosch and Parker, 2008). This alteration results from numerous mechanisms, such as bottlenecks, hybridization, polyploidy, and stress-induced modification of the genome (Lee, 2002; Prentis et al., 2008). While these mechanisms may promote rapid adaptation in invasive species, it is also important to consider that evolutionary change in invasive species may not necessarily be adaptive (Keller and Taylor, 2008).

The ways in which invasive species influence the evolution of natives is complex. Multiple species may interact with an invasive and show differential responses to the novel selection pressure (e.g. Mealor and Hild, 2007). Species in the recipient community may differ in plasticity, in the degree of preexisting adaptations, or the genetic architecture underlying traits under selection, which can all influence whether or not they experience evolutionary responses to invasive species. Species may not experience evolutionary change, due to natives already possessing preadapted traits (exaptations) that enhance coexistence with the invader (examples reviewed in Strauss et al., 2006b) or due to an 'evolutionary trap', where the traits or behavioural decision-making rules of a species are in mismatch with the new postinvasion selective environment and thus lead to reduced survival or reproduction (Schlaepfer et al., 2005). These varying degrees of adaptive ability in the invasive or in the recipient community may actually determine the ultimate impact of the invasive species.

An ecological network perspective may be necessary to understand evolutionary responses to invasive species because it has been increasingly shown that evolution in native and invasive species can depend on variation in ecological interactions with other species (i.e. Thompson, 2005). For example, *Alliaria petiolata*, a Eurasian understory plant that has invaded North American forests, has a patchy distribution in the invaded habitat and differentially invests in an allelochemical that slows the growth of competitors. This spatial variation in competitive strength has in turn led to coevolution in the degree of tolerance of understory competitors in the invaded communities (Lankau, 2012). Another plant species native to California grasslands, *Lotus wrangelianus*, demonstrated adaptive responses to compete with invasive *Medicago polymorpha*, but this adaptive response was only observed in communities that did not have the exotic insect herbivore *Hypera brunneipennis* present (Lau, 2006). Increasing trophic complexity—for example, the presence of alternate host plants for invasive parsnip webworms that attack native wild parsnips—can decrease the selective impact of the invasive species (Berenbaum and Zangerl, 2006). These studies indicate that the interaction network—the presence and magnitude of connections between invasive and invaded species—sets a context for the degree of evolution and coevolution in an invaded system. Researchers have even posited that in addition to escape from enemies (Enemy release hypothesis; Keane and Crawley, 2002) and escape from the costs of defence (evolution of increased competitive ability; Blossey and Notzold, 1995), an introduced species' escape from community complexity in their native habitat and consequent decrease in interactions with fitness-impacting species may be an additional type of 'release' in the novel habitat (Müller-Schärer et al., 2004; Strauss, 2014).

The complexity of experimental and survey designs will necessarily increase in order to separate adaptive from nonadaptive evolution (i.e. Keller and Taylor, 2008) and to take into account the effects of interactions in a trophic network. However, there are many interesting hypotheses to test. The likelihood of evolutionary rescue, where evolution rescues a declining population from extinction after an environmental change (Gomulkiewicz and Holt, 1995), may be influenced not only by the degree of maladaptation and the speed of adaptive evolution but also by interactions with other species in a community context. Theoretical studies (de Mazancourt et al., 2008) and empirical tests in experimental adaptive radiations of the bacteria *Pseudomonas fluorescens* (Fukami et al., 2007; Gómez and Buckling, 2013) have demonstrated that biodiversity can restrict evolutionary opportunities in groups of competitors. It may be that the number of links connecting resident to invasive species or the per-capita influence of the invasive species on interacting natives, via direct or indirect effects, mitigates the selection pressure exerted on species in the native habitat after an invasion. Evolutionary rescue or rapid evolution in native species can reduce the impact of invasive species. Evolution of feeding morphology in the Australian soapberry bug *Leptocoris tagalicus* occurred on a relatively rapid time scale following the spread of an invasive vine, leading to an almost doubled attack rate of the herbivore on the seeds of the invasive species (Carroll et al., 2005). Eco-evolutionary dynamics—where ecological and evolutionary processes influence one another and may lead to dynamic feedback loops where evolution alters an ecological process that in turn drives further evolutionary change (Fussmann et al., 2007)—may also be important to understand the consequences of invasive species. Shifts in community composition and diversity caused by phenotypic differentiation in a focal species may be almost as substantial as those caused by the addition of the species to the system (Bassar et al., 2010; Walsh et al., 2012), and local adaptation to environmental conditions may affect community structure as strongly as the presence or absence of those environmental conditions (Pantel et al., 2015). It is currently unknown whether the evolutionary changes that occur following invasions in ecological networks have the effect either of mitigating the ecological and functional consequences of invasion or of leading to further such ecological impacts. To understand how evolution is likely to propagate across food web networks after invasion, future studies might explore comparisons between ecological networks based on trophic interactions and evolutionary networks of fitness impacts or selection strengths.

5. DYNAMICAL CONSIDERATIONS FOR INVASION IN ECOLOGICAL NETWORKS

Question 10   How do the Impacts of Invaders on Invaded Networks Change Over Time and How does Network Structure Influence This?

Though the consequences of invasive species for ecosystem composition, structure, and function can be profound, much less attention has been paid to understanding how these consequences manifest over time. In a 2006 study, Strayer et al. (2006) reviewed 185 papers spanning a 5-year period (2001–05) that studied the effects of an invading species and found that 40% of the studies did not record the time since the invasion occurred and very few included multiple time points in the study. Given the variety and complexity of effects invasive species may have on a system—influencing abundance and biomass distributions, species interactions, ecosystem functioning, and evolution—it is important to implement the additional dimension of understanding temporal dynamics in these effects. Strayer et al. (2006) emphasize considering not only the 'acute' phase of invasions, immediately after the new species has arrived, but also the 'chronic' phase of invasions, when ecological and evolutionary processes have reacted to the incorporation of the new species. They also suggest a combination of mathematical modelling, microcosm experiments, and chronosequences (surveys that include site variation in the time since the invasion event has occurred) to better elucidate how invaded systems have changed with time.

Some attention has been paid to the existence of lags between a new species arrival and when it begins to affect the new ecosystem. The presence of such lags, whether in population growth or in range expansion, can vary because the match between the invasive species and the abiotic and biotic features of the novel habitat will vary for all invaders and invaded systems. The timing of the lag will thus depend on the timing over which a mismatch between the invasive species and the novel environment shifts to become more favourable to the invasive species. This can occur if the environment shifts to become more favourable to the alien species or if the alien species experiences genetic shifts to become better adapted to the novel landscape (Crooks and Soulé, 1999; see also Question 9: How do invasive species affect subsequent network evolutionary dynamics?). Delays in the effects of invasive species may also arise if these effects are density-dependent (i.e. an invasive canopy tree shading understory growth) or dependent on the ontogeny of the invading species (i.e. a long-lived invasive with stage-dependent impacts on other species; Simberloff et al., 2013). Two potential evolutionary mechanisms for sudden changes in impacts of invasive species in their novel habitats include: (i) hybridization of invasive with native species, leading to evolutionary changes that alter the way the invasive species participates in the ecosystem it inhabits (Hastings et al., 2005; Lee, 2002), and (ii) evolution of host shifts, for example, when parasites arrive together with their hosts and later adapt to infect resident host species (Dunn, 2009). Even if the invaded pathogens and parasites do not evolve, these species could contribute to decreasing the impact of an invasive host species on the novel ecosystem over time if they display a lag in the start of epidemics that serve to decrease any temporary (acute) abundance accumulations experienced by the invasive species (Amsellem et al., 2017; Médoc et al., 2017). The influence of network structure on the dynamics of invasive species effects through time remains an unexploited and novel area of research.

Question 11 Does Invasion Influence the Probability of Subsequent Invasion in Ecological Networks?

The North American Great Lakes are now home to approximately 180 invasive species introduced since the era of European exploration and colonization. Though one of the most important vectors of such nonnative taxa has been ballast water transport, the rate of invasion increased sharply even after the implementation of ballast water controls in 1989 (Holeck et al., 2004). Examples of this acceleration in the rate of successful invasions over time are referred to as 'invasional meltdown' (Cohen and Carlton, 1998; Simberloff and von Holle, 1999). Potential mechanisms for invasional meltdowns that are well known include the replacement of invasion-resistant native herbivores with exotic herbivores (Parker et al., 2006), habitat alteration, and introduction of the entirety of species needed to complete parasitic life cycles (Ricciardi, 2001; Ricciardi and MacIsaac, 2000). However, the invasion acceleration observed in the Great Lakes in the 1990s coincided with a profound shift in the region of origin of nonnative species, where 70% of species invading since 1985 are native to the fresh and brackish waters of the Ponto-Caspian region (Ricciardi and MacIsaac, 2000). This suggests that previously shared habitat may be another driver of invasional meltdown.

We simulated sequential species invasions to evaluate the role of shared history on establishment success. The simulations were constructed similarly to those described in Romanuk et al. (in press): briefly, we constructed food webs at each of three connectance levels ($C = 0.05, 0.15$, and 0.25) using the niche model (Williams and Martinez, 2000), computed the dynamics of the food webs for 2000 time steps using the structure-dynamical integrated model (Brose et al., 2005, 2006; Williams and Martinez, 2004b; Yodzis and Innes, 1992) to generate dynamically persistent webs, then randomly selected 50 webs from each connectance level (10 webs each with species richness $S = 26, 27, 28, 29$, and 30) to serve as native webs to be invaded. We then generated two types of invaders to serve as sources, copersistent and random invaders. For copersistent invaders, we constructed food webs with $C = 0.15$ using the niche model, computed the dynamics of the food webs for 2000 time steps using the structure-dynamical integrated model to generate dynamically persistent webs, then randomly selected webs with $S = 26, 27, 28, 29$, or 30 (150 webs total). For random invaders, instead of constructing webs, we simply assembled a list of species using niche properties (n_i, r_i, c_i; Williams and Martinez, 2000) associated with $C = 0.15$ then randomly selected 150 sets of 30 species from this list. By not constructing webs from these species, there were thus no interactions or associated biomass shifts as there were for copersistent invaders. We thus generated sets of species to be used as sources for invaders that either did or did not have a shared history prior to their arrival in the new food web. To simulate invasions, for each of the 150 native webs, we randomly chose either one of the 150 copersistent invader webs or one of the 150 random invader species groups, without replacement. We then randomly assigned the invasive species an order of introduction and sequentially added these invasive species to the native web every 200 time steps, beginning at time step 2000. The simulation ended 200 time steps after the last addition of an invasive species. Biomass of introduced species ranged between 10^{-9} and 10^{-2} and invasive and native species were eliminated from the web if their abundance decreased below an extinction threshold of 10^{-10}. Parameterization of the dynamical model and functional responses follows the methods described in Romanuk et al. (2009). Invasion success was calculated as the proportion of species added to a native web that persisted until 200 time steps following the last invasion.

Our simulations revealed that invasion success increased over time for both the copersistent and random invaders, but the slope of this increase was steeper for invaders with shared history (Fig. 3A), indicating that shared history increased invasion success. The magnitude of these differences were similar across low, medium, and high connectance webs. However, invasion success for both random and copersistent invaders decreased with increasing connectance. As invaders from basal trophic levels were almost always successful, differences in invasion success were driven primarily by consumer invaders (Fig. 3B).

The benefit of shared history that we observed matches similar advantages noted in 'invasion cartels' of species invading, for example, the Great Lakes where prior invasion of Eurasian zebra mussels, *D. polymorpha*, and amphipods, *Echinogammarus ischnus*, provided established prey to facilitate the invasion of the round goby, *Neogobius melanostomous*, and lakes in Spain where invasive North American populations of the northern pike, *Esox lucius*, are sustained by previously invasive populations of the American crayfish, *Procambarus clarkii* (Ricciardi, 2005). Our results, combined with additional recent evidence that nonnative species are accumulating at an accelerating rate in some freshwater (Reid and Orlova, 2002; Ricciardi, 2001), marine (Cohen and Carlton, 1998), and terrestrial (Parker et al., 2006) systems, suggest that hypotheses to explain the likelihood of invasion success must incorporate the facilitative mechanisms known to promote invasional meltdown. Future studies will also benefit from assessing the importance of invader facilitation relative to native biotic resistance.

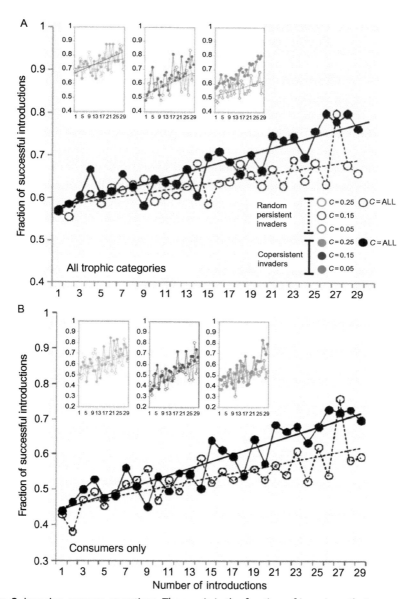

Fig. 3 Invasion success over time. The *y* axis is the fraction of invasions that were successful (e.g. the proportion of species added to a native web that persisted until 200 time steps following the last invasion), and the *x* axis shows how this changes across each sequential species invasion attempt. *Solid lines* and *filled symbols* are for copersistent invaders, with a shared history, and *dashed lines* and *unfilled symbols* are for random invaders, without a shared history. The main panels show invasion success

(Continued)

Question 12 How does the [In]Stability of the Network Facilitate or Prevent Invasions?

The stability of a system is defined as its ability to return to an equilibrium state after a disturbance. Unstable systems may therefore by definition enhance the likelihood of invasion success, since invasions represent a perturbation to the system and unstable systems enhance or propagate perturbations. However, this broad theoretical perspective has not been tested specifically for the perturbation of invasion. Some recent studies have used ecological stability concepts to understand evolutionary systems (Débarre et al., 2014; Svardal et al., 2014) and vice versa (Borrelli et al., 2015), and this cross-use of stability concepts may provide more concrete evidence that a perturbation of any type may promote a perturbation of any other type. Instead of stability inherently favouring or disfavouring invasion, unstable ecosystems may be adverse for species that have evolved in stable systems. In addition to selection for r and K strategists, c strategies may be observed in species well adapted to life in unstable, chaotic environments (Doebeli, 1995). For example, the theoretical study of Demetrius (2000) found that demographic stability of a population was positively correlated with a

Fig. 3—Cont'd   across all levels of connectance (C) for (a) all trophic categories and (b) consumer invaders only. The inset panels show invasion success within low ($C=0.05$, *green*), medium ($C=0.10$, *blue*), and high ($C=0.25$, *red*) connectance webs. For analyses conducted on invaders at all trophic levels (A), the regression equation for copersistent invaders across all C (*black solid line*) is $y=0.007x+0.5721$ ($r^2=0.802$, $P<0.001$), for random invaders across all C (*black dashed line*) is $y=0.004x+0.577$ ($r^2=0.496$, $P<0.001$), for copersistent invaders at low C (*green solid line*) is $y=0.005x+0.669$ ($r^2=0.398$, $P<0.001$), for random invaders at low C (*green dashed line*) is $y=0.003x+0.702$ ($r^2=0.137$, $P=0.048$), for copersistent invaders at medium C (*blue solid line*) is $y=0.008x+0.536$ ($r^2=0.535$, $P<0.001$), for random invaders at medium C (*blue dashed line*) is $y=0.006x+0.518$ ($r^2=0.327$, $P=0.001$), for copersistent invaders at high C (*red solid line*) is $y=0.007x+0.571$ ($r^2=0.791$, $P<0.001$), and for random invaders at high C (*red dashed line*) is $y=0.004x+0.512$ ($r^2=0.319$, $P<0.001$). For analyses conducted for consumer invaders only (B), the regression equation for copersistent invaders across all C (*black solid line*) is $y=0.010x+0.435$ ($r^2=0.828$, $P<0.001$), for random invaders across all C (*black dashed line*) is $y=0.006x+0.443$ ($r^2=0.575$, $P<0.001$), for copersistent invaders at low C (*green solid line*) is $y=0.009x+0.505$ ($r^2=0.442$, $P<0.001$), for random invaders at low C (*green dashed line*) is $y=0.00x+0.536$ ($r^2=0.225$, $P=0.009$), for copersistent invaders at medium C (*blue solid line*) is $y=0.010x+0.397$ ($r^2=0.670$, $P<0.001$), for random invaders at medium C (*blue dashed line*) is $y=0.008x+0.384$ ($r^2=0.460$, $P<0.001$), for copersistent invaders at high C (*red solid line*) is $y=0.010x+0.403$ ($r^2=0.526$, $P<0.001$), and for random invaders at high C (*red dashed line*) is $y=0.006x+0.408$ ($r^2=0.357$, $P<0.001$).

measure of the variability in the age of reproducing individuals in the populations. Therefore, instability may introduce an environmental context that favours species adapted to this instability and disfavours invasion by species from more stable habitats.

Moving from theoretical to applied instances of system instability, managed agricultural ecosystems are highly perturbed by anthropogenic interventions such as regular modifications of the cultivated plant and associated other species, the mechanical alteration of soil, nutrient input, and pesticide use and therefore present an excellent case study to understand the effects of instability for invasion success. Invasive species in agroecosystems may follow dynamics similar to the reinvasion of pests that have been previously excluded via pesticide application or host suppression. Less perturbed agricultural networks—in this instance organic farms that maintain a higher evenness of natural enemies than conventional farms—clearly demonstrated an increased resistance to pest outbreak (Crowder et al., 2010). The mechanism for this resistance to invasion may be increased connectance observed in the organic networks, a property that is often associated with decreased success of invasive species (Baiser et al., 2010). Less intensively managed agricultural systems that are less perturbed should also, in accordance with the intermediate disturbance hypothesis, experience increased diversity and system stability and decreased likelihood of invasion success (Catford et al., 2012; Connell, 1978). This prediction is consistent with agroforestry systems, which are not intensively managed and experience increased stability associated with the presence of perennial plants, and in turn demonstrate a diverse array of species interactions that allow increased control of pests (Liere et al., 2012; Morris et al., 2015). The management of associated plants—plants that grow inside the agricultural system but are not the targets of cultivation—can have important impacts on stability and consequent control of pest reestablishment as well. Increasing the diversity of cover crops can help stabilize the food web (Djigal et al., 2012) and enhance predation on invasive pests (Mollot et al., 2012). Decreasing the frequency of mowing cover crops can increase system stability and pest control (Tixier et al., 2013) via an increase in generalist-feeding species (Dassou and Tixier, 2016; Letourneau et al., 2011) and enhanced complexity of interspecific interactions (e.g. enhanced intraguild predation or trophic coherence, which is a measure of food webs associated with species that feed on species in a variety of trophic levels; Johnson et al., 2014).

6. FUTURE DIRECTIONS FOR RESEARCH IN INVASION NETWORKS

Question 13 How does Spatial Connectivity Influence Ecological Networks and What are the Implications for the Study of Invasive Species?

Invasions span multiple stages—introduction, establishment, dispersal to new sites, and subsequent spread—and any of these stages may be sensitive to properties of the landscape where the invasion occurs. For example, the potential for introduction of alien species can depend on the degree of spatial or dispersal connectivity between the native and introduced habitat. This connectivity can be modelled in a variety of ways, such as considering a very strong spatial connection where all species invade on a similar timescale (consistent with the sudden appearance of bridges connecting previously isolated habitats) or considering sequential invasion attempts by individual species at varying rates (as in the Trophic Theory of Island Biogeography, Massol et al., this issue, or in metacommunity models related to food web dynamics, Calcagno et al., 2011; Pillai et al., 2011). Establishment of invasive species can be influenced by the spatial distribution of resources, of other species, and of habitat availability and quality (Hastings et al., 2005; With, 2002). Diversity (Fridley et al., 2007; Shea and Chesson, 2002) and competitive strength of the recipient community (Hart and Gardner, 1997) influence invasibility and these factors vary widely in a landscape as well.

Spatial properties such as patch size, shape, and connectivity also influence the spread of invasive species (Harrison et al., 2001; Maheu-Giroux and de Blois, 2007). Spatial models of the spread of invasive species (Ellner et al., 1998; Fisher, 1937; Shigesada and Kawasaki, 1997) can be used to determine how invasive species might occupy a novel landscape and to determine practices to slow their spread (Sharov and Liebhold, 1998). This approach has been further developed to consider potential evolution of invading species along the advancing invasion front. Evolution of dispersal in invasive species can be important to understand their spread (Burton et al., 2010; Phillips et al., 2008) and was observed in the invasive Australian cane toad (Phillips et al., 2006). Trade-offs among life history traits have also led to the evolution of deleterious traits in advancing invasive species. In the case of the invasive Australian cane toad, evolution of increased dispersal was associated with decreased reproductive rates (Hudson et al., 2015).

Spatial locations can also be described as networks, where edges represent distance or connectivity and nodes represent a geographic location (Dale and Fortin, 2010). This network perspective was used to identify inland lakes in Ontario, Canada that served as major hubs of transport for the invasive spiny waterflea *Bythotrephes longimanus* and thus were the most promising targets for management efforts to reduce boat traffic (Muirhead and MacIsaac, 2005). Properties of the spatial networks underlying food webs in a landscape are likely to have important effects for the stability and structure of resident populations and food webs. For example, species persistence is strongly influenced by spatial network spatial configuration (Adler and Nuernberger, 1994) and topology (Gilarranz and Bascompte, 2012). Bellmore et al. (2015) used an incremental aggregation of fish and aquatic invertebrate food webs from different locations in a river floodplain into increasingly complex metafood webs to show that increasing spatial complexity led to reduced strength of consumer–resource interactions, which is an important driver of food web stability. Much work remains to incorporate the full complexity of spatial networks into ecological network dynamics and this task is not simple (see Gravel et al., 2016; Mougi and Kondoh, 2016). It is currently unknown what spatial properties of food webs in a landscape render them more or less susceptible to invasion. It is also unclear whether these properties would depend on the trophic level or other attributes of the invasive species in the ecological network. Future research may identify geographic nodes that contribute most strongly to the spread of invasive species (e.g. Muirhead and MacIsaac, 2005) and try to identify properties of those nodes that promote invasion. Controlled invasion experiments (e.g. Morel-Journel et al., 2015) and theoretical modelling will be important tools to facilitate such analyses.

Question 14 How can we Integrate HTS Tools to Monitor Networks Before and After Invasions?

Monitoring multispecies interactions as well their response to various alterations such as biological invasions is resource-demanding and methodologically challenging. However, predicting the success of management actions requires the explicit and detailed quantification of direct and indirect ecological interactions within an invaded ecosystem (Bull and Courchamp, 2009). Emerging molecular techniques based on HTS such as metabarcoding or metagenomics have the potential to alleviate current methodological hurdles in network ecology (Kamenova et al., this issue; Vacher et al., 2016). HTS techniques allow the detailed reconstruction of past and present biodiversity even from minute amounts of environmental DNA (eDNA) (Cannon et al., 2016; Hajibabaei et al., 2011; Willerslev et al., 2014; Yoccoz et al., 2012; Zaiko et al., 2015; Zhou et al., 2013). HTS methods can quantify complex trophic interactions without extensive a priori knowledge or taxonomic expertise (Bohmann et al., 2011; Boyer et al., 2013; Deagle et al., 2009; Mollot et al., 2014; Valentini et al., 2009). Finally, genomic HTS data can be used to derive direct quantitative estimates of species relative abundances or biomass (Andersen et al., 2012; Evans et al., 2016; Willerslev et al., 2014). Such quantitative biodiversity and trophic interaction data could therefore be used to quickly and inexpensively monitor weighted interaction networks across gradients of invasion or after management actions.

HTS-based food webs are scarce in the literature and studies that use this approach for understanding biological invasions have not been conducted yet. However, the application of HTS to other ecological questions can illuminate how this approach might benefit studies of invasion of networks of interacting species. HTS has been used to estimate prey attack rates at the level of complete interacting communities (Mollot et al., 2014; Vacher et al., 2016) and could similarly be used to estimate parasitism or pollination rates. Such data could be readily used to quantify the top-down impact of an invasive species or to assess the degree to which a potential biocontrol species affects an invader. HTS-based assessments of diet preference and the degree of trophic plasticity could be used to predict competition between native and invasive species or to forecast undesired cascading effects in feeding behaviours following management actions such as addition of biocontrol species or removal of target nuisance species. The simultaneous characterization of multiple types of species interactions is possible using HTS techniques (i.e. trophic or parasitic; Tiede et al., 2016), and such data could be used to simulate realistic species management scenarios and anticipate the impact of unexpected synergies or antagonisms that might increase invasion success.

Despite its vast potential, integrating HTS tools to monitor biological invasions will present technical and conceptual challenges as well. Technical challenges are inherent in setting up a universal HTS methodological framework to identify multiple types of species interactions in any type of habitat and organism in the field. Developing universal primers that amplify across a range of organisms spanning entire food webs will be challenging, and developing a bioinformatics pipeline to manage the massive amounts of data generated by HTS will require substantial time as well.

Conceptually, the use of HTS data does not provide the answer to questions that researchers must decide. Which network metrics are most suitable for efficient monitoring of invasions in natural systems? Which food web alterations are caused by the invasion in progress as opposed to other potential drivers of change in the system? One potentially interesting forward direction might be to use HTS to estimate and manage the functional role of invasive species, facilitated by techniques in the fast growing field of functional metagenomics (Chistoserdova, 2009). Invasive species monitoring could then focus on the profiles of gene or protein expression or functional activity, rather than on taxonomic identity per se. This approach could be useful to identify key genes influencing invader success and impact (Dlugosch et al., 2013; Scully et al., 2013) or to implement management strategies that seek to optimize an invasive species' integration in invaded ecosystems and minimize its impact on ecosystem functioning, rather than removal of the species.

7. CONCLUSION

Considering groups of interacting species as a network—where the links between the species represent interactions of competition, predation, parasitism, and mutualism—is a framework that may improve our understanding of and our ability to predict the dynamics of invasive species. Network properties such as connectance and modularity, which emerge only when arrays of ecosystem residents are considered, are indicative of system stability and provide important information about the system's susceptibility to invasion. Our goal here has been to address the information gaps and highlight exciting future directions of research needed to implement a network perspective in the study of invasive species. Some of the questions raised will contribute to advancing the field of ecology from a theoretical and conceptual perspective (e.g. Questions 2, 3, 5, 8, 10, and 12), while others are directly relevant to assist system management decisions (e.g. Questions 13 and 14). We end by emphasizing that it is important to consider not just the invasive organism and its strongest interaction partners but instead to identify the role the novel species plays in the context of the broader set of existing connections between resident species. Only then can researchers identify emergent signals that can be used to monitor the establishment and growth of invasive species and to manage them in ways that mitigate their harmful impacts for ecosystems.

REFERENCES

Adler, F.R., Nuernberger, B., 1994. Persistence in patchy irregular landscapes. Theor. Popul. Biol. 45, 41–75.

Aizen, M.A., Morales, C.L., Morales, J.M., 2008. Invasive mutualists erode native pollination webs. PLoS Biol. 6. e31.

Amsellem, L., Brouat, C., Duron, O., Porter, S.S., Vilcinskas, A., Facon, B., 2017. Importance of microorganisms to macroorganisms invasions: is the essential invisible to the eye? (The little prince, A. de Saint-Exupéry, 1943). Adv. Ecol. Res. 57, 99–146.

Andersen, K., Bird, K.L., Rasmussen, M., Haile, J., Breuning-Madsen, H., Kjaer, K.H., Orlando, L., Gilbert, M.T.P., Willerslev, E., 2012. Meta-barcoding of "dirt" DNA from soil reflects vertebrate biodiversity. Mol. Ecol. 21, 1966–1979.

Asner, G.P., Vitousek, P.M., 2005. Remote analysis of biological invasion and biogeochemical change. Proc. Natl. Acad. Sci. U.S.A. 102, 4383–4386.

Azzurro, E., Tuset, V.M., Lombarte, A., Maynou, F., Simberloff, D., Rodríguez-Pérez, A., Solé, R.V., 2014. External morphology explains the success of biological invasions. Ecol. Lett. 17, 1455–1463.

Baiser, B., Russell, G.J., Lockwood, J.L., 2010. Connectance determines invasion success via trophic interactions in model food webs. Oikos 119, 1970–1976.

Bartomeus, I., Vilà, M., Santamaría, L., 2008. Contrasting effects of invasive plants in plant-pollinator networks. Oecologia 155, 761–770.

Bascompte, J., Jordano, P., Melián, C.J., Olesen, J.M., 2003. The nested assembly of plant-animal mutualistic networks. Proc. Natl. Acad. Sci. U.S.A. 100, 9383–9387.

Bassar, R.D., Marshall, M.C., López-Sepulcre, A., Zandonà, E., Auer, S.K., Travis, J., Pringle, C.M., Flecker, A.S., Thomas, S.A., Fraser, D.F., Reznick, D.N., 2010. Local adaptation in Trinidadian guppies alters ecosystem processes. Proc. Natl. Acad. Sci. U.S.A. 107, 3616–3621.

Beckerman, A.P., Petchey, O.L., Warren, P.H., 2006. Foraging biology predicts food web complexity. Proc. Natl. Acad. Sci. U.S.A. 103, 13745–13749.

Bellay, S., De Oliveira, E.F., Almeida-Neto, M., Mello, M.A.R., Takemoto, R.M., Luque, J.L., 2015. Ectoparasites and endoparasites of fish form networks with different structures. Parasitology 142, 901–909.

Bellmore, J.R., Baxter, C.V., Connolly, P.J., 2015. Spatial complexity reduces interaction strengths in the meta-food web of a river floodplain mosaic. Ecology 96, 274–283.

Berenbaum, M.R., Zangerl, A.R., 2006. Parsnip webworms and host plants at home and abroad: trophic complexity in a geographic mosaic. Ecology 87, 3070–3081.

Blackburn, T.M., Cassey, P., Lockwood, J.L., 2009. The role of species traits in the establishment success of exotic birds. Glob. Chang. Biol. 15, 2852–2860.

Blossey, B., Notzold, R., 1995. Evolution of increased competitive ability in invasive nonindigenous plants: a hypothesis. J. Ecol. 83, 887–889.

Bohmann, K., Monadjem, A., Noer, C.L., Rasmussen, M., Zeale, M.R.K., Clare, E., Jones, G., Willerslev, E., Gilbert, M.T.P., 2011. Molecular diet analysis of two African free-tailed bats (Molossidae) using high throughput sequencing. PLoS One 6. e21441.

Borrelli, J.J., Allesina, S., Amarasekare, P., Arditi, R., Chase, I., Damuth, J., Holt, R.D., Logofet, D.O., Novak, M., Rohr, R.P., Rossberg, A.G., Spencer, M., Khai, J.K., Ginzburg, L.R., 2015. Selection on stability across ecological scales. Trends Ecol. Evol. 30, 417–425.

Bossdorf, O., Auge, H., Lafuma, L., Rogers, W.E., Siemann, E., Prati, D., 2005. Phenotypic and genetic differentiation between native and introduced plant populations. Oecologia 144, 1–11.

Boyer, S., Wratten, S.D., Holyoake, A., Abdelkrim, J., Cruickshank, R.H., 2013. Using next-generation sequencing to analyse the diet of a highly endangered land snail (Powelliphanta augusta) feeding on endemic earthworms. PLoS One 8. e75962.

Brooks, M.L., D'antonio, C.M., Richardson, D.M., Grace, J.B., Keeley, J.E., DiTomaso, J.M., Hobbs, R.J., Pellant, M., Pyke, D., 2004. Effects of invasive alien plants on fire regimes. BioScience 54, 677–688.

Brose, U., Berlow, E.L., Martinez, N.D., 2005. Scaling up keystone effects from simple to complex ecological networks. Ecol. Lett. 9, 1228–1236.

Brose, U., Williams, R.J., Martinez, N.D., 2006. Allometric scaling enhances stability in complex food webs. Ecol. Lett. 9, 1228–1236.

Brouwer, N.L., Hale, A.N., Kalisz, S., 2015. Mutualism-disrupting allelopathic invader drives carbon stress and vital rate decline in a forest perennial herb. AoB Plants 7, plv014.

Brown, S.P., Le Chat, L., Taddei, F., 2008. Evolution of virulence: triggering host inflammation allows invading pathogens to exclude competitors. Ecol. Lett. 11, 44–51.

Bull, L.S., Courchamp, F., 2009. Management of interacting invasives: ecosystem approaches. In: Clout, M.N., Williams, P.A. (Eds.), Invasive Species Management: A Handbook of Principles and Techniques. Oxford University Press, Oxford, pp. 232–247.

Burton, O.J., Phillips, B.L., Travis, J.M.J., 2010. Trade-offs and the evolution of life-histories during range expansion. Ecol. Lett. 13, 1210–1220.

Cagnolo, L., Salvo, A., Valladares, G., 2011. Network topology: patterns and mechanisms in plant-herbivore and host-parasitoid food webs. J. Anim. Ecol. 80, 342–351.

Calcagno, V., Massol, F., Mouquet, N., Jarne, P., David, P., 2011. Constraints on food chain length arising from regional metacommunity dynamics. Proc. R. Soc. Lond. B Biol. Sci. 278, 3042–3049. http://dx.doi.org/10.1098/rspb.2011.0112.

Cannon, M.V., Hester, J., Shalkhauser, A., Chan, E.R., Logue, K., Small, S.T., Serre, D., 2016. In silico assessment of primers for eDNA studies using PrimerTree and application to characterize the biodiversity surrounding the Cuyahoga River. Sci. Rep. 6, 22908.

Capps, K.A., Flecker, A.S., 2013. Invasive aquarium fish transform ecosystem nutrient dynamics. Proc. R. Soc. Lond. B Biol. Sci. 280, 20131520.

Carpentier, A., Gozlan, R.E., Cucherousset, J., Paillisson, J.-M., Marion, L., 2007. Is topmouth gudgeon Pseudorasbora parva responsible for the decline in sunbleak Leucaspius delineatus populations? J. Fish Biol. 71, 274–278.

Carroll, S.P., Loye, J.E., Dingle, H., Mathieson, M., Famula, T.R., Zalucki, M.P., 2005. And the beak shall inherit—evolution in response to invasion. Ecol. Lett. 8, 944–951.

Catford, J.A., Daehler, C.C., Murphy, H.T., Sheppard, A.W., Hardesty, B.D., Westcott, D.A., Rejmánek, M., Bellingham, P.J., Pergl, J., Horvitz, C.C., Hulmei, P.E., 2012. The intermediate disturbance hypothesis and plant invasions: implications for species richness and management. Perspect. Plant Ecol. Evol. Syst. 14, 231–241.

Chagnon, P.-L., U'Ren, J.M., Miadlikowska, J., Lutzoni, F., Arnold, A.E., 2016. Interaction type influences ecological network structure more than local abiotic conditions: evidence from endophytic and endolichenic fungi at a continental scale. Oecologia 180, 181–191.

Chistoserdova, L., 2009. Functional metagenomics: recent advances and future challenges. Biotechnol. Genet. Eng. Rev. 26, 335–352.

Chytrý, M., Jarošík, V., Pyšek, P., Hájek, O., Knollová, I., Tichý, L., Danihelka, J., 2008. Separating habitat invasibility by alien plants from the actual level of invasion. Ecology 89, 1541–1553.

Cohen, A.N., Carlton, J.T., 1998. Accelerating invasion rate in a highly invaded estuary. Science 279, 555–558.

Colautti, R.I., MacIsaac, H.J., 2004. A neutral terminology to define "invasive" species. Divers. Distrib. 10, 135–141.

Colautti, R.I., Ricciardi, A., Grigorovich, I.A., MacIsaac, H.J., 2004. Is invasion success explained by the enemy release hypothesis? Ecol. Lett. 7, 721–733.

Colautti, R.I., Grigorovich, I.A., MacIsaac, H.J., 2006. Propagule pressure: a null model for biological invasions. Biol. Invasions 8, 1023–1037.

Connell, J.H., 1978. Diversity in tropical rain forests and coral reefs. Science 199, 1302–1310.

Correia, A.M., 2001. Seasonal and interspecific evaluation of predation by mammals and birds on the introduced red swamp crayfish Procambarus clarkii (Crustacea, Cambaridae) in a freshwater marsh (Portugal). J. Zool. 255, 533–541.

Croll, D.A., Maron, J.L., Estes, J.A., Danner, E.M., Byrd, G.V., 2005. Introduced predators transform subarctic islands from grassland to tundra. Science 307, 1959–1961.

Crooks, J.A., Soulé, M.E., 1999. Lag times in population explosions of invasive species: causes and implications. In: Sandlund, O.T., Schei, P.J., Viken, Å. (Eds.), Invasive Species and Biodiversity Management. Kluwer Academic Publishers, Dordrecht, pp. 103–125.

Crowder, D.W., Northfield, T.D., Strand, M.R., Snyder, W.E., 2010. Organic agriculture promotes evenness and natural pest control. Nature 466, 109–112.

Dale, M.R.T., Fortin, M.J., 2010. From graphs to spatial graphs. Annu. Rev. Ecol. Evol. Syst. 41, 21–38.

Darwin, C., 1859. On the Origin of Species. John Murray, London.

Dassou, A.G., Tixier, P., 2016. Response of pest control by generalist predators to local-scale plant diversity: a meta-analysis. Ecol. Evol. 6, 1143–1153.

David, P., Thébault, E., Anneville, O., Duyck, P.-F., Chapuis, E., Loeuille, N., 2017. Impacts of invasive species on food webs: a review of empirical data. Adv. Ecol. Res. 56, 1–60.

Davis, M.A., Grime, J.P., Thompson, K., 2000. Fluctuating resources in plant communities: a general theory of invasibility. J. Ecol. 88, 528–534.

de Gruchy, M.A., Reader, R.J., Larson, D.W., 2005. Biomass, productivity, and dominance of alien plants: a multihabitat study in a national park. Ecology 86, 1259–1266.

de Mazancourt, C., Johnson, E., Barraclough, T.G., 2008. Biodiversity inhibits species' evolutionary responses to changing environments. Ecol. Lett. 11, 380–388.

Deagle, B.E., Kirkwood, R., Jarman, S.N., 2009. Analysis of Australian fur seal diet by pyrosequencing prey DNA in faeces. Mol. Ecol. 18, 2022–2038.

Débarre, F., Nuismer, S.L., Doebeli, M., 2014. Multidimensional (co)evolutionary stability. Am. Nat. 184, 158–171.

DeMelo, R., France, R., McQueen, D.J., 1992. Biomanipulation: hit or myth? Limnol. Oceanogr. 37, 192–207.

Demetrius, L., 2000. Directionality theory and the evolution of body size. Proc. R. Soc. Lond. B Biol. Sci. 267, 2385–2391.

Diez, J.M., Sullivan, J.J., Hulme, P.E., Edwards, G., Duncan, R.P., 2008. Darwin's naturalization conundrum: dissecting taxonomic patterns of species invasions. Ecol. Lett. 11, 674–681.

Djigal, D., Chabrier, C., Duyck, P.-F., Achard, R., Quénéhervé, P., Tixier, P., 2012. Cover crops alter the soil nematode food web in banana agroecosystems. Soil Biol. Biochem. 48, 142–150.

Dlugosch, K.M., Parker, I.M., 2008. Founding events in species invasions: genetic variation, adaptive evolution, and the role of multiple introductions. Mol. Ecol. 17, 431–449.

Dlugosch, K.M., Lai, Z., Bonin, A., Hierro, J., Rieseberg, L.H., 2013. Allele identification for transcriptome-based population genomics in the invasive plant Centaurea solstitialis. G3: Genes Genomes. Genet. 3, 359–367.

Doebeli, M., 1995. Evolutionary predictions from invariant physical measures of dynamic processes. J. Theor. Biol. 173, 377–387.

Dorcas, M.E., Willson, J.D., Reed, R.N., Snow, R.W., Rochford, M.R., Miller, M.A., Meshaka, W.E., Andreadis, P.T., Mazzotti, F.J., Romagosa, C.M., Hart, K.M., 2012. Severe mammal declines coincide with proliferation of invasive Burmese pythons in Everglades National Park. Proc. Natl. Acad. Sci. U.S.A. 109, 2418–2422.

Dukes, J.S., Mooney, H.A., 2004. Disruption of ecosystem processes in western North America by invasive species. Rev. Chil. Hist. Nat. 77, 411–437.

Duncan, R.P., Williams, P.A., 2002. Ecology: Darwin's naturalization hypothesis challenged. Nature 417, 608–609.

Dunn, A.M., 2009. Parasites and biological invasions. In: Advances in parasitology, vol. 68. Academic Press, pp. 161–184.

Dunne, J.A., Williams, R.J., Martinez, N.D., 2002a. Network structure and biodiversity loss in food webs: robustness increases with connectance. Ecol. Lett. 5, 558–567.

Dunne, J.A., Williams, R.J., Martinez, N.D., 2002b. Food-web structure and network theory: the role of connectance and size. Proc. Natl. Acad. Sci. U.S.A. 99, 12917–12922.

Duyck, P.-F., David, P., Junod, G., Brunel, C., Dupont, R., Quilici, S., 2006. Importance of competition mechanisms in successive invasions by polyphagous tephritids in La Réunion. Ecology 87, 1770–1780.

Duyck, P.-F., David, P., Quilici, S., 2007. Can more K-selected species be better invaders? A case study of fruit flies in La Réunion. Divers. Distrib. 13, 535–543.

Eastwood, M.M., Donahue, M.J., Fowler, A.E., 2007. Reconstructing past biological invasions: niche shifts in response to invasive predators and competitors. Biol. Invasions 9, 397–407.

Ehrenfeld, J.G., 2010. Ecosystem consequences of biological invasions. Annu. Rev. Ecol. Evol. Syst. 41, 59–80.

Elias, M., Fontaine, C., van Veen, F.J.F., 2013. Evolutionary history and ecological processes shape a local multilevel antagonistic network. Curr. Biol. 23, 1355–1359.

Ellner, P.S., Sasaki, A., Haraguchi, Y., Matsuda, H., 1998. Speed of invasion in lattice population models: pair-edge approximation. J. Math. Biol. 36, 469–484.

Elton, C.S., 1958. The Ecology of Invasions by Animals and Plants. Methuen and Co., London.

Evans, N.T., Olds, B.P., Renshaw, M.A., Turner, C.R., Li, Y., Jerde, C.L., Mahon, A.R., Pfrender, M.E., Lamberti, G.A., Lodge, D.M., 2016. Quantification of mesocosm fish and amphibian species diversity via environmental DNA metabarcoding. Mol. Ecol. Resour. 16, 29–41.

Ezenwa, V.O., Gerardo, N.M., Inouye, D.W., Medina, M., Xavier, J.B., 2012. Animal behavior and the microbiome. Science 338, 198–199.

Facon, B., Genton, B.J., Shykoff, J., Jarne, P., Estoup, A., David, P., 2006. A general eco-evolutionary framework for understanding bioinvasions. Trends Ecol. Evol. 21, 130–135.

Fargione, J., Brown, C.S., Tilman, D., 2003. Community assembly and invasion: an experimental test of neutral versus niche processes. Proc. Natl. Acad. Sci. U.S.A. 100, 8916–8920.

Fisher, R.A., 1937. The wave of advance of advantageous genes. Ann. Eugen. 7, 355–369.

Fontaine, C., Thébault, E., 2015. Comparing the conservatism of ecological interactions in plant-pollinator and plant-herbivore networks. Popul. Ecol. 57, 29–36.

Fontaine, C., Guimarães, P.R., Kéfi, S., Loeuille, N., Memmott, J., van der Putten, W.H., van Veen, F.J.F., Thébault, E., 2011. The ecological and evolutionary implications of merging different types of networks. Ecol. Lett. 14, 1170–1181.

Fortunato, S., 2010. Community detection in graphs. Phys. Rep. 486, 75–174.

Fridley, J., Stachowicz, J., Naeem, S., Sax, D., Seabloom, E., Smith, M., Stohlgren, T., Tilman, D., Von Holle, B., 2007. The invasion paradox: reconciling pattern and process in species invasions. Ecology 88, 3–17.

Fukami, T., Beaumont, H.J.E.E., Zhang, X.-X., Rainey, P.B., 2007. Immigration history controls diversification in experimental adaptive radiation. Nature 446, 436–439.

Fussmann, G.F., Loreau, M., Abrams, P.A., 2007. Eco-evolutionary dynamics of communities and ecosystems. Funct. Ecol. 21, 465–477.

Gerhold, P., Cahill, J.F., Winter, M., Bartish, I.V., Prinzing, A., 2015. Phylogenetic patterns are not proxies of community assembly mechanisms (they are far better). Funct. Ecol. 29, 600–614.

Gerla, D.J., Vos, M., Kooi, B.W., Mooij, W.M., 2009. Effects of resources and predation on the predictability of community composition. Oikos 118, 1044–1052.

Gianuca, A.T., Pantel, J.H., De Meester, L., 2016. Disentangling the effect of body size and phylogenetic distances on zooplankton top-down control of algae. Proc. R. Soc. B 283, 20160487.

Gilarranz, L.J., Bascompte, J., 2012. Spatial network structure and metapopulation persistence. J. Theor. Biol. 297, 11–16.

Gómez, P., Buckling, A., 2013. Real-time microbial adaptive diversification in soil. Ecol. Lett. 16, 650–655.

Gomulkiewicz, R., Holt, R.D., 1995. When does evolution by natural selection prevent extinction? Evolution 49, 201–207.

González, A.L., Kominoski, J.S., Danger, M., Ishida, S., Iwai, N., Rubach, A., 2010. Can ecological stoichiometry help explain patterns of biological invasions? Oikos 119, 779–790.

Gounand, I., Mouquet, N., Canard, E., Guichard, F., Hauzy, C., Gravel, D., 2014. The paradox of enrichment in metaecosystems. Am. Nat. 184, 752–763.

Gravel, D., Guichard, F., Loreau, M., Mouquet, N., 2010. Source and sink dynamics in meta-ecosystems. Ecology 91, 2172–2184.

Gravel, D., Massol, F., Canard, E., Mouillot, D., Mouquet, N., 2011. Trophic theory of island biogeography. Ecol. Lett. 14, 1010–1016.

Gravel, D., Massol, F., Leibold, M.A., 2016. Stability and complexity in model meta-ecosystems. Nat. Commun. 7, 12457.

Green, P.T., O'Dowd, D.J., Abbott, K.L., Jeffery, M., Retallick, K., Mac Nally, R., 2011. Invasional meltdown: invader–invader mutualism facilitates a secondary invasion. Ecology 92, 1758–1768.

Grove, S., Parker, I.M., Haubensak, K.A., 2015. Persistence of a soil legacy following removal of a nitrogen-fixing invader. Biol. Invasions 17, 2621–2631.

Gurevitch, J., Padilla, D.K., 2004. Response to Ricciardi. Assessing species invasions as a cause of extinction. Trends Ecol. Evol. 19, 620.

Hajibabaei, M., Shokralla, S., Zhou, X., Singer, G.A.C., Baird, D.J., 2011. Environmental barcoding: a next-generation sequencing approach for biomonitoring applications using river benthos. PLoS One 6. e17497.

Harrison, S., Rice, K., Maron, J., 2001. Habitat patchiness promotes invasion by alien grasses on serpentine soil. Biol. Conserv. 100, 45–53.

Hart, D.R., Gardner, R.H., 1997. A spatial model for the spread of invading organisms subject to competition. J. Math. Biol. 35, 935–948.

Hastings, A., Cuddington, K., Davies, K.F., Dugaw, C.J., Elmendorf, S., Freestone, A., Harrison, S., Holland, M., Lambrinos, J., Malvadkar, U., Melbourne, B.A., Moore, K., Taylor, C., Thomson, D., 2005. The spatial spread of invasions: new developments in theory and evidence. Ecol. Lett. 8, 91–101.

Hecht, A.L., Casterline, B.W., Earley, Z.M., Goo, Y.A., Goodlett, D.R., Wardenburg, J.B., 2016. Strain competition restricts colonization of an enteric pathogen and prevents colitis. EMBO Rep. 17, 1281–1291.

Hejda, M., Pyšek, P., Jarošík, V., 2009. Impact of invasive plants on the species richness, diversity and composition of invaded communities. J. Ecol. 97, 393–403.

Higgins, S.N., Vander Zanden, M.J., 2010. What a difference a species makes: a meta-analysis of dreissenid mussel impacts on freshwater ecosystems. Ecol. Monogr. 80, 179–196.

Holeck, K.T., Mills, E.L., MacIsaac, H.J., Dochoda, M.R., Colautti, R.I., Ricciardi, A., 2004. Bridging troubled waters: biological invasions, transoceanic shipping, and the Laurentian Great Lakes. BioScience 54, 919–929.

Hudson, C.M., Phillips, B.L., Brown, G.P., Shine, R., 2015. Virgins in the vanguard: low reproductive frequency in invasion-front cane toads. Biol. J. Linn. Soc. 116, 743–747.

Hui, C., Richardson, D.M., Landi, P., Minoarivelo, H.O., Garnas, J., Roy, H.E., 2016. Defining invasiveness and invasibility in ecological networks. Biol. Invasions 18, 971–983.

Inderjit, van der Putten, W.H., 2010. Impacts of soil microbial communities on exotic plant invasions. Trends Ecol. Evol. 25, 512–519.

Ives, A.R., Godfray, H.C.J., 2006. Phylogenetic analysis of trophic associations. Am. Nat. 168, E1–E14.

Jeschke, J.M., Strayer, D.L., 2006. Determinants of vertebrate invasion success in Europe and North America. Glob. Chang. Biol. 12, 1608–1619.

Johnson, S., Domínguez-García, V., Donetti, L., Muñoz, M.A., 2014. Trophic coherence determines food-web stability. Proc. Natl. Acad. Sci. U.S.A. 111, 17923–17928.

Jones, C.G., Lawton, J.H., Shachak, M., 1994. Organisms as ecosystem engineers. Oikos 69, 373–386.

Jones, E.I., Nuismer, S.L., Gomulkiewicz, R., 2013. Revisiting Darwin's conundrum reveals a twist on the relationship between phylogenetic distance and invasibility. Proc. Natl. Acad. Sci. U.S.A. 110, 20627–20632.

Jordano, P., Bascompte, J., Olesen, J.M., 2003. Invariant properties in coevolutionary networks of plant–animal interactions. Ecol. Lett. 6, 69–81.

Kamenova, S., Bartley, T.J., Bohan, D.A., Boutain, J.R., Colautti, R.I., Domaizon, I., Fontaine, C., Lemainque, A., Le Viol, I., Mollot, G., Perga, M.-E., Ravigné, V., Massol, F., 2017. Invasions toolkit: current methods for tracking the spread and impact of invasive species. Adv. Ecol. Res. 56, 85–182.

Karatayev, A.Y., Burlakova, L.E., Padilla, D.K., 2002. Impacts of zebra mussels on aquatic communities and their role as ecosystem engineers. In: Leppäkoski, E., Gollasch, S., Olenin, S. (Eds.), Invasive Aquatic Species of Europe. Distribution, Impacts and Management. Springer Science + Business Media, Dordrecht, The Netherlands, pp. 433–446.

Keane, R.M., Crawley, M.J., 2002. Exotic plant invasions and the enemy release hypothesis. Trends Ecol. Evol. 17, 164–170.

Kéfi, S., Berlow, E.L., Wieters, E.A., Navarrete, S.A., Petchey, O.L., Wood, S.A., Boit, A., Joppa, L.N., Lafferty, K.D., Williams, R.J., Martinez, N.D., Menge, B.A., Blanchette, C.A., Iles, A.C., Brose, U., 2012. More than a meal... integrating non-feeding interactions into food webs. Ecol. Lett. 15, 291–300.

Kéfi, S., Berlow, E.L., Wieters, E.A., Joppa, L.N., Wood, S.A., Brose, U., Navarrete, S.A., 2015. Network structure beyond food webs: mapping non-trophic and trophic interactions on Chilean rocky shores. Ecology 96, 291–303.

Kéfi, S., Miele, V., Wieters, E.A., Navarrete, S.A., Berlow, E.L., 2016. How structured is the entangled bank? The surprisingly simple organization of multiplex ecological networks leads to increased persistence and resilience. PLoS Biol. 14. e1002527.

Keller, S.R., Taylor, D.R., 2008. History, chance and adaptation during biological invasion: separating stochastic phenotypic evolution from response to selection. Ecol. Lett. 11, 852–866.

Khosravi, A., Mazmanian, S.K., 2013. Disruption of the gut microbiome as a risk factor for microbial infections. Curr. Opin. Microbiol. 16, 221–227.

King, S.A., Buckney, R.T., 2001. Exotic plants in the soil-stored seed bank of urban bushland. Aust. J. Bot. 49, 717–720.

Kolar, C.S., Lodge, D.M., 2001. Progress in invasion biology: predicting invaders. Trends Ecol. Evol. 16, 199–204.

Krause, A.E., Frank, K.A., Mason, D.M., Ulanowicz, R.E., Taylor, W.W., 2003. Compartments revealed in food-web structure. Nature 426, 282–285.

Lankau, R.A., 2012. Coevolution between invasive and native plants driven by chemical competition and soil biota. Proc. Natl. Acad. Sci. U.S.A. 109, 11240–11245.

Lau, J.A., 2006. Evolutionary responses of native plants to novel community members. Evolution 60, 56–63.

Lee, C.E., 2002. Evolutionary genetics of invasive species. Trends Ecol. Evol. 17, 386–391.

Lee, K.-M., Lee, S.Y., Connolly, R.M., 2011. Combining stable isotope enrichment, compartmental modelling and ecological network analysis for quantitative measurement of food web dynamics. Methods Ecol. Evol. 2, 56–65.

León, J.A., Tumpson, D.B., 1975. Competition between two species for two complementary or substitutable resources. J. Theor. Biol. 50, 185–201.

Letourneau, D.K., Armbrecht, I., Rivera, B.S., Lerma, J.M., Carmona, E.J., Daza, M.C., Escobar, S., Galindo, V., Gutiérrez, C., López, S.D., Mejía, J.L., Rangel, A.M.A., Rangel, J.H., Rivera, L., Saavedra, C.A., Torres, A.M., Trujillo, A.R., 2011. Does plant diversity benefit agroecosystems? A synthetic review. Ecol. Appl. 21, 9–21.

Levine, J.M., D'Antonio, C.M., 1999. Elton revisited: a review of evidence linking diversity and invasibility. Oikos 87, 15–26.

Liere, H., Jackson, D., Vandermeer, J., 2012. Ecological complexity in a coffee agroecosystem: spatial heterogeneity, population persistence and biological control. PLoS One 7. e45508.

Lockwood, J.L., Cassey, P., Blackburn, T., 2005. The role of propagule pressure in explaining species invasions. Trends Ecol. Evol. 20, 223–228.

Loeuille, N., Leibold, M.A., 2008. Ecological consequences of evolution in plant defenses in a metacommunity. Theor. Popul. Biol. 74, 34–45.

Loeuille, N., Loreau, M., 2004. Nutrient enrichment and food chains: can evolution buffer top-down control? Theor. Popul. Biol. 65, 285–298.

Loeuille, N., Loreau, M., 2010. Emergence of complex food web structure in community evolution models. In: Verhoef, H.A., Morin, P.J. (Eds.), Community Ecology: Processes, Models, and Applications. Oxford University Press, Oxford, pp. 163–178.

Lonsdale, W.M., 1999. Global patterns of plant invasions and the concept of invasibility. Ecology 80, 1522–1536.

Loreau, M., Mouquet, N., Holt, R.D., 2003. Meta-ecosystems: a theoretical framework for a spatial ecosystem ecology. Ecol. Lett. 6, 673–679.

MacArthur, R.H., Wilson, E.O., 1967. Theory of Island Biogeography. Princeton University Press, Princeton, New Jersey.

Maheu-Giroux, M., de Blois, S., 2007. Landscape ecology of Phragmites australis invasion in networks of linear wetlands. Landsc. Ecol. 22, 285–301.

Marchetti, M.P., Moyle, P.B., Levine, R., 2004. Invasive species profiling? Exploring the characteristics of non-native fishes across invasion stages in California. Freshw. Biol. 49, 646–661.

Martin, F.-P.J., Wang, Y., Sprenger, N., Yap, I.K.S., Lundstedt, T., Lek, P., Rezzi, S., Ramadan, Z., van Bladeren, P., Fay, L.B., Kochhar, S., Lindon, J.C., Holmes, E., Nicholson, J.K., 2008. Probiotic modulation of symbiotic gut microbial–host metabolic interactions in a humanized microbiome mouse model. Mol. Syst. Biol. 4, 157.

Massol, F., Dubart, M., Calcagno, V., Cazelles, K., Jacquet, C., Kéfi, S., Gravel, D., 2017. Island biogeography of food webs. Adv. Ecol. Res. 56, 183–262.

McGill, B.J., Etienne, R.S., Gray, J.S., Alonso, D., Anderson, M.J., Benecha, H.K., Dornelas, M., Enquist, B.J., Green, J.L., He, F., Hurlbert, A.H., Magurran, A.E., Marquet, P.A., Maurer, B.A., Ostling, A., Soykan, C.U., Ugland, K.I., White, E.P., 2007. Species abundance distributions: moving beyond single prediction theories to integration within an ecological framework. Ecol. Lett. 10, 995–1015.

Mealor, B.A., Hild, A.L., 2007. Post-invasion evolution of native plant populations: a test of biological resilience. Oikos 116, 1493–1500.

Médoc, V., Firmat, C., Sheath, D.J., Pegg, J., Andreou, D., Britton, J.R., 2017. Parasites and biological invasions: predicting ecological alterations at levels from individual hosts to whole networks. Adv. Ecol. Res. 57, 1–54.

Melián, C.J., Bascompte, J., Jordano, P., Krivan, V., 2009. Diversity in a complex ecological network with two interaction types. Oikos 118, 122–130.

Memmott, J., Waser, N.M., Price, M.V., 2004. Tolerance of pollination networks to species extinctions. Proc. R. Soc. Lond. B Biol. Sci. 271, 2605–2611.

Mitchell, C.E., Agrawal, A.A., Bever, J.D., Gilbert, G.S., Hufbauer, R.A., Klironomos, J.N., Maron, J.L., Morris, W.F., Parker, I.M., Power, A.G., Seabloom, E.W., Torchin, M.E., Vázquez, D.P., 2006. Biotic interactions and plant invasions. Ecol. Lett. 9, 726–740.

Moe, S.J., Stelzer, R.S., Forman, M.R., Harpole, W.S., Daufresne, T., Yoshida, T., 2005. Recent advances in ecological stoichiometry: insights for population and community ecology. Oikos 109, 29–39.

Mollot, G., Tixier, P., Lescourret, F., Quilici, S., Duyck, P.-F., 2012. New primary resource increases predation on a pest in a banana agroecosystem. Agric. For. Entomol. 14, 317–323.

Mollot, G., Duyck, P.-F., Lefeuvre, P., Lescourret, F., Martin, J.-F., Piry, S., Canard, E., Tixier, P., 2014. Cover cropping alters the diet of arthropods in a banana plantation: a metabarcoding approach. PLoS One 9. e93740.

Morel-Journel, T., Girod, P., Mailleret, L., Auguste, A., Blin, A., Vercken, E., 2015. The highs and lows of dispersal: how connectivity and initial population size jointly shape establishment dynamics in discrete landscapes. Oikos 125, 769–777.

Morris, J.R., Vandermeer, J., Perfecto, I., 2015. A keystone ant species provides robust biological control of the coffee berry borer under varying pest densities. PLoS One 10. e0142850.

Mougi, A., Kondoh, M., 2016. Food-web complexity, meta-community complexity and community stability. Sci. Rep. 6, 24478.

Mucha, P.J., Richardson, T., Macon, K., Porter, M.A., Onnela, J.-P., 2010. Community structure in time-dependent, multiscale, and multiplex networks. Science 328, 876–878.

Muirhead, J.R., MacIsaac, H.J., 2005. Development of inland lakes as hubs in an invasion network. J. Appl. Ecol. 42, 80–90.

Müller-Schärer, H., Schaffner, U., Steinger, T., 2004. Evolution in invasive plants: implications for biological control. Trends Ecol. Evol. 19, 417–422.

Murall, C.L., Abbate, J.L., Puelma Touzel, M., Allen-Vercoe, E., Alizon, S., Froissart, R., McCann, K., 2017. Invasions of host-associated microbiome networks. Adv. Ecol. Res. 57, 201–281.

Naisbit, R.E., Rohr, R.P., Rossberg, A.G., Kehrli, P., Bersier, L.-F., 2012. Phylogeny versus body size as determinants of food web structure. Proc. R. Soc. Lond. B Biol. Sci. 279, 3291–3297. rspb20120327.

Nuismer, S.L., Jordano, P., Bascompte, J., 2013. Coevolution and the architecture of mutualistic networks. Evolution 67, 338–354.

Nyberg, C.D., Wallentinus, I., 2005. Can species traits be used to predict marine macroalgal introductions? Biol. Invasions 7, 265–279.

Ødegaard, F., Diserud, O.H., Østbye, K., 2005. The importance of plant relatedness for host utilization among phytophagous insects. Ecol. Lett. 8, 612–617.

Olesen, J.M., Eskildsen, L.I., Venkatasamy, S., 2002. Invasion of pollination networks on oceanic islands: importance of invader complexes and endemic super generalists. Divers. Distrib. 8, 181–192.

Pantel, J.H., Duvivier, C., De Meester, L., 2015. Rapid local adaptation mediates zooplankton community assembly in experimental mesocosms. Ecol. Lett. 18, 992–1000.

Parker, J.D., Burkepile, D.E., Hay, M.E., 2006. Opposing effects of native and exotic herbivores on plant invasions. Science 311, 1459–1461.

Pearse, I.S., Altermatt, F., 2013. Predicting novel trophic interactions in a non-native world. Ecol. Lett. 16, 1088–1094.

Petchey, O.L., Brose, U., Rall, B.C., 2010. Predicting the effects of temperature on food web connectance. Philos. Trans. R. Soc. Lond. B Biol. Sci. 365, 2081–2091.

Phillips, B.L., Brown, G.P., Webb, J.K., Shine, R., 2006. Invasion and the evolution of speed in toads. Nature 439, 803.

Phillips, B.L., Brown, G.P., Travis, J.M.J., Shine, R., 2008. Reid's paradox revisited: the evolution of dispersal kernels during range expansion. Am. Nat. 172, S34–S48.

Pillai, P., Gonzalez, A., Loreau, M., 2011. Metacommunity theory explains the emergence of food web complexity. Proc. Natl. Acad. Sci. U.S.A. 108, 19293–19298.

Pintor, L.M., Byers, J.E., 2015. Do native predators benefit from non-native prey? Ecol. Lett. 18, 1174–1180.

Pocock, M.J.O., Evans, D.M., Memmott, J., 2012. The robustness and restoration of a network of ecological networks. Science 335, 973–977.

Prentis, P.J., Wilson, J.R.U., Dormontt, E.E., Richardson, D.M., Lowe, A.J., 2008. Adaptive evolution in invasive species. Trends Plant Sci. 13, 288–294.

Reid, D.F., Orlova, M.I., 2002. Geological and evolutionary underpinnings for the success of Ponto-Caspian species invasions in the Baltic Sea and North American Great Lakes. Can. J. Fish. Aquat. Sci. 59, 1144–1158.

Rejmánek, M., Richardson, D.M., 1996. What attributes make some plant species more invasive? Ecology 77, 1655–1661.

Rezende, E.L., Albert, E.M., Fortuna, M.A., Bascompte, J., 2009. Compartments in a marine food web associated with phylogeny, body mass, and habitat structure. Ecol. Lett. 12, 779–788.

Ricciardi, A., 2001. Facilitative interactions among aquatic invaders: is an "invasional meltdown" occurring in the Great Lakes? Can. J. Fish. Aquat. Sci. 58, 2513–2525.

Ricciardi, A., 2004. Assessing species invasions as a cause of extinction. Trends Ecol. Evol. 19, 619.

Ricciardi, A., 2005. Facilitation and synergistic interactions between introduced aquatic species. In: Mooney, H.A., Mack, R.N., McNeely, J.A., Neville, L.E., Schei, P.J., Waage, J.K. (Eds.), Invasive Alien Species: A New Synthesis. Island Press, Washington, D.C., pp. 162–178.

Ricciardi, A., Atkinson, S.K., 2004. Distinctiveness magnifies the impact of biological invaders in aquatic ecosystems. Ecol. Lett. 7, 781–784.

Ricciardi, A., MacIsaac, H.J., 2000. Recent mass invasion of the North American Great Lakes by Ponto–Caspian species. Trends Ecol. Evol. 15, 62–65.

Richardson, D.M., Pyšek, P., 2006. Plant invasions: merging the concepts of species invasiveness and community invasibility. Prog. Phys. Geogr. 30, 409–431.

Rodda, G.H., Fritts, T.H., Chiszar, D., 1997. The disappearance of Guam's wildlife. BioScience 47, 565–574.

Rodríguez-Echeverría, S., Traveset, A., 2015. Putative linkages between below-and aboveground mutualisms during alien plant invasions. AoB Plants 7. plv062.

Romanuk, T.N., Zhou, Y., Brose, U., Berlow, E.L., Williams, R.J., Martinez, N.D., 2009. Predicting invasion success in complex ecological networks. Philos. Trans. R. Soc. Lond. B Biol. Sci. 364, 1743–1754.

Romanuk, T.N., Zhou Y., Valdovinos F.S., Martinez N.D., 2017. Robustness trade-offs in model food webs: invasion probability decreases while invasion consequences increase with connectance. Adv. Ecol. Res. 56, 263–291.

Roy, H.E., Brown, P.M.J., Rothery, P., Ware, R.L., Majerus, M.E.N., 2008. Interactions between the fungal pathogen Beauveria bassiana and three species of coccinellid: Harmonia axyridis, Coccinella septempunctata and Adalia bipunctata. BioControl 53, 265–276.

Roy, H.E., Rhule, E., Harding, S., Handley, L.-J.L., Poland, R.L., Riddick, E.W., Steenberg, T., 2011. Living with the enemy: parasites and pathogens of the ladybird Harmonia axyridis. BioControl 56, 663–679.

Sakai, A.K., Allendorf, F.W., Holt, J.S., Lodge, D.M., Molofsky, J., With, K.A., Baughman, S., Cabin, R.J., Cohen, J.E., Ellstrand, N.C., McCauley, D.E.,

O'Neil, P., Parker, I.M., Thompson, J.N., Weller, S.G., 2001. The population biology of invasive species. Annu. Rev. Ecol. Syst. 32, 305–332.

Sauve, A.M.C., Fontaine, C., Thébault, E., 2014. Structure–stability relationships in networks combining mutualistic and antagonistic interactions. Oikos 123, 378–384.

Sauve, A.M.C., Thébault, E., Pocock, M.J.O., Fontaine, C., 2016. How plants connect pollination and herbivory networks and their contribution to community stability. Ecology 97, 908–917.

Schlaepfer, M.A., Sherman, P.W., Blossey, B., Runge, M.C., 2005. Introduced species as evolutionary traps. Ecol. Lett. 8, 241–246.

Scully, E.D., Geib, S.M., Hoover, K., Tien, M., Tringe, S.G., Barry, K.W., del Rio, T.G., Chovatia, M., Herr, J.R., Carlson, J.E., 2013. Metagenomic profiling reveals lignocellulose degrading system in a microbial community associated with a wood-feeding beetle. PLoS One 8. e73827.

Seabloom, E.W., Borer, E.T., Buckley, Y.M., Cleland, E.E., Davies, K.F., Firn, J., Harpole, W.S., Hautier, Y., Lind, E.M., MacDougall, A.S., et al., 2015. Plant species' origin predicts dominance and response to nutrient enrichment and herbivores in global grasslands. Nat. Commun. 6, 7710.

Sharov, A.A., Liebhold, A.M., 1998. Model of slowing the spread of gypsy moth (Lepidoptera: Lymantriidae) with a barrier zone. Ecol. Appl. 8, 1170–1179.

Shea, K., Chesson, P., 2002. Community ecology theory as a framework for biological invasions. Trends Ecol. Evol. 17, 170–176.

Sheath, D.J., Williams, C.F., Reading, A.J., Britton, J.R., 2015. Parasites of non-native freshwater fishes introduced into England and Wales suggest enemy release and parasite acquisition. Biol. Invasions 17, 2235–2246.

Shigesada, N., Kawasaki, K., 1997. Biological Invasions: Theory and Practice. Oxford University Press, United Kingdom.

Shine, C., Williams, N., Gündling, L., 2000. A Guide to Designing Legal and Institutional Frameworks on Alien Invasive Species. IUCN, Gland, Switzerland/Cambridge/Bonn.

Simberloff, D., 2011. How common are invasion-induced ecosystem impacts? Biol. Invasions 13, 1255–1268.

Simberloff, D., Martin, J.L., Genovesi, P., Maris, V., Wardle, D.A., Aronson, J., Courchamp, F., Galil, B., García-Berthou, E., Pascal, M., Pyšek, P., 2013. Impacts of biological invasions: what's what and the way forward. Trends in Ecology & Evolution 28 (1), 58–66.

Simberloff, D., Von Holle, B., 1999. Positive interactions of nonindigenous species: invasional meltdown? Biol. Invasions 1, 21–32.

Staniczenko, P.P.A., Kopp, J.C., Allesina, S., 2013. The ghost of nestedness in ecological networks. Nat. Commun. 4, 1391.

Stinson, K.A., Campbell, S.A., Powell, J.R., Wolfe, B.E., Callaway, R.M., Thelen, G.C., Hallett, S.G., Prati, D., Klironomos, J.N., 2006. Invasive plant suppresses the growth of native tree seedlings by disrupting belowground mutualisms. PLoS Biol. 4. e140.

Stouffer, D.B., Bascompte, J., 2011. Compartmentalization increases food-web persistence. Proc. Natl. Acad. Sci. U.S.A. 108, 3648–3652.

Strauss, S.Y., 2014. Ecological and evolutionary responses in complex communities: implications for invasions and eco-evolutionary feedbacks. Oikos 123, 257–266.

Strauss, S.Y., Webb, C.O., Salamin, N., 2006a. Exotic taxa less related to native species are more invasive. Proc. Natl. Acad. Sci. U.S.A. 103, 5841–5845.

Strauss, S.Y., Lau, J.A., Carroll, S.P., 2006b. Evolutionary responses of natives to introduced species: what do introductions tell us about natural communities? Ecol. Lett. 9, 357–374.

Strayer, D.L., 2012. Eight questions about invasions and ecosystem functioning. Ecol. Lett. 15, 1199–1210.

Strayer, D.L., Eviner, V.T., Jeschke, J.M., Pace, M.L., 2006. Understanding the long-term effects of species invasions. Trends Ecol. Evol. 21, 645–651.

Strong, D.R., 1992. Are trophic cascades all wet? Differentiation and donor-control in speciose ecosystems. Ecology 73, 747–754.

Svardal, H., Rueffler, C., Doebeli, M., 2014. Organismal complexity and the potential for evolutionary diversification. Evolution 68, 3248–3259.

Thébault, E., Fontaine, C., 2010. Stability of ecological communities and the architecture of mutualistic and trophic networks. Science 329, 853–856.

Thompson, J.N., 2005. The Geographic Mosaic of Coevolution. University of Chicago Press, Chicago.

Thuiller, W., Richardson, D.M., Rouget, M., Procheş, Ş., Wilson, J.R.U., 2006. Interactions between environment, species traits, and human uses describe patterns of plant invasions. Ecology 87, 1755–1769.

Tiede, J., Wemheuer, B., Traugott, M., Daniel, R., Tscharntke, T., Ebeling, A., Scherber, C., 2016. Trophic and non-trophic interactions in a biodiversity experiment assessed by next-generation sequencing. PLoS One 11. e0148781.

Tilman, D., 1980. Resources: a graphical-mechanistic approach to competition and predation. Am. Nat. 116, 362–393.

Tilman, D., 1982. Resource Competition and Community Structure. Princeton University Press, Princeton, New Jersey.

Tixier, P., Peyrard, N., Aubertot, J.-N., Gaba, S., Radoszycki, J., Caron-Lormier, G., Vinatier, F., Mollot, G., Sabbadin, R., 2013. Modelling interaction networks for enhanced ecosystem services in agroecosystems. Adv. Ecol. Res. 49, 437–480.

Traveset, A., Richardson, D.M., 2014. Mutualistic interactions and biological invasions. Annu. Rev. Ecol. Evol. Syst. 45, 89–113.

Turner, P.J., Virtue, J.G., 2006. An eight year removal experiment measuring the impact of bridal creeper (Asparagus asparagoides (L.) Druce) and the potential benefit from its control. Plant Prot. Q. 21, 79–84.

Turner, P.J., Scott, J.K., Spafford, H., 2008. The ecological barriers to the recovery of bridal creeper (Asparagus asparagoides (L.) Druce) infested sites: impacts on vegetation and potential increase in other exotic species. Aust. Ecol. 33, 713–722.

Vacher, C., Tamaddoni-Nezhad, A., Kamenova, S., Peyrard, N., Moalic, Y., Sabbadin, R., Schwaller, L., Chiquet, J., Smith, M.A., Vallance, J., Fievet, V., Jakuschkin, B., Bohan, D.A., 2016. Learning ecological networks from next-generation sequencing data. Adv. Ecol. Res. 54, 1–39.

Valentini, A., Miquel, C., Nawaz, M.A., Bellemain, E.V.A., Coissac, E., Pompanon, F., Gielly, L., Cruaud, C., Nascetti, G., Wincker, P., Swenson, J.E., Taberlet, P., 2009. New perspectives in diet analysis based on DNA barcoding and parallel pyrosequencing: the trnL approach. Mol. Ecol. Resour. 9, 51–60.

Valéry, L., Fritz, H., Lefeuvre, J.-C., Simberloff, D., 2008. In search of a real definition of the biological invasion phenomenon itself. Biol. Invasions 10, 1345–1351.

Van Kleunen, M., Weber, E., Fischer, M., 2010. A meta-analysis of trait differences between invasive and non-invasive plant species. Ecol. Lett. 13, 235–245.

Vilà, M., Espinar, J.L., Hejda, M., Hulme, P.E., Jarošik, V., Maron, J.L., Pergl, J., Schaffner, U., Sun, Y., Pyšek, P., 2011. Ecological impacts of invasive alien plants: a meta-analysis of their effects on species, communities and ecosystems. Ecol. Lett. 14, 702–708.

Vivanco, J.M., Bais, H.P., Stermitz, F.R., Thelen, G.C., Callaway, R.M., 2004. Biogeographical variation in community response to root allelochemistry: novel weapons and exotic invasion. Ecol. Lett. 7, 285–292.

Wainwright, C.E., Cleland, E.E., 2013. Exotic species display greater germination plasticity and higher germination rates than native species across multiple cues. Biol. Invasions 15, 2253–2264.

Walsh, M.R., DeLong, J.P., Hanley, T.C., Post, D.M., 2012. A cascade of evolutionary change alters consumer-resource dynamics and ecosystem function. Proc. R. Soc. B 279, 3184–3192.

Walsh, J.R., Carpenter, S.R., Vander Zanden, M.J., 2016. Invasive species triggers a massive loss of ecosystem services through a trophic cascade. Proc. Natl. Acad. Sci. U.S.A. 113, 4081–4085.

Weiblen, G.D., Webb, C.O., Novotny, V., Basset, Y., Miller, S.E., 2006. Phylogenetic dispersion of host use in a tropical insect herbivore community. Ecology 87, S62–S75.

Wexler, A.G., Bao, Y., Whitney, J.C., Bobay, L.-M., Xavier, J.B., Schofield, W.B., Barry, N.A., Russell, A.B., Tran, B.Q., Goo, Y.A., Goodlett, D.R., Ochman, H., Mougous, J.D., Goodman, A.L., 2016. Human symbionts inject and neutralize antibacterial toxins to persist in the gut. Proc. Natl. Acad. Sci. U.S.A. 113, 3639–3644.

White, T.C.R., 2005. Why Does the World Stay Green? Nutrition and Survival of Plant-Eaters. CSIRO Publishing, Collingwood, Australia.

Whitney, K.D., Gabler, C.A., 2008. Rapid evolution in introduced species, "invasive traits" and recipient communities: challenges for predicting invasive potential. Divers. Distrib. 14, 569–580.

Wiles, G.J., Bart, J., Beck, R.E., Aguon, C.F., 2003. Impacts of the brown tree snake: patterns of decline and species persistence in Guam's avifauna. Conserv. Biol. 17, 1350–1360.

Willerslev, E., Davison, J., Moora, M., Zobel, M., Coissac, E., Edwards, M.E., Lorenzen, E.D., Vestergard, M., Gussarova, G., Haile, J., Craine, J., Gielly, L., Boessenkool, S., Epp, L.S., Pearman, P.B., Cheddadi, R., Murray, D., Brathen, K.A., Yoccoz, N., Binney, H., Cruaud, C., Wincker, P., Goslar, T., Alsos, I.G., Bellemain, E., Brysting, A.K., Elven, R., Sonstebo, J.H., Murton, J., Sher, A., Rasmussen, M., Ronn, R., Mourier, T., Cooper, A., Austin, J., Moller, P., Froese, D., Zazula, G., Pompanon, F., Rioux, D., Niderkorn, V., Tikhonov, A., Savvinov, G., Roberts, R.G., MacPhee, R.D.E., Gilbert, M.T.P., Kjaer, K.H., Orlando, L., Brochmann, C., Taberlet, P., 2014. Fifty thousand years of Arctic vegetation and megafaunal diet. Nature 506, 47–51.

Williams, R.J., Martinez, N.D., 2000. Simple rules yield complex food webs. Nature 404, 180–183.

Williams, R.J., Martinez, N.D., 2004a. Limits to trophic levels and omnivory in complex food webs: theory and data. Am. Nat. 163, 458–468.

Williams, R.J., Martinez, N.D., 2004b. Stabilization of chaotic and non-permanent food-web dynamics. Eur. Phys. J. B 38, 297–303.

Wilsey, B.J., Teaschner, T.B., Daneshgar, P.P., Isbell, F.I., Polley, H.W., 2009. Biodiversity maintenance mechanisms differ between native and novel exotic-dominated communities. Ecol. Lett. 12, 432–442.

With, K.A., 2002. The landscape ecology of invasive spread. Conserv. Biol. 16, 1192–1203.

Woodward, G., Hildrew, A.G., 2002. Body-size determinants of niche overlap and intraguild predation within a complex food web. J. Anim. Ecol. 71, 1063–1074.

Woodward, G., Ebenman, B., Emmerson, M., Montoya, J.M., Olesen, J.M., Valido, A., Warren, P.H., 2005. Body size in ecological networks. Trends Ecol. Evol. 20, 402–409.

Yoccoz, N.G., Bråthen, K.A., Gielly, L., Haile, J., Edwards, M.E., Goslar, T., Von Stedingk, H., Brysting, A.K., Coissac, E., Pompanon, F., SøNstebø, J.H., Miquel, C., Valentini, A., de Bello, F., Chave, J., Thuiller, W., Wincker, P., Cruaud, C., Gavory, F., Rasmussen, M., Gilbert, M.T.P., Orlando, L., Brochmann, C., Willerslev, E., Taberlet, P., 2012. DNA from soil mirrors plant taxonomic and growth form diversity. Mol. Ecol. 21, 3647–3655.

Yodzis, P., Innes, S., 1992. Body size and consumer-resource dynamics. Am. Nat. 139, 1151–1175.

Zaiko, A., Samuiloviene, A., Ardura, A., Garcia-Vazquez, E., 2015. Metabarcoding approach for nonindigenous species surveillance in marine coastal waters. Mar. Pollut. Bull. 100, 53–59.

Zhou, X., Li, Y., Liu, S., Yang, Q., Su, X., Zhou, L., Tang, M., Fu, R., Li, J., Huang, Q., 2013. Ultra-deep sequencing enables high-fidelity recovery of biodiversity for bulk arthropod samples without PCR amplification. Gigascience 2, 1.

INDEX

Note: Page numbers followed by "*f*" indicate figures, "*t*" indicate tables, and "*b*" indicate boxes.

ADVANCES IN ECOLOGICAL RESEARCH
VOLUME 1-56

 CUMULATIVE LIST OF TITLES

Aerial heavy metal pollution and terrestrial ecosystems, **11**, 218

Age determination and growth of Baikal seals (*Phoca sibirica*), **31**, 449

Age-related decline in forest productivity: pattern and process, **27**, 213

Allometry of body size and abundance in 166 food webs, **41**, 1

Analysis and interpretation of long-term studies investigating responses to climate change, **35**, 111

Analysis of processes involved in the natural control of insects, **2**, 1

Ancient Lake Pennon and its endemic molluscan faun (Central Europe; Mio-Pliocene), **31**, 463

Ant-plant-homopteran interactions, **16**, 53

Anthropogenic impacts on litter decomposition and soil organic matter, **38**, 263

Arctic climate and climate change with a focus on Greenland, **40**, 13

Arrival and departure dates, **35**, 1

Assessing the contribution of micro-organisms and macrofauna to biodiversity-ecosystem functioning relationships in freshwater microcosms, **43**, 151

A belowground perspective on Dutch agroecosystems: how soil organisms interact to support ecosystem services, **44**, 277

The benthic invertebrates of Lake Khubsugul, Mongolia, **31**, 97

Big data and ecosystem research programmes, **51**, 41

Biodiversity, species interactions and ecological networks in a fragmented world **46**, 89

Biogeography and species diversity of diatoms in the northern basin of Lake Tanganyika, **31**, 115

Biological strategies of nutrient cycling in soil systems, **13**, 1

Biomanipulation as a restoration tool to combat eutrophication: recent advances and future challenges, **47**, 411

Biomonitoring of human impacts in freshwater ecosystems: the good, the bad and the ugly, **44**, 1

Bray-Curtis ordination: an effective strategy for analysis of multivariate ecological data, **14**, 1

Ecology, systematics and evolution of Australian frogs, **5**, 37

Ecophysiology of trees of seasonally dry Tropics: comparison among phonologies, **32**, 113

Ecosystems and their services in a changing world: an ecological perspective, **48**, 1

Effect of flooding on the occurrence of infectious disease, **39**, 107

Effects of food availability, snow, and predation on breeding performance of waders at Zackenberg, **40**, 325

Effect of hydrological cycles on planktonic primary production in Lake Malawi Niassa, **31**, 421

Effects of climatic change on the population dynamics of crop pests, **22**, 117

Effects of floods on distribution and reproduction of aquatic birds, **39**, 63

The effects of invasive species on the decline in species richness: a global meta-analysis, **56**, 61

The effects of modern agriculture nest predation and game management on the population ecology of partridges (*Perdix perdix* and *Alectoris rufa*), **11**, 2

Effective river restoration in the 21st century: From trial and error to novel evidence-based approaches, **55**, 529

El Niño effects on Southern California kelp forest communities, **17**, 243

Empirically characterising trophic networks: What emerging DNA-based methods, stable isotope and fatty acid analyses can offer, **49**, 177

Empirical evidences of density-dependence in populations of large herbivores, **41**, 313

Endemism in the Ponto-Caspian fauna, with special emphasis on the Oncychopoda (Crustacea), **31**, 179

Energetics, terrestrial field studies and animal productivity, **3**, 73

Energy in animal ecology, **1**, 69

Environmental warming in shallow lakes: a review of potential changes in community structure as evidenced from space-for-time substitution approaches, **46**, 259

Environmental warming and biodiversity-ecosystem functioning in freshwater microcosms: partitioning the effects of species identity, richness and metabolism, **43**, 177

Estimates of the annual net carbon and water exchange of forests: the EUROFLUX methodology, **30**, 113

Estimating forest growth and efficiency in relation to canopy leaf area, **13**, 327

Edwards Brothers Malloy
Ann Arbor MI. USA
January 30, 2017